Lecture Notes in Mathematics

Volume 2266

This series reports on new developments in all areas of mathematics and their applications - quickly, informally and at a high level. Mathematical texts analysing new developments in modelling and numerical simulation are welcome. The type of material considered for publication includes:

1. Research monographs 2. Lectures on a new field or presentations of a new angle in a classical field 3. Summer schools and intensive courses on topics of current research.

Texts which are out of print but still in demand may also be considered if they fall within these categories. The timeliness of a manuscript is sometimes more important than its form, which may be preliminary or tentative.

More information about this series at http://www.springer.com/series/304

Bo'az Klartag • Emanuel Milman
Editors

Geometric Aspects of Functional Analysis

Israel Seminar (GAFA) 2017-2019
Volume II

 Springer

Editors
Bo'az Klartag
School of Mathematical Sciences
Tel Aviv University
Tel Aviv, Israel

Emanuel Milman
Department of Mathematics
Technion – Israel Institute of Technology
Haifa, Israel

Department of Mathematics
Weizmann Institute of Science
Rehovot, Israel

ISSN 0075-8434   ISSN 1617-9692   (electronic)
Lecture Notes in Mathematics
ISBN 978-3-030-46761-6   ISBN 978-3-030-46762-3   (eBook)
https://doi.org/10.1007/978-3-030-46762-3

Mathematics Subject Classification (2010): Primary: 46-XX; Secondary: 52-XX

This Springer imprint is published by the registered company Springer Nature Switzerland AG.
The registered company address is: Gewerbestrasse 11, 6330 Cham, Switzerland

Preface

Since the mid-1980s, the following volumes containing collections of papers reflecting the activity of the Israel Seminar in Geometric Aspects of Functional Analysis have appeared:

1983–1984 Published privately by Tel Aviv University
1985–1986 Springer Lecture Notes in Mathematics, vol. 1267
1986–1987 Springer Lecture Notes in Mathematics, vol. 1317
1987–1988 Springer Lecture Notes in Mathematics, vol. 1376
1989–1990 Springer Lecture Notes in Mathematics, vol. 1469
1992–1994 Operator Theory: Advances and Applications, vol. 77, Birkhäuser
1994–1996 MSRI Publications, vol. 34, Cambridge University Press
1996–2000 Springer Lecture Notes in Mathematics, vol. 1745
2001–2002 Springer Lecture Notes in Mathematics, vol. 1807
2002–2003 Springer Lecture Notes in Mathematics, vol. 1850
2004–2005 Springer Lecture Notes in Mathematics, vol. 1910
2006–2010 Springer Lecture Notes in Mathematics, vol. 2050
2011–2013 Springer Lecture Notes in Mathematics, vol. 2116
2014–2016 Springer Lecture Notes in Mathematics, vol. 2169

The first six were edited by Lindenstrauss and Milman, the seventh by Ball and Milman, the subsequent four by Milman and Schechtman, the subsequent one by Klartag, Mendelson, and Milman, and the last two by the present editors.

This is the second of two volumes from the years 2017–2019, the first volume is published in Springer Lecture Notes in Mathematics, vol. 2256. As in the previous Seminar Notes, these two volumes reflect general trends in the study of Geometric Aspects of Functional Analysis, understood in a broad sense. Two classical topics represented are the Concentration of Measure Phenomenon in the Local Theory of Banach Spaces, which has recently had triumphs in Random Matrix Theory, and the Central Limit Theorem, one of the earliest examples of regularity and order in high dimensions. Central to the text is the study of the Poincaré and log-Sobolev functional inequalities, their reverses, and other inequalities, in which a crucial role is often played by convexity assumptions such as log-concavity. The concept and properties of entropy form an important subject, with Bourgain's slicing problem and its variants drawing much attention. Constructions related to convexity theory are proposed and revisited, as well as inequalities that go beyond the Brunn–Minkowski theory. One of the major current research directions addressed is the identification of lower-dimensional structures with remarkable properties in rather arbitrary high-dimensional objects. In addition to functional analytic results, connections to computer science and to differential geometry are also discussed. All contributions are original research papers and were subject to the usual refereeing standards.

We are grateful to Vitali Milman for his help and guidance in preparing and editing these two volumes.

Tel Aviv/Rehovot, Israel   Bo'az Klartag
Haifa, Israel   Emanuel Milman

Jean Bourgain: In Memoriam

Our friend and mentor Jean Bourgain passed away on December 22, 2018, at the age of 64.

Jean Bourgain by Jan Rauchwerger. Courtesy of Vitali Milman

Jean Bourgain was one of the most outstanding mathematicians of our time. Bourgain changed the face of analysis; he revolutionized our understanding of analysis. He has introduced, mastered, and perfected many different methods in every corner of analysis, including a dozen of neighboring fields, and has left his mark in each of these directions. His achievements, vision, and insight united many distant and very diverse directions of mathematics into one enormously powerful and broad entity. When we say "Analysis" today we mean, besides classical directions, also ergodic theory, PDE, several directions of analytical number theory, geometry, and combinatorics (including complexity). This is undoubtedly the result of Bourgain's activity, unprecedented in its strength and diversity. The torrent of his achievements is difficult to grasp, the number of very long-standing problems Bourgain has solved can be counted in tens, perhaps approaching a hundred, and this would take a whole book to describe. It is almost impossible to believe: ∼550 hard analysis papers written over less than 40 years.

Jean's passing is a terrible loss for his family and friends, and a terrible loss to the mathematical world. We shudder at the thought of how many more theorems he could have proved and open problems he could have solved. He leaves behind an unbelievable legacy of results carrying his name, whose breadth is matched only by their depth.

In addition to being a mathematical giant, Jean has personally influenced us all. His influence was also well felt on the GAFA seminar notes. He has published a total of 50 papers (3 in the present volume) in **every** volume of the GAFA seminar notes since its inception (45 in Springer Math. Notes series, 2 in Birkhauser series, and 3 in Mathematical Sciences Research Institute (MSRI), Berkeley). We are proud that some of his last papers are published in these GAFA Seminar Notes (Volume I).

May he rest in peace.

Tel Aviv/Rehovot, Israel   Bo'az Klartag (Editor)
Haifa, Israel   Emanuel Milman (Editor)
Tel Aviv, Israel   Vitali Milman (Founding Editor)

Contents Overview for Volume I

Contents

Chapter 1
A Generalized Central Limit Conjecture for Convex Bodies

Haotian Jiang, Yin Tat Lee, and Santosh S. Vempala

Abstract The central limit theorem for convex bodies says that with high probability the marginal of an isotropic log-concave distribution along a random direction is close to a Gaussian, with the quantitative difference determined asymptotically by the Cheeger/Poincare/KLS constant. Here we propose a generalized CLT for marginals along random directions drawn from any isotropic log-concave distribution; namely, for x, y drawn independently from isotropic log-concave densities p, q, the random variable $\langle x, y \rangle$ is close to Gaussian. Our main result is that this generalized CLT is quantitatively equivalent (up to a small factor) to the KLS conjecture. Any polynomial improvement in the current KLS bound of $n^{1/4}$ in $\mathbb{R}^n$ implies the generalized CLT, and vice versa. This tight connection suggests that the generalized CLT might provide insight into basic open questions in asymptotic convex geometry.

This research of the author "Yin Tat Lee" was supported in part by NSF Awards CCF-1740551, CCF-1749609, and DMS-1839116.

This research of the author "Santosh S. Vempala" was supported in part by NSF Awards CCF-1563838, CCF-1717349, DMS-1839323 and E2CDA-1640081.

H. Jiang
University of Washington, Seattle, WA, USA
e-mail: jhtdavid@uw.edu

Y. T. Lee
University of Washington, Seattle, WA, USA

Microsoft Research, Redmond, WA, USA
e-mail: yintat@uw.edu

S. S. Vempala (✉)
Georgia Tech, Atlanta, GA, USA
e-mail: vempala@gatech.edu

© Springer Nature Switzerland AG 2020
B. Klartag, E. Milman (eds.), *Geometric Aspects of Functional Analysis*,
Lecture Notes in Mathematics 2266,
https://doi.org/10.1007/978-3-030-46762-3_1

1.1   Introduction

Convex bodies in high dimensions exhibit surprising asymptotic properties, i.e., phenomena that become sharper as the dimension increases. As an elementary example, most of the measure of a sphere or ball in $\mathbb{R}^n$ lies within distance $O(1/\sqrt{n})$ of any bisecting hyperplane, and a one-dimensional marginal is close to a Gaussian, i.e., its total variation distance to a Gaussian of the same variance is $O(1/\sqrt{n})$. A striking generalization of this is the central limit theorem for convex bodies in Theorem 1.1, originally due to Klartag [16]. A function $h : \mathbb{R}^n \to \mathbb{R}_+$ is called *log-concave* if it takes the form $h = \exp(-f)$ for a convex function $f : \mathbb{R}^n \to \mathbb{R} \cup \{\infty\}$. A probability measure is log-concave if it has a log-concave density. A measure is said to be *isotropic* if it has zero mean and identity covariance.

Theorem 1.1 (Central Limit Theorem) *Let p be an isotropic log-concave measure in $\mathbb{R}^n$ and $y \sim p$. Then we have*

$$\mathbb{P}_{x \sim S^{n-1}} \left[d_{\mathrm{TV}} \left(\langle x, y \rangle, \mathcal{N}(0, 1) \right) \geq c_n \right] \leq c_n,$$

for some constants c_n that tends to 0 as $n \to +\infty$.

The central limit theorem is closely related to the *thin-shell* conjecture (also known as the *variance hypothesis*) [2, 4]. Let $\sigma_n \geq 0$ satisfy

$$\sigma_n^2 = \sup_p \mathbb{E}_{x \sim p} \left[\left(\|x\| - \sqrt{n} \right)^2 \right],$$

where the supremum is taken over all isotropic, log-concave measures p in $\mathbb{R}^n$. The thin-shell conjecture [2, 4] asserts the existence of a universal constant C such that $\sigma_n^2 < C$ for all $n \in \mathbb{N}$. It is closely connected to the CLT: by a direct calculation, the CLT implies a bound on σ_n (and the conjectured CLT parameter implies the thin-shell conjecture); Moreover, $c_n = O(\sigma_n \log n / \sqrt{n})$ [2, 10]. The first non-trivial bound on σ_n, which gives the first non-trivial bound on c_n in Theorem 1.1, was due to Klartag [16]. This was followed by several improvements and refinements [12, 14, 17, 27]. The current best bound is $\sigma_n = O(n^{1/4})$ which implies $c_n = O(n^{-1/4} \log n)$ [18]. This follows from the well-known fact that $\sigma_n = O(\psi_n)$, where ψ_n is the KLS constant (also known as the inverse Cheeger constant) defined as follows.

Definition 1.2 (KLS Constant) For a log-concave density p in $\mathbb{R}^n$ with induced measure μ_p, the KLS constant ψ_p is defined as

$$\frac{1}{\psi_p} = \inf_{S \subset \mathbb{R}^n, \mu_p(S) \leq 1/2} \frac{\mu_p(\partial S)}{\mu_p(S)}.$$

We define ψ_n be the supremum of ψ_p over all isotropic log-concave densities p in $\mathbb{R}^n$.

Theorem 1.3 ([18]) *The KLS constant of any isotropic log-concave density in $\mathbb{R}^n$ is $O(n^{1/4})$.*

For other connections and implications of the KLS conjecture, including its equivalence to spectral gap and its implication of the slicing conjecture, the reader is referred to recent surveys [13, 20] and this comprehensive book [5].

A key fact used in the above theorem is the following elementary lemma about log-concave densities.

Lemma 1.4 (Third Moment) *For x, y drawn independently from an isotropic log-concave density p, we have $\mathbb{E}(\langle x, y \rangle^3) = O(n^{1.5})$.*

We remark that the third moment bound in Lemma 1.4, holds even if x, y are drawn independently from different measures.

If the KLS conjecture is true, then the expression above is $O(n)$. It is shown in an earlier version of [18] that any polynomial improvement in the third moment bound to $n^{1.5-\epsilon}$ for some $\epsilon > 0$ would lead to an improvement in the bound on the KLS constant to $n^{1/4-\epsilon'}$ for some $\epsilon' > 0$. (The techniques used in the corresponding part of the preprint [18] are formally included in this paper.)

Motivated by the above connection, we propose a generalized CLT in this paper. To formally state our generalized CLT, we need the definition of L_p Wasserstein distance.

Definition 1.5 (L_p Wasserstein Distance or W_p Distance) The L_p Wasserstein distance between two probability measures μ and ν in $\mathbb{R}$ for $p \geq 1$ is defined by

$$W_p(\mu, \nu) \overset{\text{def}}{=} \inf_{\pi} \left[\int |x - y|^p d\pi(x, y) \right]^{\frac{1}{p}},$$

where the infimum is over all couplings of μ and ν, i.e. probability measures π in $\mathbb{R}^2$ that have marginals μ and ν.

When convenient we will denote $W_p(\mu, \nu)$ also be $W_p(x, y)$ where $x \sim \mu$, $y \sim \nu$. Our generalized CLT is stated using the W_2 distance, which is a natural choice, also used in related work on CLT's [11, 28].

The content of the conjecture is that one can replace the uniform distribution on the sphere (or Gaussian) with any isotropic log-concave density, i.e., along most directions with respect to any isotropic log-concave measure, the marginal of an isotropic log-concave measure is approximately Gaussian.

Conjecture 1.6 (Generalized CLT) Let x, y be independent random vectors drawn from isotropic log-concave densities p, q respectively and $G \sim \mathcal{N}(0, n)$. Then,

$$W_2(\langle x, y \rangle, G) = O(1). \tag{1.1.1}$$

The current best upper bound on the W_2 distance in Eq. (1.1.1) is the trivial bound of $O(\sqrt{n})$. As we will see later, a third moment bound of order $O(n)$ in Lemma 1.4 would be implied if Conjecture 1.6 holds.

Our main result is that this Generalized CLT is equivalent (up to a small factor) to the KLS conjecture, and any polynomial improvement in one leads to a similar improvement in the other.

Theorem 1.7 (Generalized CLT Equivalent to KLS) *Fix $\epsilon \in (0, 1/2)$. If for every isotropic log-concave measure p in $\mathbb{R}^n$ and independent vectors $x, y \sim p$ and $g \sim \mathcal{N}(0, n)$, we have $W_2(\langle x, y \rangle, g) = O\left(n^{1/2-\epsilon}\right)$, then for any $\delta > 0$, we have $\psi_n = O\left(n^{1/4-\epsilon/2+\delta}\right)$.*

On the other hand, if we have $\psi_n = O\left(n^{1/4-\epsilon/2}\right)$, then for any isotropic log-concave measures p, q in $\mathbb{R}^n$, independent vectors $x \sim p$, $y \sim q$ and $\delta > 0$, we have $W_2(\langle x, y \rangle, G) = O\left(n^{1/2-\epsilon+\delta}\right)$.

Remark 1.8 We emphasize that the equivalence between Generalized CLT and the KLS conjecture in Theorem 1.7 does not hold in a pointwise sense, i.e. the Generalized CLT for a specific isotropic log-concave measure p in $\mathbb{R}^n$ alone does not imply the corresponding bound for ψ_p and vice versa. One needs to establish the Generalized CLT for all isotropic log-concave measures in $\mathbb{R}^n$ in order to deduce the KLS conjecture.

The proof of Theorem 1.7 proceeds in three steps: (1) in Theorem 1.9 below, we show that an improved third moment bound implies an improved bound on the KLS constant (an earlier version of this part of the proof is implicit in the preprint [18]), (2) in Theorem 1.27, we show that an improved bound for Generalized CLT implies an improved third moment bound, and (3) in Theorem 1.49, we show that an improved bound on the KLS constant implies an improved bound for Generalized CLT. While all three parts are new and unpublished (except on the arXiv), the proof of (3) is via a coupling with Brownian motion (we discuss the similarity to existing literature [11]), (2) is relatively straightforward, and (1) is the most technical, based on a carefully chosen potential function and several properties of an associated tensor.

The main intermediate result in our proof that the Generalized CLT implies the KLS conjecture is the following theorem.

Theorem 1.9 *Fix $\epsilon \in (0, 1/2)$. If for every isotropic log-concave distribution p in $\mathbb{R}^n$ and independent vectors $x, y \sim p$, we have*

$$\mathbb{E}_{x,y\sim p}\left(\langle x, y \rangle^3\right) = O\left(n^{1.5-\epsilon}\right), \tag{1.1.2}$$

then for any $\delta > 0$, we have $\psi_n = O\left(n^{1/4-\epsilon/2+\delta}\right)$.

In fact what we show is that the KLS constant ψ_n can be bounded in terms of the third moment.

Theorem 1.10 *Let p range over all isotropic log-concave distributions in $\mathbb{R}^n$. Then,*

$$\psi_n^2 \leq \frac{\widetilde{O}(1)}{n} \cdot \sup_p \mathbb{E}_{x,y\sim p}\left(\langle x, y\rangle^3\right) = \widetilde{O}(1) \cdot \sup_p \mathbb{E}_{\theta\sim S^{n-1}} \left\|\mathbb{E}_{x\sim p}\left(\langle x, \theta\rangle xx^T\right)\right\|_F. \tag{1.1.3}$$

This intermediate result might be of independent interest and is in fact a refinement of the following bound on the KLS constant given by Eldan [10].

$$\psi_n^2 \leq \widetilde{O}(1) \cdot \sup_p \sup_{\theta\in S^{n-1}} \left\|\mathbb{E}_{x\sim p}\left(\langle x, \theta\rangle xx^T\right)\right\|_F. \tag{1.1.4}$$

We replace the supremum over $\theta \in S^{n-1}$ on the RHS by the expectation over S^{n-1}. Here $\|\cdot\|_F$ stands for the *Frobenius norm* (see Sect. 1.2.1). To see how (1.1.3) refines (1.1.4), let $x, y \sim p$ be independent vectors and σ be the uniform measure on S^{n-1}. Then,

$$\int_{S^{n-1}} \left\|\mathbb{E}_{x\sim p}\left(\langle x, \theta\rangle xx^T\right)\right\|_F d\sigma(\theta) = \int_{S^{n-1}} \mathbb{E}_{x,y\sim p}\left(\langle x, \theta\rangle \cdot \langle y, \theta\rangle \cdot \langle x, y\rangle^2\right) d\sigma(\theta)$$

$$= \frac{1}{n}\mathbb{E}_{x,y\sim p}\left(\langle x, y\rangle^3\right).$$

1.2 Preliminaries

In this section, we review background definitions.

1.2.1 Notation and Definitions

A function $h : \mathbb{R}^n \rightarrow \mathbb{R}_+$ is called *log-concave* if it takes the form $h(x) = \exp(-f(x))$ for a convex function $f : \mathbb{R}^n \rightarrow \mathbb{R}\cup\{\infty\}$. It is *t-strongly log-concave* if it takes the form $h(x) = h'(x)e^{-\frac{t}{2}\|x\|_2^2}$ where $h'(x) : \mathbb{R}^n \rightarrow \mathbb{R}_+$ is an integrable log-concave function. A probability measure is log-concave (*t*-strongly log-concave) if it has a log-concave (resp. *t*-strongly log-concave) density function.

Given a matrix $A \in \mathbb{R}^{m\times n}$, we define its *Frobenius norm* (also known as Hilbert-Schmidt norm), denoted as $\|A\|_F$, to be

$$\|A\|_F = \sqrt{\sum_{i=1}^{m}\sum_{j=1}^{n}|A_{i,j}|^2} = \text{Tr}\left(A^T A\right).$$

The *operator norm* (also known as spectral norm) of A, denoted $\|A\|_{\mathrm{op}}$, is defined as

$$\|A\|_{\mathrm{op}} = \sqrt{\lambda_{\max}\left(A^T A\right)},$$

where $\lambda_{\max}(\cdot)$ stands for the maximum eigenvalue.

1.2.2 Stochastic Calculus

Given real-valued stochastic processes x_t and y_t, the quadratic variations $[x]_t$ and $[x, y]_t$ are real-valued stochastic processes defined by

$$[x]_t = \lim_{|P| \to 0} \sum_{n=1}^{\infty} \left(x_{\tau_n} - x_{\tau_{n-1}}\right)^2 \quad \text{and}$$

$$[x, y]_t = \lim_{|P| \to 0} \sum_{n=1}^{\infty} \left(x_{\tau_n} - x_{\tau_{n-1}}\right)\left(y_{\tau_n} - y_{\tau_{n-1}}\right),$$

where $P = \{0 = \tau_0 \leq \tau_1 \leq \tau_2 \leq \cdots \uparrow t\}$ is a stochastic partition of the non-negative real numbers, $|P| = \max_n (\tau_n - \tau_{n-1})$ is called the *mesh* of P and the limit is defined using convergence in probability. Note that $[x]_t$ is non-decreasing with t and $[x, y]_t$ can be defined as

$$[x, y]_t = \frac{1}{4}\left([x + y]_t - [x - y]_t\right).$$

For example, if the processes x_t and y_t satisfy the SDEs $dx_t = \mu(x_t)dt + \sigma(x_t)dW_t$ and $dy_t = v(y_t)dt + \eta(y_t)dW_t$ where W_t is a Wiener process, we have

$$[x]_t = \int_0^t \sigma^2(x_s)ds \quad [x, y]_t = \int_0^t \sigma(x_s)\eta(y_s)ds \text{ and } d[x, y]_t = \sigma(x_t)\eta(y_t)dt.$$

For vector-valued SDEs

$$dx_t = \mu(x_t)dt + \Sigma(x_t)dW_t \text{ and } dy_t = v(y_t)dt + M(y_t)dW_t,$$

we have that

$$[x^i, x^j]_t = \int_0^t \left(\Sigma(x_s)\Sigma^T(x_s)\right)_{ij} ds \text{ and } d[x^i, y^j]_t = \left(\Sigma(x_t)M^T(y_t)\right)_{ij} dt.$$

Lemma 1.11 (Itô's Formula) *Itô [15] Let x be a semimartingale and f be a twice continuously differentiable function, then*

$$df(x_t) = \sum_i \frac{df(x_t)}{dx^i} dx^i + \frac{1}{2} \sum_{i,j} \frac{d^2 f(x_t)}{dx^i dx^j} d[x^i, x^j]_t.$$

The next two lemmas are well-known facts about Wiener processes.

Lemma 1.12 (Reflection Principle) *Given a Wiener process W_t and $a, t \geq 0$, then we have that*

$$\mathbb{P}\left(\sup_{0 \leq s \leq t} W_s \geq a \right) = 2\mathbb{P}(W_t \geq a).$$

Theorem 1.13 (Dambis, Dubins-Schwarz Theorem) *[8, 9] Every continuous local martingale M_t is of the form*

$$M_t = M_0 + W_{[M]_t} \text{ for all } t \geq 0,$$

where W_s is a Wiener process.

1.2.3 Log-Concave Functions

Theorem 1.14 (Dinghas, Prékopa, and Leindler) *The convolution of two log-concave functions is log-concave; in particular, any marginal of a log-concave density is log-concave.*

The next lemma is a "reverse" Hölder's inequality (see e.g., [23]).

Lemma 1.15 (Log-Concave Moments) *For any log-concave density p in $\mathbb{R}^n$ and any positive integer k,*

$$\mathbb{E}_{x \sim p} \|x\|^k \leq (2k)^k \cdot \left(\mathbb{E}_{x \sim p} \|x\|^2 \right)^{k/2}.$$

The following inequality bounding the small ball probability is from [3].

Theorem 1.16 ([3, Thm. 10.4.7]) *For any isotropic log-concave density p and any $\epsilon < \epsilon_0$,*

$$\mathbb{P}_{x \sim p}\left(\|x\|_2 \leq \epsilon \sqrt{n} \right) \leq \epsilon^{c\sqrt{n}},$$

where ϵ_0, c are absolute constants.

The following theorem from [6, 24] states that the Poincaré constant is bounded by the KLS constant.

Theorem 1.17 (Poincaré Constant [6, 24]) *For any isotropic log-concave density p in $\mathbb{R}^n$ and any smooth function g, we have*

$$\text{Var}_{x \sim p} g(x) \leq O\left(\psi_n^2\right) \cdot \mathbb{E}_{x \sim p} \|\nabla g(x)\|^2 .$$

An immediate consequence of the above theorem is the following lemma which is central to our analysis. We give a proof of this central lemma for completeness.

Lemma 1.18 *For any matrix A and any isotropic log-concave density p,*

$$\text{Var}_{x \sim p}\left(x^T A x\right) \leq O\left(\psi_r^2\right) \cdot \|A\|_F^2 ,$$

where $r = \text{rank}(A + A^T)$.

Proof Since $x^T A x = x^T A^T x$, we have $\text{Var}_{x \sim p}(x^T A x) = \text{Var}_{x \sim p}(x^T (A + A^T) x)/4$. Now applying Theorem 1.17 to the projection of p onto the orthogonal complement of the null space of matrix A finishes the proof. $\qquad\square$

To prove a upper bound on the KLS constant, it suffices to consider subsets of measure $1/2$. We quote a theorem from [26, Thm 1.8].

Theorem 1.19 *The KLS constant of any log-concave density is achieved by a subset of measure $1/2$.*

The next theorem is an essentially best possible tail bound on large deviations for log-concave densities, due to Paouris [27].

Theorem 1.20 *There exists a universal constant c such that for any isotropic log-concave density p in $\mathbb{R}^n$ and any $t > 1$, $\mathbb{P}_{x \sim p}\left(\|x\| > c \cdot t \sqrt{n}\right) \leq e^{-t\sqrt{n}}$.*

1.2.4 Distance Between Probability Measures

The total variation distance is used in the statement of classical central limit theorem (e.g. [16]).

Definition 1.21 The total variation distance between two probability measures μ and ν in $\mathbb{R}$ is defined by

$$d_{\text{TV}}(\mu, \nu) \overset{\text{def}}{=} \sup_{A \subseteq \mathbb{R}} |\mu(A) - \nu(A)| .$$

The following lemma relates total variation distance to L_1-Wasserstein distance (see Definition 1.5) for isotropic log-concave distributions.

Lemma 1.22 ([25, Prop 1]) *Let μ and ν be isotropic log-concave distributions in $\mathbb{R}$, then we have*

$$d_{\mathrm{TV}}(\mu, \nu) = O(1) \cdot \sqrt{W_1(\mu, \nu)}.$$

Now we relate L_s Wasserstein distance to L_t Wasserstein distance for $1 \leq s \neq t$. By Hölder's inequality, one can show that for any $s \leq t$, we have $W_s(\mu, \nu) \leq W_t(\mu, \nu)$. In the special case where both μ and ν are isotropic log-concave distributions in $\mathbb{R}$, it is shown in [25, Prop 5] that

$$W_t(\mu, \nu)^t \leq O(1) \cdot W_s(\mu, \nu)^s \log^{t-s}\left(\frac{t^t}{W_s(\mu, \nu)^s}\right).$$

In the following, we generalize this result to cases where μ or ν might be the measure of the inner product of two independent isotropic log-concave vectors. This generalization might be useful for future applications. The proof is essentially the same as that in [25] as is therefore postponed to Appendix 1.

Lemma 1.23 *Let μ and ν be two probability measures in $\mathbb{R}$. Suppose one of the following holds:*

1. *Both μ and ν are isotropic log-concave distributions.*
2. *The distribution μ is isotropic log-concave, while ν is the measure of the random variable $\frac{1}{\sqrt{n}}\langle x, y \rangle$ where $x \sim p$ and $y \sim q$ are independent random vectors and p, q are isotropic log-concave distributions in $\mathbb{R}^n$.*
3. *There exist isotropic log-concave distributions p_μ, q_μ, p_ν and q_ν in $\mathbb{R}^n$ such that μ is the measure of the random variable $\frac{1}{\sqrt{n}}\langle x_\mu, y_\mu \rangle$ and ν is the measure of the random variable $\frac{1}{\sqrt{n}}\langle x_\nu, y_\nu \rangle$, where $x_\mu \sim p_\mu$, $y_\mu \sim q_\mu$, $x_\nu \sim p_\nu$ and $y_\nu \sim q_\nu$ are independent random vectors.*

Then there exists a universal constant $c > 0$ such that for any $1 \leq s < t$, we have

$$W_t(\mu, \nu)^t \leq c W_s(\mu, \nu)^s \log^{t-s}\left(\frac{c^t t^{2t}}{W_s(\mu, \nu)^s}\right) + c^t t^{2t} \exp(-c\sqrt{n}).$$

Moreover, the above bound is valid even when the coupling (μ, ν) on the left-hand side is taken to be the best coupling for $W_s(\mu, \nu)$ instead of the best coupling for $W_t(\mu, \nu)$.

1.2.5   Matrix Inequalities

For any symmetric matrix B, we define $|B| = \sqrt{B^2}$, namely, the matrix formed by taking absolute value of all eigenvalues of B.

Lemma 1.24 (Matrix Hölder Inequality) *Given a symmetric matrices A and B and any $s, t \geq 1$ with $s^{-1} + t^{-1} = 1$, we have*

$$\mathrm{Tr}(AB) \leq \left(\mathrm{Tr}\,|A|^s\right)^{1/s} \left(\mathrm{Tr}\,|B|^t\right)^{1/t}.$$

Lemma 1.25 (Lieb-Thirring Inequality [22]) *Given positive semi-definite matrices A and B and $r \geq 1$, we have*

$$\mathrm{Tr}\left(\left(B^{1/2}AB^{1/2}\right)^r\right) \leq \mathrm{Tr}\left(B^{r/2}A^r B^{r/2}\right).$$

Lemma 1.26 ([1, 10]) *Given a symmetric matrix B, a positive semi-definite matrix A and $\alpha \in [0, 1]$, we have*

$$\mathrm{Tr}\left(A^\alpha B A^{1-\alpha} B\right) \leq \mathrm{Tr}\left(AB^2\right).$$

1.2.6 From Generalized CLT to Third Moment Bound

In this subsection, we prove that an improved bound for Generalized CLT implies an improved third moment bound.

Theorem 1.27 *Fix $\epsilon \in (0, 1/2)$. Let p be any isotropic log-concave distribution in $\mathbb{R}^n$, x, y be independent random vectors drawn from p and $G \sim \mathcal{N}(0, n)$. If we have*

$$W_2(\langle x, y \rangle, G)^2 = O\left(n^{1-2\epsilon}\right), \tag{1.2.1}$$

then it follows that

$$\mathbb{E}_{x,y \sim p}\left(\langle x, y \rangle^3\right) = O\left(n^{1.5-\epsilon}\right).$$

We remark that while the equivalence between Generalized CLT and the KLS conjecture in our main theorem (Theorem 1.7) does not hold in a point-wise sense, the result in Theorem 1.27 holds for every isotropic log-concave p.

Proof Let π_2 be the best coupling between $\langle x, y \rangle$ and G in (1.2.1). In the rest of the proof, we use $\mathbb{E}_{\pi_2}$ to denote the expectation where $\langle x, y \rangle$ and G satisfies the coupling π_2. Applying Lemma 1.23, we have

$$\mathbb{E}_{\pi_2}|\langle x, y \rangle, G|^3 = O\left(n^{\frac{3}{2}-2\epsilon}\log n\right).$$

Now we can bound $\mathbb{E}_{x,y\sim p}\langle x, y \rangle^3$ using the coupling π_2 as

$$\mathbb{E}_{x,y\sim p}\langle x, y \rangle^3 = \mathbb{E}_{\pi_2} \left(\langle x, y \rangle - G + G\right)^3$$
$$= \mathbb{E}_{\pi_2} \left(G^3 + 3G^2(\langle x, y \rangle - G) + 3G(\langle x, y \rangle - G)^2 + (\langle x, y \rangle - G)^3\right).$$

The first term is zero due to symmetry. For the second term, we have

$$\mathbb{E}_{\pi_2} G^2(\langle x, y \rangle - G) \leq \sqrt{\mathbb{E}_{G\sim N(0,n)} G^4} \cdot \sqrt{\mathbb{E}_{\pi_2}(\langle x, y \rangle - G)^2}$$
$$= O(n) \cdot O\left(n^{0.5-\epsilon}\right) = O\left(n^{1.5-\epsilon}\right).$$

The last two terms can be bounded similarly as

$$\mathbb{E}_{\pi_2} G(\langle x, y \rangle - G)^2 \leq \left(\mathbb{E}_{G\sim N(0,n)} |G|^3\right)^{\frac{1}{3}} \cdot \left(\mathbb{E}_{\pi_2} |\langle x, y \rangle - G|^3\right)^{\frac{2}{3}}$$
$$= O\left(\sqrt{n}\right) \cdot O\left(n^{1-\frac{4}{3}\epsilon} \log^{\frac{2}{3}} n\right) = O\left(n^{1.5-\epsilon}\right),$$

and

$$\mathbb{E}_{\pi_2}(\langle x, y \rangle - G)^3 \leq \mathbb{E}_{\pi_2} |\langle x, y \rangle - G|^3 = O\left(n^{1.5-2\epsilon} \log n\right) = O\left(n^{1.5-\epsilon}\right).$$

This completes the proof of Theorem 1.27. □

1.3 Stochastic Localization

The key technique used in part of our proofs is the stochastic localization scheme introduced in [10]. The idea is to transform a given log-concave density into one that is proportional to a Gaussian times the original density. This is achieved by a martingale process by modifying the current density infinitesimally according to an exponential in a random direction. By having a martingale, the measures of subsets are maintained in expectation, and the challenge is to control how close they remain to their expectations over time. We now define a simple version of the process we will use, which is the same as in [18].

1.3.1 The Process and Its Basic Properties

Given a distribution with a log-concave density $p(x)$, we start at time $t = 0$ with this distribution and at each time $t > 0$, we apply an infinitesimal change to the density. This is done by picking a random direction from a standard Gaussian.

Definition 1.28 Given a log-concave distribution p, we define the following stochastic differential equation:

$$c_0 = 0, \quad dc_t = dW_t + \mu_t dt, \tag{1.3.1}$$

where the probability distribution p_t, the mean μ_t and the covariance A_t are defined by

$$p_t(x) = \frac{e^{c_t^T x - \frac{t}{2}\|x\|_2^2} p(x)}{\int_{\mathbb{R}^n} e^{c_t^T y - \frac{t}{2}\|y\|_2^2} p(y) dy}, \quad \mu_t = \mathbb{E}_{x \sim p_t} x, \quad A_t = \mathbb{E}_{x \sim p_t} (x - \mu_t)(x - \mu_t)^T.$$

The following basic lemmas will be used in the analysis. For a more rigorous account of the construction and further details of the process, the reader is referred to [11, 18, 20]

Lemma 1.29 *For any $x \in \mathbb{R}^n$, we have $dp_t(x) = (x - \mu_t)^T dW_t \, p_t(x)$.*

Next we state the change of the mean and the covariance matrix.

Lemma 1.30 $d\mu_t = A_t dW_t$ and $dA_t = \int_{\mathbb{R}^n} (x - \mu_t)(x - \mu_t)^T \left((x - \mu_t)^T dW_t\right) p_t$ $(x) dx - A_t^2 dt$.

1.3.2 Bounding the KLS Constant

The following lemmas from [18] are used to bound the KLS constant by the spectral norm of the covariance matrix at time t. First, we bound the measure of a set of initial measure $\frac{1}{2}$.

Lemma 1.31 *For any set $E \subset \mathbb{R}^n$ with $\int_E p(x) dx = \frac{1}{2}$ and $t \geq 0$, we have that*

$$\mathbb{P}\left(\frac{1}{4} \leq \int_E p_t(x) dx \leq \frac{3}{4}\right) \geq \frac{9}{10} - \mathbb{P}\left(\int_0^t \|A_s\|_{\text{op}} \, ds \geq \frac{1}{64}\right).$$

At time t, the distribution is t-strongly log-concave and it is known that it has KLS constant $O\left(t^{-1/2}\right)$. The following isoperimetric inequality was proved in [7] and was also used in [10].

Theorem 1.32 *Let $h(x) = f(x)e^{-\frac{t}{2}\|x\|_2^2} / \int f(y)e^{-\frac{t}{2}\|y\|_2^2} dy$ where $f : \mathbb{R}^n \to \mathbb{R}_+$ is an integrable log-concave function. Then h is log-concave and for any measurable subset S of $\mathbb{R}^n$,*

$$\int_{\partial S} h(x) dx = \Omega\left(\sqrt{t}\right) \cdot \min\left\{\int_S h(x) dx, \int_{\mathbb{R}^n \setminus S} h(x) dx\right\}.$$

In other words, the KLS constant of h is $O\left(t^{-1/2}\right)$.

This gives a bound on the KLS constant.

Lemma 1.33 *Given a log-concave distribution* p, *let* A_t *be given by Definition 1.28 using initial distribution* p. *Suppose that there is* $T > 0$ *such that*

$$\mathbb{P}\left(\int_0^T \|A_s\|_{\mathrm{op}}\, ds \le \frac{1}{64}\right) \ge \frac{3}{4},$$

then we have $\psi_p = O\left(T^{-1/2}\right)$.

Thus to prove a bound on ψ_p, it suffices to give an upper bound on $\|A_t\|_{\mathrm{op}}$. The potential function we will use to bound $\|A_t\|_{\mathrm{op}}$ is $\Phi_t = \mathrm{Tr}((A_t - I)^q)$ for some even integer q. We give the detailed analysis in Sect. 1.4.

The following result from [18] will be useful. It shows that the operator norm stays bounded up to a certain time with probability close to 1.

Lemma 1.34 ([18], Lemma 58) *Assume for* $k \ge 1$, $\psi_p = O(n^{1/2k})$ *for any isotropic log-concave distribution* p *in* $\mathbb{R}^n$. *There is a constant* $c \ge 0$ *s.t. for any*

$$0 \le T \le \frac{1}{c \cdot k \cdot (\log n)^{1-\frac{1}{k}} \cdot n^{1/k}},$$

we have

$$\mathbb{P}\left[\max_{t \in [0,T]} \|A_t\|_{\mathrm{op}} \ge 2\right] \le 2\exp\left(-\frac{1}{cT}\right). \tag{1.3.2}$$

1.3.3 Bounding the Potential

In order to bound the potential $\Phi_t = \mathrm{Tr}((A_t - I)^q)$, we bound its derivative. We go from the derivative to the potential itself via the following lemma, which might also be useful in future applications.

Lemma 1.35 *Let* $\{\Phi_t\}_{t\ge 0}$ *be an* n-*dimensional Itô process with* $\Phi_0 \le \frac{U}{2}$ *and* $d\Phi_t = \delta_t dt + v_t^T dW_t$. *Let* $T > 0$ *be some fixed time,* $U > 0$ *be some target upper bound, and* f *and* g *be some auxiliary functions such that for all* $0 \le t \le T$

1. $\delta_t \le f(\Phi_t)$ *and* $\|v_t\|_2 \le g(\Phi_t)$,
2. *Both* $f(\cdot)$ *and* $g(\cdot)$ *are non-negative non-decreasing functions,*
3. $f(U) \cdot T \le \frac{U}{8}$ *and* $g(U) \cdot \sqrt{T} \le \frac{U}{8}$.

Then, we have the following upper bound on Φ_t:

$$\mathbb{P}\left[\max_{t \in [0,T]} \Phi_t \ge U\right] \le 0.01.$$

Proof We denote the Itô process formed by the martingale term as $\{Y_t\}_{t\geq 0}$, i.e. $Y_0 = 0$ and $dY_t = v_t^T dW_t$. We first show that in order to control Φ_t, it suffices to control Y_t.

Claim 1.36 For any $0 \leq t_0 \leq T$, if $\max_{t \in [0,t_0]} Y_t \leq \frac{U}{3}$, then we have

$$\max_{t \in [0,t_0]} \Phi_t \leq U.$$

Proof of Claim 1.36 Assume for the purpose of contradiction that $\max_{t \in [0,t_0]} \Phi_t > U$. Denote $t' = \inf\{t \in [0,t_0] | \Phi_t \geq U\}$. It follows that for any $t \in [0, t']$, we have $\Phi_t \leq U$ and $f(\Phi_t) \cdot t' \leq f(U) \cdot T \leq \frac{U}{8}$. It follows that

$$\Phi_t \leq \Phi_0 + \frac{U}{8} + Y_t < U,$$

which leads to a contradiction.   □

Since Y_t is a martingale, it follows from Theorem 1.13 that there exists a Wiener process $\{B_t\}_{t\geq 0}$ such that $Y_t = B_{[Y]_t}$, for all $t \geq 0$. The next claim bounds Y_t using B_t.

Claim 1.37 If $\max_{t \in [0,U^2/64]} B_t \leq \frac{U}{3}$, then we have

$$\max_{t \in [0,T]} Y_t \leq U/3,$$

Proof of Claim 1.37 Assume for the purpose of contradiction that $\max_{t \in [0,T]} Y_t \geq \frac{U}{3}$. Define t_0 as the first time when Y_t becomes at least $\frac{U}{3}$. By definition, for any $t \in [0, t_0]$, $Y_t \leq \frac{U}{3}$. Using Claim 1.36, we have $\max_{t \in [0,t_0]} \Phi_t \leq U$. It follows that

$$[Y]_{t_0} = \int_0^{t_0} \|v_t\|_2^2 \, dt \leq T \cdot g^2(U) \leq \frac{U^2}{64}.$$

This implies that

$$Y_{t_0} = B_{[Y]_{t_0}} \leq \max_{t \in [0,U^2/64]} B_t \leq \frac{U}{3},$$

which leads to a contradiction.   □

Now it suffices to bound the probability that the Wiener process $\{B_t\}_{t\geq 0}$ exceeds $U/3$ in the time period $[0, U^2/64]$. Using the reflection principle in Lemma 1.12, we have

$$\Pr\left[\max_{t \in [0,T]} \Phi_t \geq U\right] \leq \Pr\left[\max_{t \in [0,U^2/64]} B_t > U/3\right] = 2\Pr\left[B_{U^2/64} > U/3\right] \leq 0.01.$$

□

1.4   From Third Moment Bound to KLS

In this section, we show that an improved third moment bound implies an improved bound on the KLS constant. Theorems 1.9 and 1.27 together imply the first part of Theorem 1.7.

Theorem 1.9 *Fix $\epsilon \in (0, 1/2)$. If for every isotropic log-concave distribution p in $\mathbb{R}^n$ and independent vectors $x, y \sim p$, we have*

$$\mathbb{E}_{x,y \sim p}\left(\langle x, y \rangle^3\right) = O\left(n^{1.5-\epsilon}\right), \tag{1.1.2}$$

then for any $\delta > 0$, we have $\psi_n = O\left(n^{1/4-\epsilon/2+\delta}\right)$.

The rest of this section is devoted to proving Theorem 1.9. Throughout this section, we assume the condition in Theorem 1.9 holds, i.e. for every isotropic log-concave distribution p in $\mathbb{R}^n$ and independent vectors $x, y \sim p$, one has

$$\mathbb{E}_{x,y \sim p}\left(\langle x, y \rangle^3\right) = O\left(n^{1.5-\epsilon}\right). \tag{1.4.1}$$

1.4.1   Tensor Inequalities

The proof of Theorem 1.9 is based on the potential function $\Phi_t = \mathrm{Tr}\left((A_t - I)^q\right)$ for some even integer q. This potential is the one of the key technical differences between this paper and previous work using stochastic localization, which used $\mathrm{Tr}(A_t^q)$ [10, 19]. The proof of a tight log-Sobolev inequality [21] used a Stieltjes-type potential function, $\mathrm{Tr}((uI - A)^{-q})$ to avoid logarithmic factors. The potential we use here, $\mathrm{Tr}\left((A_t - I)^q\right)$ allows us to track how close A_t is to I (not just bounding how large A_t is). For example, in Lemma 1.43, we bound the derivative of the potential Φ_t by some powers of Φ_t. Since Φ_t is 0 initially, this gives a significantly tighter bound around $t = 0$ (compared to $\mathrm{Tr}(A_t^q)$). We will discuss this again in the course of the proof.

For the analysis we define the following tensor and derive some of its properties.

Definition 1.38 (3-Tensor) For an isotropic log-concave distribution p in $\mathbb{R}^n$ and symmetric matrices A, B and C, define

$$T_p(A, B, C) = \mathbb{E}_{x,y \sim p}\left(x^T A y\right)\left(x^T B y\right)\left(x^T C y\right)$$

We drop the subscript p to indicate the worst case bound over all isotropic log-concave distributions

$$T(A, B, C) \overset{\text{def}}{=} \sup_{\text{isotropic log-concave } p} \mathbb{E}_{x,y \sim p}\left(x^T A y\right)\left(x^T B y\right)\left(x^T C y\right)$$

It is clear from the definition that T is invariant under permutation of A, B and C. In the rest of this subsection, we give a few tensor inequalities that will be used throughout the rest of our proofs. The proofs of these tensor inequalities are postponed to Appendix 2.

Lemma 1.39 *For any $A_1, A_2, A_3 \succeq 0$, we have that $T(A_1, A_2, A_3) \geq 0$ and for any symmetric matrices B_1, B_2, B_3, we have that*

$$T(B_1, B_2, B_3) \leq T\left(|B_1|, |B_2|, |B_3|\right).$$

In the next lemma, we collect tensor inequalities that will be useful for later proofs.

Lemma 1.40 *Suppose that $\psi_k \leq \alpha k^\beta$ for all $k \leq n$ for some fixed $0 \leq \beta \leq \frac{1}{2}$ and $\alpha \geq 1$. For any isotropic log-concave distribution p in $\mathbb{R}^n$ and symmetric matrices A and B, we have that*

1. $T(A, I, I) \leq T(I, I, I) \cdot \|A\|_{\mathrm{op}}$.
2. $T(A, I, I) \leq O\left(\psi_n^2\right) \cdot \mathrm{Tr}\,|A|$.
3. $T(A, B, I) \leq O\left(\psi_r^2\right) \cdot \|B\|_{\mathrm{op}} \,\mathrm{Tr}\,|A|$ *where* $r = \min(2 \cdot \mathrm{rank}(B), n)$.
4. $T(A, B, I) \leq O\left(\alpha^2 \log n\right) \cdot \left(\mathrm{Tr}\,|B|^{1/(2\beta)}\right)^{2\beta} \mathrm{Tr}\,|A|$.
5. $T(A, B, I) \leq \left(T\left(|A|^s, I, I\right)\right)^{1/s} \cdot \left(T\left(|B|^t, I, I\right)\right)^{1/t}$, *for any $s, t \geq 1$ with* $s^{-1} + t^{-1} = 1$.

Lemma 1.41 *For any positive semi-definite matrices A, B, C and any $\alpha \in [0, 1]$, then*

$$T\left(B^{1/2} A^\alpha B^{1/2}, B^{1/2} A^{1-\alpha} B^{1/2}, C\right) \leq T\left(B^{1/2} A B^{1/2}, B, C\right).$$

1.4.2 Derivatives of the Potential

The next lemma computes the derivative of $\Phi_t = \mathrm{Tr}((A_t - I)^q)$, as done in [18]. For the reader's convenience, we include a proof here.

Lemma 1.42 *Let A_t be defined by Definition 1.28. For any integer $q \geq 2$, we have that*

$$d\mathrm{Tr}\left((A_t - I)^q\right) = q \cdot \mathbb{E}_{x \sim p_t} (x - \mu_t)^T (A_t - I)^{q-1} (x - \mu_t)(x - \mu_t)^T dW_t$$

$$- q \cdot \mathrm{Tr}\left((A_t - I)^{q-1} A_t^2\right) dt$$

$$+ \frac{q}{2} \cdot \sum_{\alpha+\beta=q-2} \mathbb{E}_{x,y \sim p_t} (x - \mu_t)^T (A_t - I)^\alpha (y - \mu_t)$$

$$\times (x - \mu_t)^T (A_t - I)^\beta (y - \mu_t)(x - \mu_t)^T (y - \mu_t) dt.$$

Proof Let $\Phi(X) = \mathrm{Tr}((X - I)^q)$. Then the first and second-order directional derivatives of Φ at X is given by

$$\left.\frac{\partial \Phi}{\partial X}\right|_H = q \cdot \mathrm{Tr}\left((X - I)^{q-1} H\right) \quad \text{and}$$

$$\left.\frac{\partial^2 \Phi}{\partial X \partial X}\right|_{H_1, H_2} = q \cdot \sum_{k=0}^{q-2} \mathrm{Tr}\left((X - I)^k H_2 (X - I)^{q-2-k} H_1\right).$$

Using these and Itô's formula, we have that

$$d\mathrm{Tr}((A_t - I)^q) = q \cdot \mathrm{Tr}\left((A_t - I)^{q-1} dA_t\right)$$

$$+ \frac{q}{2} \cdot \sum_{\alpha+\beta=q-2} \sum_{ijkl} \mathrm{Tr}\left((A_t - I)^\alpha e_{ij}(A_t - I)^\beta e_{kl}\right) d[A_{ij}, A_{kl}]_t,$$

where e_{ij} is the matrix that is 1 in the entry (i, j) and 0 otherwise, and A_{ij} is the real-valued stochastic process defined by the (i, j)th entry of A_t.

Using Lemmas 1.30 and 1.29, we have that

$$dA_t = \mathbb{E}_{x \sim p_t}(x - \mu_t)(x - \mu_t)^T (x - \mu_t)^T dW_t - A_t A_t dt$$

$$= \mathbb{E}_{x \sim p_t}(x - \mu_t)(x - \mu_t)^T (x - \mu_t)^T e_z dW_{t,z} - A_t A_t dt, \tag{1.4.2}$$

where $W_{t,z}$ is the zth coordinate of W_t. Therefore,

$$d[A_{ij}, A_{kl}]_t = \sum_z \left(\mathbb{E}_{x \sim p_t}(x - \mu_t)_i (x - \mu_t)_j (x - \mu_t)^T e_z\right)$$

$$\times \left(\mathbb{E}_{x \sim p_t}(x - \mu_t)_k (x - \mu_t)_l (x - \mu_t)^T e_z\right) dt$$

$$= \mathbb{E}_{x, y \sim p_t}(x - \mu_t)_i (x - \mu_t)_j (y - \mu_t)_k (y - \mu_t)_l (x - \mu_t)^T$$

$$\times (y - \mu_t) dt. \tag{1.4.3}$$

Using the formula for dA_t (1.4.2) and $d[A_{ij}, A_{kl}]_t$ (1.4.3), we have that

$$d\mathrm{Tr}\left((A_t - I)^q\right) = q \cdot \mathbb{E}_{x \sim p_t}(x - \mu_t)^T (A_t - I)^{q-1}(x - \mu_t)(x - \mu_t)^T dW_t$$

$$- q \cdot \mathrm{Tr}\left((A_t - I)^{q-1} A_t^2\right) dt$$

$$+ \frac{q}{2} \cdot \sum_{\alpha+\beta=q-2} \sum_{ijkl} \mathrm{Tr}\left((A_t - I)^\alpha e_{ij}(A_t - I)^\beta e_{kl}\right) \mathbb{E}_{x, y \sim p_t}$$

$$\times (x - \mu_t)_i (x - \mu_t)_j (y - \mu_t)_k (y - \mu_t)_l (x - \mu_t)^T (y - \mu_t) dt$$

$$= q \cdot \mathbb{E}_{x \sim p_t} (x - \mu_t)^T (A_t - I)^{q-1} (x - \mu_t)(x - \mu_t)^T dW_t$$

$$- q \cdot \mathrm{Tr}\left((A_t - I)^{q-1} A_t^2\right) dt$$

$$+ \frac{q}{2} \cdot \sum_{\alpha + \beta = q-2} \mathbb{E}_{x, y \sim p_t} (x - \mu_t)^T (A_t - I)^{\alpha} (y - \mu_t)$$

$$\times (x - \mu_t)^T (A_t - I)^{\beta} (y - \mu_t)(x - \mu_t)^T (y - \mu_t) dt.$$

$\square$

1.4.3 Bounding the Potential

The derivative of the potential has drift (dt) and stochastic/Martingale (dW_t) terms. The next lemma bounds the drift and Martingale parts of the change in the potential by tensor quantities. We will then bound each one separately.

Lemma 1.43 *Let A_t and p_t be defined as in Definition 1.28. Let $\Phi_t = \mathrm{Tr}((A_t - I)^q)$ for some even integer $q \geq 2$, then we have that $d\Phi_t = \delta_t dt + v_t^T dW_t$ with*

$$\delta_t \leq \frac{1}{2} q(q-1) \cdot T\left(A_t(A_t - I)^{q-2}, A_t, A_t\right) + 2q \cdot \left(\Phi_t^{1+\frac{1}{q}} + \Phi_t^{1-\frac{1}{q}} n^{\frac{1}{q}}\right)$$

and

$$\|v_t\|_2 \leq q \cdot \left\|\mathbb{E}_{x \sim p}(x - \mu_t)^T (A - I)^{q-1}(x - \mu_t)(x - \mu_t)^T\right\|_2.$$

Proof By Lemma 1.42, we have

$$d\Phi_t = q \cdot \mathbb{E}_{x \sim p_t} (x - \mu_t)^T (A_t - I)^{q-1} (x - \mu_t)(x - \mu_t)^T dW_t$$

$$- q \cdot \mathrm{Tr}\left((A_t - I)^{q-1} A_t^2\right) dt$$

$$+ \frac{q}{2} \cdot \sum_{\alpha + \beta = q-2} \mathbb{E}_{x, y \sim p_t} (x - \mu_t)^T (A_t - I)^{\alpha} (y - \mu_t)(x - \mu_t)^T$$

$$\times (A_t - I)^{\beta} (y - \mu_t)(x - \mu_t)^T (y - \mu_t) dt$$

$$= q \cdot \mathbb{E}_{x \sim p}(x - \mu_t)^T (A - I)^{q-1} (x - \mu_t)(x - \mu_t)^T dW_t$$

$$- q \cdot \mathrm{Tr}\left((A_t - I)^{q-1} A_t^2\right) dt$$

$$+ \frac{q}{2} \cdot \sum_{\alpha + \beta = q-2} \mathbb{E}_{x, y \sim \tilde{p}_t} x^T A_t (A_t - I)^{\alpha} y x^T A_t (A_t - I)^{\beta} y x^T$$

$$\times A_t y dt \stackrel{\mathrm{def}}{=} \delta_t dt + v_t^T dW_t.$$

where $\tilde{p}_t$ is the isotropic correspondence of p_t defined by $\tilde{p}_t(x) = p\left(A_t^{1/2}x + \mu_t\right)$, $\delta_t dt$ is the drift term in $d\Phi_t$ and $v_t^T dW_t$ is the martingale term in $d\Phi_t$.

For the drift term $\alpha_t dt$, we have

$$\delta_t \leq \frac{q}{2} \cdot \sum_{\alpha+\beta=q-2} T\left(A_t(A_t - I)^\alpha, A_t(A_t - I)^\beta, A_t\right) - q \cdot \text{Tr}\left((A_t - I)^{q-1}A_t^2\right).$$

The first drift term is

$$\frac{q}{2} \cdot \sum_{\alpha+\beta=q-2} T\left(A_t(A_t - I)^\alpha, A_t(A_t - I)^\beta, A_t\right)$$

$$\leq \frac{q}{2} \cdot \sum_{\alpha+\beta=q-2} T\left(A_t |A_t - I|^\alpha, A_t |A_t - I|^\beta, A_t\right) \qquad \text{(Lem 1.39)}$$

$$\leq \frac{q}{2} \cdot \sum_{\alpha+\beta=q-2} T\left(A_t |A_t - I|^{q-2}, A_t, A_t\right) \qquad \text{(Lem 1.41)}$$

$$= \frac{q(q-1)}{2} \cdot T\left(A_t(A_t - I)^{q-2}, A_t, A_t\right).$$

For the second drift term, since q is even, we have that

$$-q \cdot \text{Tr}\left((A_t - I)^{q-1}A_t^2\right) \leq q \cdot \text{Tr}\left(|A_t - I|^{q-1}(A_t - I + I)^2\right)$$

$$\leq 2q \cdot \text{Tr}\left(|A_t - I|^{q+1}\right) + 2q \cdot \text{Tr}\left(|A_t - I|^{q-1}\right)$$

$$\leq 2q \cdot \Phi_t^{1+\frac{1}{q}} + 2q \cdot \Phi_t^{1-\frac{1}{q}} n^{\frac{1}{q}}.$$

For the Martingale term $v_t^T dW_t$, we note that

$$\|v_t\|_2 = q \cdot \left\| \mathbb{E}_{x \sim p}(x - \mu_t)^T (A - I)^{q-1}(x - \mu_t)(x - \mu_t)^T \right\|_2.$$

$\square$

The Martingale term is relatively straightforward to bound. We use the following lemma from [18] in our analysis.

Lemma 1.44 ([18, Lem 25]) *Given a log-concave distribution p with mean μ and covariance A. For any positive semi-definite matrix C, we have that*

$$\left\| \mathbb{E}_{x \sim p}(x - \mu)(x - \mu)^T C(x - \mu) \right\|_2 = O\left(\|A\|_{\text{op}}^{1/2} \cdot \text{Tr}\left(A^{1/2}CA^{1/2}\right)\right).$$

Lemma 1.45 *Let p_t be the log-concave distribution at time t with covariance matrix A_t. Let $\Phi_t = \text{Tr}((A_t - I)^q)$ for some even integer $q \geq 2$ and $d\Phi_t = $*

$\delta_t dt + v_t^T dW_t$. Assume $\Phi_t \le n$. Then,

$$\|v_t\|_2 \le q \cdot \left\| \mathbb{E}_{x \sim p_t} (x - \mu_t)^T (A - I)^{q-1} (x - \mu_t)(x - \mu_t)^T \right\|_2$$

$$\le O(q) \cdot \left(\Phi_t^{1 - \frac{1}{2q}} n^{\frac{1}{q}} + n^{\frac{1}{q}} \right).$$

Proof Note that

$$\left\| \mathbb{E}_{x \sim p} (x - \mu_t)^T (A_t - I)^{q-1} (x - \mu_t)(x - \mu_t)^T \right\|_2$$

$$\le O(1) \cdot \|A_t\|_{op}^{1/2} \operatorname{Tr} \left| A_t^{1/2} (A_t - I)^{q-1} A_t^{1/2} \right| \qquad \text{(Lem 1.44)}$$

$$\le O(1) \cdot \|A_t\|_{op}^{1/2} \operatorname{Tr} |A_t - I|^{q-1} + O(1) \cdot \|A_t\|_{op}^{1/2} \operatorname{Tr} |A_t - I|^q$$

$$\le O\left(1 + \Phi_t^{\frac{1}{2q}} \right) \cdot \Phi_t^{1 - \frac{1}{q}} n^{\frac{1}{q}} + O\left(1 + \Phi_t^{\frac{1}{2q}} \right) \cdot \Phi_t$$

$$\le O\left(\Phi_t^{1 - \frac{1}{2q}} n^{\frac{1}{q}} + \Phi_t^{1 + \frac{1}{2q}} + n^{\frac{1}{q}} \right).$$

$\square$

Next we bound the drift term. This takes more work. We write

$$\delta_t \le \frac{1}{2} q(q - 1) \delta_t^{(1)} + q \delta_t^{(2)},$$

where

$$\delta_t^{(1)} = T\left(A_t (A_t - I)^{q-2}, A_t, A_t \right) \quad \text{and} \quad \delta_t^{(2)} = \Phi_t^{1 + \frac{1}{q}} + \Phi_t^{1 - \frac{1}{q}} n^{\frac{1}{q}}.$$

We bound $\delta_t^{(1)}$ in the following lemma. This is the core lemma which needs several tensor properties and bounds. It is also the reason we use $\operatorname{Tr}((A_t - I)^q)$ as the potential. Specifically, using this potential lets us write $A - I$ as the sum of two matrices one with small eigenvalues and the other of low rank, by choosing the threshold for "small" eigenvalue appropriately.

Lemma 1.46 *Suppose that $\psi_k \le \alpha k^\beta$ for all $k \le n$ for some $\alpha \ge 1$ and β s.t. $1/4 - \epsilon/2 \le \beta \le 1/4$. Let $\Phi = \operatorname{Tr}((A - I)^q)$ for some even integer $q \ge \frac{1}{2\beta}$ and $\Lambda = 4\beta + 2\epsilon - 1$. Assume $\Phi \le n$. Then*

$$\delta^{(1)} \le O(\alpha^2) \cdot \Phi n^{2\beta} \cdot \left[n^{-\frac{1}{q}} \Phi^{\frac{1}{q}} \log n + n^{-\frac{\Lambda}{4q}} \cdot n^{\frac{2}{q}} \Phi^{-\frac{2}{q}} \right].$$

Proof We have that

$$
\begin{aligned}
\delta^{(1)} &= T\left(A(A-I)^{q-2}, A, A\right) \\
&= T\left((A-I)^{q-1} + (A-I)^{q-2}, A-I+I, A-I+I\right) \\
&\le T\left(|A-I|^{q-1}, |A-I|, |A-I|\right) + 2T\left(|A-I|^{q-1}, |A-I|, I\right) \\
&\quad + T\left(|A-I|^{q-1}, I, I\right) \hspace{4cm} \text{(Lem 1.39)} \\
&\quad + T\left((A-I)^{q-2}, |A-I|, |A-I|\right) + 2T\left((A-I)^{q-2}, |A-I|, I\right) \\
&\quad + T\left((A-I)^{q-2}, I, I\right) \\
&\le T\left(|A-I|^{q-1}, |A-I|, |A-I|\right) + 3T\left(|A-I|^{q-1}, |A-I|, I\right) \\
&\quad + 3T\left(|A-I|^{q-1}, I, I\right) + T\left((A-I)^{q-2}, I, I\right) \hspace{1.5cm} \text{(Lem 1.41)} \\
&\overset{\Delta}{=} \delta_1^{(1)} + 3\delta_2^{(1)} + 3\delta_3^{(1)} + \delta_4^{(1)}.
\end{aligned}
$$

We first bound $\delta_1^{(1)}$ as follows

$$
\begin{aligned}
\delta_1^{(1)} &= T\left(|A-I|^{q-1}, |A-I|, |A-I|\right) \\
&\le T\left(|A-I|^q, |A-I|, I\right) \hspace{4cm} \text{(Lem 1.41)} \\
&\le O(\alpha^2 \log n) \cdot \Phi\left(\mathrm{Tr}|A-I|^{1/2\beta}\right)^{2\beta} \hspace{2cm} \text{(Lem 1.40.4)} \\
&\le O(\alpha^2 \log n) \cdot \Phi\left(\left(\mathrm{Tr}|A-I|^q\right)^{\frac{1}{2\beta q}} n^{1-\frac{1}{2\beta q}}\right)^{2\beta} \hspace{1cm} \text{(Lem 1.24)} \\
&\le O(\alpha^2 \log n) \cdot n^{2\beta - \frac{1}{q}} \Phi^{1+\frac{1}{q}}.
\end{aligned}
$$

For $\delta_2^{(1)}$, we write

$$
|A-I| = B_1 + B_2,
$$

where B_1 consists of the eigen-components of $|A-I|$ with eigenvalues at most η and B_2 is the remaining part. Then we can bound $\delta_2^{(1)}$ as follows

$$
\begin{aligned}
\delta_2^{(1)} &= T\left(B_1^{q-1}, B_1, I\right) + T\left(B_1^{q-1}, B_2, I\right) + T\left(B_2^{q-1}, B_1, I\right) \\
&\quad + T\left(B_2^{q-1}, B_2, I\right). \hspace{5cm} \text{(1.4.4)}
\end{aligned}
$$

The first term in Eq. (1.4.4) can be bounded as

$$T\left(B_1^{q-1}, B_1, I\right) \le T\left(B_1^q, I, I\right) \qquad \text{(Lem 1.41)}$$

$$\le T\left(I, I, I\right) \cdot \|B_1\|^q \qquad \text{(Lem 1.40.1)}$$

$$\le O\left(\eta^q n^{1.5-\epsilon}\right).$$

The second term in Eq. (1.4.4) is bounded as

$$T\left(B_1^{q-1}, B_2, I\right) \le T\left(B_1^q, I, I\right)^{\frac{q-1}{q}} \cdot T\left(B_2^q, I, I\right)^{\frac{1}{q}} \qquad \text{(Lem 1.40.5)}$$

$$\le O\left(\eta^q n^{1.5-\epsilon}\right)^{\frac{q-1}{q}} \cdot O\left(\psi_n^2 \Phi\right)^{\frac{1}{q}}$$

$$\text{(Lem 1.40.1 and Lem 1.40.2)}$$

$$= O(1) \cdot \alpha^{\frac{2}{q}} \eta^{q-1} n^{\frac{(1.5-\epsilon)(q-1)}{q}+\frac{2\beta}{q}} \Phi^{\frac{1}{q}},$$

where we used $\mathrm{Tr}\left(B_2^q\right) \le \mathrm{Tr}((A-I)^q) \le \Phi$ in the last line. For the third term in Eq. (1.4.4), we have

$$T\left(B_2^{q-1}, B_1, I\right) \le T\left(B_2^q, I, I\right)^{\frac{q-1}{q}} \cdot T\left(B_1^q, I, I\right)^{\frac{1}{q}} \qquad \text{(Lem 1.40.5)}$$

$$\le O\left(\psi_n^2 \Phi\right)^{\frac{q-1}{q}} \cdot O\left(\eta^q n^{1.5-\epsilon}\right)^{\frac{1}{q}}$$

$$\text{(Lem 1.40.1 and Lem 1.40.2)}$$

$$= O(1) \cdot \alpha^{\frac{2(q-1)}{q}} \eta n^{2\beta-\frac{2\beta}{q}+(1.5-\epsilon)\cdot\frac{1}{q}} \Phi^{\frac{q-1}{q}}.$$

For the last term in Eq. (1.4.4), let P be the orthogonal projection from $\mathbb{R}^n$ to the range of B_2. Notice that $\mathrm{rank}(B_2) \le \frac{\Phi}{\eta^q}$ because each positive eigenvalue of B_2 is at least η. We have

$$T\left(B_2^{q-1}, B_2, I\right) = T\left(PB_2^{q-1}P, PB_2P, I\right)$$

$$\le T\left(PB_2^q P, P, I\right) \qquad \text{(Lem 1.41)}$$

$$\le O\left(\psi_{2\cdot\mathrm{rank}(B_2)}^2\right) \cdot \Phi \qquad \text{(Lem 1.40.3)}$$

$$= O\left(\frac{\alpha^2 \Phi^{1+2\beta}}{\eta^{2\beta q}}\right).$$

Summing up these four terms, we get

$$\delta_2^{(1)} \leq O(1) \cdot \left[\eta^q n^{1.5-\epsilon} + \alpha^{\frac{2}{q}} \eta^{q-1} n^{-\frac{(1.5-\epsilon)(q-1)}{q} + \frac{2\beta}{q}} \Phi^{\frac{1}{q}} \right.$$

$$+ \alpha^{\frac{2(q-1)}{q}} \eta n^{2\beta - \frac{2\beta}{q} + (1.5-\epsilon) \cdot \frac{1}{q}} \Phi^{\frac{q-1}{q}} + \left. \frac{\alpha^2 \Phi^{1+2\beta}}{\eta^{2\beta q}} \right]$$

$$\leq O(\alpha^2) \cdot \left[\eta^q n^{1.5-\epsilon} + \eta^{q-1} n^{\frac{2\beta + (1.5-\epsilon)(q-1)}{q}} \Phi^{\frac{1}{q}} + \eta n^{2\beta - \frac{2\beta}{q} + (1.5-\epsilon) \cdot \frac{1}{q}} \right.$$

$$\times \left. \Phi^{\frac{q-1}{q}} + \frac{\Phi^{1+2\beta}}{\eta^{2\beta q}} \right].$$

It turns out that when $1/4 - \epsilon/2 \leq \beta \leq 1/4$, the last two terms dominate the first two terms (which is justified shortly). Balancing the last two terms, we choose $\eta = \Phi^{\frac{1}{q}} n^{-\frac{2\beta(q-1)+1.5-\epsilon}{q(1+2\beta q)}}$, and this gives

$$\delta_2^{(1)} \leq O(\alpha^2) \cdot \left[\Phi n^{2\beta} \cdot n^{\frac{\beta(1-4\beta-2\epsilon)q}{1+2\beta q}} + \Phi n^{2\beta} \cdot n^{\frac{\beta(1-4\beta-2\epsilon)(q-1)}{1+2\beta q}} \right.$$

$$+ \left. \Phi n^{2\beta} \cdot n^{\frac{\beta(1-4\beta-2\epsilon)}{1+2\beta q}} + \Phi n^{2\beta} \cdot n^{\frac{\beta(1-4\beta-2\epsilon)}{1+2\beta q}} \right].$$

Since $\beta \geq 1/4 - \epsilon/2$, $\beta(1 - 4\beta - 2\epsilon) \leq 0$ which implies that the last two terms dominate the first two terms in this case. We therefore have

$$\delta_2^{(1)} \leq O(\alpha^2) \cdot \Phi n^{2\beta} \cdot n^{\frac{\beta(1-4\beta-2\epsilon)}{1+2\beta q}}.$$

The third term $\delta_3^{(1)}$ is bounded as

$$\delta_3^{(1)} = T\left(|A - I|^{q-1}, I, I \right)$$

$$= T\left(B_1^{q-1}, I, I \right) + T\left(B_2^{q-1}, I, I \right)$$

$$\leq O(1) \cdot \left(\eta^{q-1} n^{1.5-\epsilon} + \alpha^2 n^{2\beta} \Phi / \eta \right) \qquad \text{(Lem 1.40.1 and Lem 1.40.2)}$$

$$\leq O(\alpha^2) \cdot n^{\frac{2\beta(q-1)+1.5-\epsilon}{q}} \Phi^{\frac{q-1}{q}},$$

where the last line is by choosing $\eta = \left(n^{2\beta-1.5+\epsilon} \Phi \right)^{1/q}$. The final term $\delta_4^{(1)}$ is bounded as

$$\delta_4^{(1)} = T\left(|A - I|^{q-2}, I, I \right)$$

$$= T\left(B_1^{q-2}, I, I\right) + T\left(B_2^{q-2}, I, I\right)$$

$$\leq O(1) \cdot \left(\eta^{q-2} n^{1.5-\epsilon} + \alpha^2 n^{2\beta} \Phi/\eta^2\right) \qquad \text{(Lem 1.40.1 and Lem 1.40.2)}$$

$$\leq O(\alpha^2) \cdot n^{\frac{2\beta(q-2)+2(1.5-\epsilon)}{q}} \Phi^{\frac{q-2}{q}}.$$

Combining all the terms we have

$$\delta^{(1)} \leq O(\alpha^2) \cdot \Phi n^{2\beta} \cdot \left[n^{-\frac{1}{q}} \Phi^{\frac{1}{q}} \log n + n^{-\frac{\beta}{1+2\beta q}} \cdot \Lambda + n^{-\frac{\Lambda}{2q}} n^{\frac{1}{q}} \Phi^{-\frac{1}{q}} + n^{-\frac{\Lambda}{q}} n^{\frac{2}{q}} \Phi^{-\frac{2}{q}}\right].$$

Simplifying the above with the assumptions $\Phi \leq n$ and $q \geq \frac{1}{2\beta}$ finishes the proof of the lemma. $\qquad\square$

1.4.4 Proof of Theorem 1.9

We note that $\Phi_0 = 0$. Using the bounds we have, we will show that when q is taken as the smallest even integer greater than $\max\{8, \lceil 1/\delta \rceil\}$, with probability close to 1, we can write

$$\Phi_t \leq O\left(n^{1-\frac{\Lambda}{12}} \log^{-q} n\right),$$

for all $t \in [0, T]$ where $T = O\left(\frac{n^{-2\beta+\frac{\Lambda}{24q}}}{\alpha^2}\right)$.

Intuitively, when $\Phi_t \leq O\left(n^{1-\frac{\Lambda}{12}} \log^{-q} n\right)$ and $T = O\left(\frac{n^{-2\beta+\frac{\Lambda}{24q}}}{\alpha^2}\right)$, we have, using the analysis of the previous section,

$$\delta_t T \leq O\left(n^{1-\frac{\Lambda}{12}} \log^{-q} n\right) \quad \text{and} \quad \|v_t\|_2 \sqrt{T} \leq O\left(n^{1-\frac{\Lambda}{12}} \log^{-q} n\right).$$

This suggests that Φ_t stays at most $O\left(n^{1-\frac{\Lambda}{12}} \log^{-q} n\right)$ during a period of length T. Formally, we prove the following lemma to get an improved bound on ψ_n. Our proof applies Lemma 1.35.

Lemma 1.47 *Suppose that $\psi_k \leq \alpha k^\beta, \forall k \leq n$ for some $\alpha \geq 1$ and $1/4 - \epsilon/2 < \beta \leq 1/4$. Let p be any isotropic log-concave distribution. Let $\Phi_t = \mathrm{Tr}((A_t - I)^q)$ with $q = 2\lceil 1/\beta \rceil$. Then for n large enough such that $n^{\frac{\Lambda}{48q}} > \log n$ where $\Lambda = 4\beta + 2\epsilon - 1$, there exists a universal constant C s.t.*

$$\mathbb{P}\left[\max_{t \in [0,T]} \Phi_t \geq n^{1-\frac{\Lambda}{12}} \log^{-q} n\right] \leq 0.01 \quad \text{with} \quad T = \frac{C n^{-2\beta+\frac{\Lambda}{24q}}}{\alpha^2}.$$

Proof We use Lemma 1.35 with the bounds from Lemma 1.45 and 1.46. Recall we have the following bound on the potential change.

$$d\Phi_t = \delta_t dt + v_t^T dW_t,$$

with $||v_t||_2 \leq g(\Phi_t)$ where $g(\Phi_t)$ is defined to be $+\infty$ when $\Phi_t > n$ and $O(q) \cdot \left(\Phi_t^{1-\frac{1}{2q}} n^{\frac{1}{q}} + n^{\frac{1}{q}}\right)$ otherwise, and $\delta_t \leq f(\Phi_t)$ where $f(\Phi_t)$ is defined to be $+\infty$ when $\Phi_t > n$ and $\frac{1}{2}q(q-1)\delta^{(1)}(\Phi_t) + q\delta^{(2)}(\Phi_t)$ otherwise where

$$\delta^{(1)}(\Phi_t) = O(\alpha^2) \cdot \Phi_t n^{2\beta} \cdot \left[n^{-\frac{1}{q}} \Phi_t^{\frac{1}{q}} \log n + n^{-\frac{\Lambda}{4q}} \cdot n^{\frac{2}{q}} \Phi_t^{-\frac{2}{q}} \right],$$

and

$$\delta^{(2)}(\Phi_t) = \Phi_t^{1+\frac{1}{q}} + \Phi_t^{1-\frac{1}{q}} n^{\frac{1}{q}}.$$

We show that the conditions in Lemma 1.35 are met with $U = n^{1-\frac{\Lambda}{12}} \log^{-q} n$ and $T = \frac{Cn^{-2\beta+\frac{\Lambda}{24q}}}{\alpha^2}$ for some small enough constant C. It is easy to see that $f(\Phi_t)$ and $g(\Phi_t)$ are non-negative and non-decreasing functions of Φ_t by our choice of q, so we only need to check that the last condition of Lemma 1.35 holds.

We first consider the martingale term. For $1 \leq U \leq n$, we have

$$g(U) \cdot \sqrt{T} = O(q) \cdot \left(U^{1-\frac{1}{2q}} n^{\frac{1}{q}} + n^{\frac{1}{q}} \right) \cdot \frac{\sqrt{C} n^{-\beta+\frac{\Lambda}{48q}}}{\alpha^2}$$

$$\leq O(q) \cdot U \cdot U^{-\frac{1}{2q}} n^{\frac{1}{q}} \cdot \frac{\sqrt{C} n^{-\beta+\frac{\Lambda}{48q}}}{\alpha^2}$$

$$\leq U \cdot O(q) \cdot \sqrt{C} \cdot n^{-\beta+\frac{1}{q}+\frac{\Lambda}{48q}}.$$

Note that $q \geq 2/\beta$ and $\Lambda \leq 1$. Thus,

$$g(U) \cdot \sqrt{T} \leq U \cdot O(q)\sqrt{C}.$$

which is bounded by $U/8$ when C is small enough.

Now we verify that $f(U) \cdot T \leq U/8$ for some suitably small constant C. We first verify this for $\delta^{(2)}(\Phi_t)$.

$$\delta^{(2)}(U) \cdot T \leq U \cdot \left(U^{\frac{1}{q}} + U^{-\frac{1}{q}} n^{\frac{1}{q}} \right) Cn^{-2\beta+\frac{\Lambda}{24q}}$$

$$= U \cdot C \left(n^{\frac{1}{q}-\frac{\Lambda}{12q}} \log^{-1} n + n^{\frac{\Lambda}{12q}} \log n \right) n^{-2\beta+\frac{\Lambda}{24q}}$$

$$\le UCn^{-2\beta+\frac{1}{q}-\frac{\Lambda}{24q}}\log n$$

$$\le UC,$$

where in the last line we used $q \ge 2/\beta$, $\Lambda \le 1$ and $n^\beta > \log n$. Now we consider $\delta^{(1)}(\Phi_t)$. We denote the two terms in $\delta^{(1)}(\Phi_t)$ as $\delta_i^{(1)}(\Phi_t)$, where $i = 1, 2$. For the first term $\delta_1^{(1)}(\Phi_t)$ we have

$$\delta_1^{(1)}(U) \cdot T = O(\alpha^2) \cdot U n^{2\beta} (\log n) n^{-\frac{1}{q}} U^{\frac{1}{q}} \cdot \frac{Cn^{-2\beta+\frac{\Lambda}{24q}}}{\alpha^2}$$

$$= O(1) \cdot UC n^{-\frac{\Lambda}{24q}}$$

$$\le O(1) \cdot UC.$$

For the second term $\delta_2^{(1)}(\Phi_t)$ we have

$$\delta_2^{(1)}(U) \cdot T = O(\alpha^2) \cdot U n^{2\beta} \cdot n^{-\frac{\Lambda}{4q}} \cdot n^{\frac{2}{q}} U^{-\frac{2}{q}} \cdot \frac{Cn^{-2\beta+\frac{\Lambda}{24q}}}{\alpha^2}$$

$$= O(1) \cdot UC n^{-\frac{\Lambda}{24q}} \log^2 n$$

$$\le O(1) \cdot UC.$$

This shows that

$$\delta^{(1)}(U)T \le O(1)UC.$$

Thus, for some suitably small C, we have $f(U) \cdot T \le U/8$. Applying Lemma 1.35 completes the proof of the lemma. □

When $1/4 - \epsilon/2 < \beta \le 1/4$, we get a better bound on ψ_n.

Lemma 1.48 *Suppose that $\psi_k \le \alpha k^\beta$, for all $k \le n$ for some $\alpha \ge 1$ and $1/4 - \epsilon/2 < \beta \le 1/4$. Let p be an isotropic log-concave distribution in $\mathbb{R}^n$. Then for n large enough such that $n^{\frac{\Lambda}{48q}} > \log n$, there exists a universal constant $C > 0$ s.t.*

$$\psi_n \le C\alpha n^{\beta - \frac{\Lambda}{48q}},$$

where $\Lambda = 4\beta + 2\epsilon - 1$ and $q = 2\lceil 1/\beta \rceil$.

Proof Using Lemma 1.47, with probability at least 0.99, for any $t \le T = \frac{Cn^{-2\beta+\frac{\Lambda}{24q}}}{\alpha^2}$ where C is some universal constant and $q = 2\lceil 1/\beta \rceil$, we have

$$\Phi_t \le n^{1-\frac{\Lambda}{12}} \log^{-q} n.$$

Assuming this event, we have

$$\int_0^T \|A_t\|_{\mathrm{op}} dt \leq \int_0^T \left(1 + \Phi_t^{1/q}\right) \leq T\left(1 + n^{\frac{1}{q} - \frac{\Lambda}{12q}} \log^{-1} n\right) \leq 1/64.$$

Now applying Lemma 1.33, we get

$$\psi_p \leq O(\alpha) \cdot n^{\beta - \frac{\Lambda}{48q}},$$

where C is some universal constant. Since p is arbitrary, we have the result.   □

Now we are finally ready to prove Theorem 1.9.

Proof of Theorem 1.9 We start with the known bound $\psi_n \leq \alpha_0 n^{\beta_0}$ for $\beta_0 = 1/4$ and some constant α_0. We construct a sequence of better and better bounds for ψ_n which hold for any n large enough such that $n^{\frac{\Lambda}{48q}} > \log n$, where $q = \Theta(1/\beta) = O(1/(1 - 2\epsilon + 4\delta))$. (Note that if $\Lambda \leq 4\delta$, then we are done by Lemma 1.48. So we can assume without loss of generality that $\Lambda > 4\delta$). Since q is fixed, one can find a fixed n_0 such that for any $n \geq n_0$, the requirement $n^{\frac{\Lambda}{48q}} > \log n$ is satisfied whenever $\Lambda > 4\delta$, regardless of the current bound on ψ_n.

Suppose $\psi_n \leq \alpha_i n^{\beta_i}$ is the current bound. If $\beta_i \leq 1/4 - \epsilon/2 + \delta$, then we are done. Otherwise, applying Lemma 1.48 gives the better bound

$$\psi_n \leq \alpha_{i+1} n^{\beta_{i+1}},$$

where $\alpha_{i+1} = C\alpha_i$ and $\beta_{i+1} = \beta_i - \frac{\Lambda}{48q} \leq \beta_i - \frac{\delta}{12q}$ (since $\Lambda \geq 4\delta$). Therefore, starting from $\beta_0 = 1/4$ and repeating the procedure at most $M = \lceil \frac{6\epsilon q}{\delta} \rceil$ times, we will get some $m \leq M$ such that $\psi_n \leq \alpha_m n^{\beta_m}$ where $\beta_m \leq 1/4 - \epsilon/2 + \delta$ and $\alpha_m \leq C^{\lceil \frac{3q}{\delta} \rceil} \alpha_0$. This holds for any large n such that $n^{\frac{\delta}{12q}} > \log n$. For small n that doesn't satisfy the requirement $n^{\frac{\delta}{12q}} > \log n$, we simply bound them by some constant. We conclude that $\psi_n \leq O\left(n^{1/4 - \epsilon/2 + \delta}\right)$ for any n. We note that in fact the bound we get is $n^{1/4 - \epsilon/2 + \delta + q/(\delta \log n)}$ and since $q = O(1/\beta)$, we can set $\delta = O(1/\sqrt{\beta \log n})$ so that the bound on β is $1/4 - \epsilon/2 + o(1)$.   □

1.5   From KLS to Generalized CLT

Theorem 1.49 *Assume* $\psi_n = O(n^{1/4 - \epsilon/2})$ *for some* $0 < \epsilon < 1/2$ *and some dimension* n. *Let* p, q *be any isotropic log-concave distributions in* $\mathbb{R}^n$, x, y *be independent random vectors drawn from* p *and* q *and* $G \sim N(0, n)$. *It follows that*

$$W_2(\langle x, y \rangle, G)^2 = O\left(n^{1 - 2\epsilon} (\log n)^{1/2 + \epsilon}\right). \tag{1.5.1}$$

This gives exactly the condition in Theorem 1.27 (up to a small polynomial factor in $\log n$). The remainder of this section is devoted to proving Theorem 1.49. We start by relating $\langle x, y \rangle$ with $\langle x, g \rangle$, where $x \sim p$, $y \sim q$ are independent vectors drawn from isotropic log-concave distributions p, q in $\mathbb{R}^n$ and $g \sim \mathcal{N}(0, I)$ is a standard Gaussian vector in $\mathbb{R}^n$.

Lemma 1.50 *Assume the conditions of Theorem 1.49. Let $g \sim \mathcal{N}(0, I)$ be independent from x and y, then we have*

$$W_2(\langle x, y \rangle, \langle x, g \rangle)^2 = O\left(n^{1-2\epsilon}(\log n)^{1/2+\epsilon}\right).$$

Before we prove Lemma 1.50, we show how to use the lemma to prove Theorem 1.49. The intuition is the following. Lemma 1.50 allows us to relate $\langle x, y \rangle$ to $\langle x, g \rangle$. Notice for fixed x, the random variable $\langle x, g \rangle$ has a Gaussian law with variance $\|x\|_2$. Since $\|x\|_2$ is concentrated around $\sqrt{n}$, it follows that $\langle x, g \rangle$ is close to the Gaussian distribution $\mathcal{N}(0, n)$.

Proof of Theorem 1.49 Using Lemma 1.50 Let g be a random vector drawn from a standard n-dimensional normal distribution $\mathcal{N}(0, I)$. By Lemma 1.50, we have

$$W_2(\langle x, y \rangle, \langle x, g \rangle)^2 = O\left(n^{1-2\epsilon}(\log n)^{1/2+\epsilon}\right). \tag{1.5.2}$$

For fixed sample x, the random variable $\langle x, g \rangle$ has the same law as $\|x\|_2 \cdot g_1$ where $g_1 \sim \mathcal{N}(0, 1)$. Notice that G has the same law as $\sqrt{n} \cdot g_2$, where $g_2 \sim \mathcal{N}(0, 1)$. When x is fixed, we obtain a coupling between $\langle x, g \rangle$ and G by identifying g_1 with g_2. It follows that

$$
\begin{aligned}
W_2(\langle x, g \rangle, G) &\leq \mathbb{E}_{x \sim p}\left(\|x\|_2 - \sqrt{n}\right)^2 \cdot \mathbb{E}_{g_1 \sim \mathcal{N}(0,1)} g_1^2 \\
&= \mathbb{E}_{x \sim p}\left(\|x\|_2 - \sqrt{n}\right)^2 \\
&\leq \mathbb{E}_{x \sim p}\left(\frac{\left(\|x\|_2^2 - n\right)^2}{\left(\|x\|_2 + \sqrt{n}\right)^2}\right) \\
&\leq \frac{1}{n} \cdot \mathrm{Var}\left(\|x\|_2^2\right) \\
&\leq \frac{1}{n} \cdot O\left(\psi_n^2\right) = O\left(n^{1-2\epsilon}\right),
\end{aligned}
$$

where the last line uses Lemma 1.18 with the matrix A being the identity matrix in $\mathbb{R}^n$. This combined with (1.5.2) finishes the proof of Theorem 1.49. $\qquad\square$

Now we are left to prove Lemma 1.50. For this we turn to the stochastic localization technique introduced in Sect. 1.3. In the proof, we make use of Lemma 1.34. Our proof here bears structural similarities to that in [11], in that both proofs use stochastic localization specifically by viewing random variables as Brownian motion.

Proof of Lemma 1.50 We apply the stochastic construction in Sect. 1.3 with initial probability distribution $p_0 = p$. Since p_t is a martingale and p_∞ is a point mass at μ_∞, we have that

$$x \sim \mu_\infty = \int_0^\infty d\mu_t = \int_0^\infty A_t dW_t^{(n)},$$

where we used Lemma 1.30 and $W_t^{(n)}$ is a standard n-dimensional Brownian motion. The inner product $\langle x, y \rangle$ can be written similarly as

$$\langle x, y \rangle = \int_0^\infty y^T A_t dW_t^{(n)}.$$

Notice that $y^T A_t dW_t^{(n)}$ is a martingale whose quadratic variation has derivative $y^T A_t^2 y$ at time t. It follows that the process $W_t^{(1)}$ defined by $dW_t^{(1)} = y^T A_t dW_t^{(n)}/\sqrt{y^T A_t^2 y}$ is a one-dimensional standard Brownian motion. We therefore have

$$\langle x, y \rangle = \int_0^\infty \sqrt{y^T A_t^2 y} \cdot dW_t^{(1)}.$$

Note that $\sqrt{y^T A_t^2 y}$ is concentrated near $\sqrt{\mathbb{E}_{y \sim q} y^T A_t^2 y} = \sqrt{\text{Tr}\left(A_t^2\right)}$. It is therefore natural to couple $\langle x, y \rangle$ with the random variable $L = \int_0^\infty \sqrt{\text{Tr}\left(A_t^2\right)} dW_t^{(1)}$. We will show that this coupling gives an upper bound on $W_2(\langle x, y \rangle, L)^2$. Notice that the first random variable $\langle x, y \rangle$ depends on both x and y but the second random variable L depends only on x. So why would this coupling work? The intuition behind the coupling is the following: as one takes the expectation over y, the random variable $\sqrt{y^T A_t^2 y}$ is concentrated around $\sqrt{\text{Tr}\left(A_t^2\right)}$ and the deviation depends on the variable $\|A_t\|_{\text{op}}^2$. In the stochastic construction in Sect. 1.3, A_t starts from identity and ends up being 0. This allows good bounds on $\|A_t\|_{\text{op}}^2$.

We use $\mathbb{E}_x$ to denote the expectation taken with respect to the randomness of $W_t^{(n)}$ (notice that both A_t and $W_t^{(1)}$ adapt to $W_t^{(n)}$). It follows that

$$W_2(\langle x, y \rangle, L)^2 \leq \mathbb{E}_{x,y} \left[\int_0^\infty \left(\sqrt{y^T A_t^2 y} - \sqrt{\text{Tr}\left(A_t^2\right)} \right) \cdot dW_t^{(1)} \right]^2$$

$$= \mathbb{E}_{x,y} \left[\int_0^\infty \left(\sqrt{y^T A_t^2 y} - \sqrt{\text{Tr}\left(A_t^2\right)} \right)^2 dt \right]$$

$$= \int_0^\infty \mathbb{E}_{x,y} \left[\left(\sqrt{y^T A_t^2 y} - \sqrt{\mathrm{Tr}\left(A_t^2\right)} \right)^2 \right] dt$$

$$= \int_0^\infty \mathbb{E}_{x,y} \left[\left(\frac{y^T A_t^2 y - \mathrm{Tr}\left(A_t^2\right)}{\sqrt{y^T A_t^2 y} + \sqrt{\mathrm{Tr}\left(A_t^2\right)}} \right)^2 \right] dt$$

$$\leq \int_0^\infty \mathbb{E}_x \left[\frac{\mathbb{E}_y \left(y^T A_t^2 y - \mathrm{Tr}\left(A_t^2\right) \right)^2}{\mathrm{Tr}\left(A_t^2\right)} \right] dt$$

$$= \int_0^\infty \mathbb{E}_x \left[\frac{\mathrm{Var}\left(y^T A_t^2 y \right)}{\mathrm{Tr}\left(A_t^2\right)} \right] dt$$

$$\leq \int_0^\infty \mathbb{E}_x \left[\frac{O\left(\psi_n^2\right) \cdot \mathrm{Tr}\left(A_t^4\right)}{\mathrm{Tr}\left(A_t^2\right)} \right] dt$$

$$\leq O\left(\psi_n^2\right) \cdot \int_0^\infty \mathbb{E}_x \left[\|A_t\|_{\mathrm{op}}^2 \right] dt,$$

where the first equality uses Ito's isometry and the last two lines follow from Lemma 1.18. The remaining thing is to bound $\|A_t\|_{\mathrm{op}}^2$.

The covariance matrix A_t corresponds to a density proportional to the log-concave density $p(x)$ multiplied by a Gaussian density $e^{-c_t^T x - \frac{t}{2}\|x\|_2^2}$. It is well known that the operator norm of such A_t is dominated by the Gaussian term (e.g. [10], Proposition 2.6), i.e.

$$\|A_t\|_{\mathrm{op}} \leq O(1/t).$$

We also need an upper bound for $\mathbb{E}_x[\|A_t\|_{\mathrm{op}}^2]$ when t is close to 0. For this take $k = \frac{1}{1/2-\epsilon}$ in Lemma 1.34, we have for any $0 \leq t \leq \frac{1/2-\epsilon}{cn^{1/2-\epsilon}(\log n)^{1/2+\epsilon}}$,

$$\mathbb{P}\left[\|A_t\|_{\mathrm{op}} \geq 2\right] \leq 2\exp\left(-\frac{1}{ct}\right). \tag{1.5.3}$$

We can therefore bound $\mathbb{E}[\|A_t\|_{\mathrm{op}}^2]$ as

$$\mathbb{E}[\|A_t\|_{\mathrm{op}}^2] \leq 4 \cdot \mathbb{P}\left[\|A_t\|_{\mathrm{op}} < 2\right] + \frac{1}{t^2} \cdot \mathbb{P}\left[\|A_t\|_{\mathrm{op}} \geq 2\right] \leq 4 + \frac{1}{t^2} \cdot 2\exp\left(-\frac{1}{ct}\right).$$

Since $t \leq \frac{1/2-\epsilon}{cn^{1/2-\epsilon}(\log n)^{1/2+\epsilon}}$, $1/t \geq \frac{cn^{1/2-\epsilon}(\log n)^{1/2+\epsilon}}{1/2-\epsilon}$. For fixed $0 < \epsilon < 1/2$, the last term $\frac{1}{t^2} \cdot 2\exp\left(-\frac{1}{ct}\right)$ becomes negligible when n is sufficiently large so $\mathbb{E}[\|A_t\|_{\mathrm{op}}^2]$

is bounded by some constant C_ϵ (that depends on ϵ) for any $t \leq \frac{1/2-\epsilon}{cn^{1/2-\epsilon}(\log n)^{1/2+\epsilon}} = T \leq 1$. It follows that

$$W_2(\langle x, y \rangle, L)^2 \leq O\left(\psi_n^2\right) \cdot \int_0^\infty \mathbb{E}_x \left[\|A_t\|_{\text{op}}^2 \right] dt$$

$$\leq O\left(\psi_n^2\right) \cdot \left(\int_0^T C_\epsilon dt + \int_T^\infty \frac{1}{t^2} dt \right)$$

$$\leq O\left(\psi_n^2\right) \cdot \frac{1}{T} = O\left(n^{1-2\epsilon}(\log n)^{1/2+\epsilon}\right).$$

We note that L is defined using only the isotropic log-concave distribution p. One can therefore prove a similar bound when q is the n-dimensional standard normal distribution, i.e.

$$W_2(\langle x, g \rangle, L)^2 = O\left(n^{1-2\epsilon}(\log n)^{1/2+\epsilon}\right).$$

Combining these two bounds, we have the desired result.

$$W_2(\langle x, y \rangle, \langle x, g \rangle)^2 = O\left(n^{1-2\epsilon}(\log n)^{1/2+\epsilon}\right).$$

$\square$

1.5.1 Connection to Classical CLT for Convex Sets

Using exactly the same approach, we prove the following theorem which is easier to compare with classical results on central limit theorem for convex sets. Here we replace the W_2 distance in Theorem 1.5.1 by the total variation distance.

Theorem 1.51 *Assume $\psi_n = O\left(n^{1/4-\epsilon/2}\right)$ for some $0 < \epsilon < 1/2$ and some dimension n. Let p, q be any isotropic log-concave distributions in $\mathbb{R}^n$. For fixed vector $x \sim p$, denote $\langle x, y \rangle$ the random variable formed by the inner product of x and y, when $y \sim q$ is independently drawn from x. Let $g \sim \mathcal{N}(0, 1)$ be a standard normal distribution. Then we have*

$$\mathbb{P}_{x \sim p}\left[d_{\text{TV}}\left(\frac{\langle x, y \rangle}{\|x\|_2}, g \right) \geq Cn^{-\epsilon/2} \right] \leq \exp\left(-cn^{\frac{1}{2}-\epsilon}(\log n)^{1/2+\epsilon} \right),$$

for some constants c and C that depend on ϵ.

The following lemma can be proved by using a similar approach as in the proof of Lemma 1.50.

Lemma 1.52 *Assume $\psi_n = O\left(n^{1/4-\epsilon/2}\right)$ for some $0 < \epsilon < 1/2$ and some dimension n. Let p, q be any isotropic log-concave distributions in $\mathbb{R}^n$ and let $x \sim p$, $y \sim q$ and $g \sim \mathcal{N}(0, I)$ be independent samples. Then with probability at least $1 - \exp\left(-cn^{\frac{1}{2}-\epsilon}(\log n)^{1/2+\epsilon}\right)$ over the random choice of x, we have*

$$W_2(\langle x, y\rangle, \langle x, g\rangle) = O\left(n^{\frac{1}{2}-\epsilon}\right),$$

where the constant c depends on ϵ.

Proof of Theorem 1.51 Using Lemma 1.52 By Lemma 1.16, we have with probability at least $1 - \exp(-\Omega(\sqrt{n}))$, $\|x\|_2 \geq C\sqrt{n}$ for some universal constant $C > 0$. We condition on this event and the event in Lemma 1.52 such that

$$W_2(\langle x, y\rangle, \langle x, g\rangle) = O\left(n^{\frac{1}{2}-\epsilon}\right).$$

The probability that these events hold at the same time is at least

$$1 - \exp\left(-\Omega\left(n^{\frac{1}{2}-\epsilon}(\log n)^{1/2+\epsilon}\right)\right).$$

In this case we have

$$W_2\left(\langle x, y\rangle/\|x\|_2, \langle x, g\rangle/\|x\|_2\right) = O\left(n^{-\epsilon}\right).$$

Notice that for a fixed x, $\langle x, y\rangle/\|x\|_2$ follows a one-dimensional isotropic log-concave distribution and $\langle x, g\rangle/\|x\|_2$ follows a standard normal distribution. Applying Lemma 1.22 finishes the proof of the theorem. □

Appendix 1: Missing Proofs in Sect. 1.2.4

We restate Lemma 1.23 below for reference.

Lemma 1.23 *Let μ and ν be two probability measures in $\mathbb{R}$. Suppose one of the following holds:*

1. *Both μ and ν are isotropic log-concave distributions.*
2. *The distribution μ is isotropic log-concave, while ν is the measure of the random variable $\frac{1}{\sqrt{n}}\langle x, y\rangle$ where $x \sim p$ and $y \sim q$ are independent random vectors and p, q are isotropic log-concave distributions in $\mathbb{R}^n$.*
3. *There exist isotropic log-concave distributions p_μ, q_μ, p_ν and q_ν in $\mathbb{R}^n$ such that μ is the measure of the random variable $\frac{1}{\sqrt{n}}\langle x_\mu, y_\mu\rangle$ and ν is the measure of the random variable $\frac{1}{\sqrt{n}}\langle x_\nu, y_\nu\rangle$, where $x_\mu \sim p_\mu$, $y_\mu \sim q_\mu$, $x_\nu \sim p_\nu$ and $y_\nu \sim q_\nu$ are independent random vectors.*

Then there exists a universal constant $c > 0$ such that for any $1 \leq s < t$, we have

$$W_t(\mu, \nu)^t \leq c W_s(\mu, \nu)^s \log^{t-s}\left(\frac{c^t t^{2t}}{W_s(\mu, \nu)^s}\right) + c^t t^{2t} \exp(-c\sqrt{n}).$$

Moreover, the above bound is valid even when the coupling (μ, ν) on the left-hand side is taken to be the best coupling for $W_s(\mu, \nu)$ instead of the best coupling for $W_t(\mu, \nu)$.

Proof of Lemma 1.23 The result for Case 1 is given by Meckes and Meckes [25, Prop 5]. Here we use the same idea to prove the result for Case 2. The proof for Case 3 is almost the same and is omitted.

We denote the random variable drawn from ν as z and the best coupling for $W_s(\mu, \nu)$ as $\left(\frac{1}{\sqrt{n}}\langle x, y\rangle, z\right)$. We use the coupling $\left(\frac{1}{\sqrt{n}}\langle x, y\rangle, z\right)$ in the rest of the proof whenever we write expectations. Denote $\mathbf{1}_{\{\cdot\}}$ the indicator function of an event. For any $R > 0$, we have

$$W_t\left(\frac{1}{\sqrt{n}}\langle x, y\rangle, z\right)^t \leq \mathbb{E}\left|\frac{1}{\sqrt{n}}\langle x, y\rangle - z\right|^t$$

$$\leq R^{t-s} \cdot \mathbb{E}\left|\frac{1}{\sqrt{n}}\langle x, y\rangle - z\right|^s$$

$$+ \mathbb{E}\left|\frac{1}{\sqrt{n}}\langle x, y\rangle - z\right|^t \mathbf{1}_{\left\{\left|\frac{1}{\sqrt{n}}\langle x, y\rangle - z\right| \geq R\right\}}$$

$$\leq R^{t-s} \cdot W_s\left(\frac{1}{\sqrt{n}}\langle x, y\rangle, z\right)^s$$

$$+ \sqrt{\mathbb{P}\left[\left|\frac{1}{\sqrt{n}}\langle x, y\rangle - z\right| \geq R\right] \cdot \mathbb{E}\left(\frac{1}{\sqrt{n}}\langle x, y\rangle - z\right)^{2t}},$$

where the last step is by Cauchy-Schwarz. Now we bound the second term in the above expression. Using Minkowski's inequality, we have

$$\left(\mathbb{E}\left(\frac{1}{\sqrt{n}}\langle x, y\rangle - z\right)^{2t}\right)^{1/2t} \leq \left(\mathbb{E}z^{2t}\right)^{1/2t} + \left(\mathbb{E}\left(\frac{1}{\sqrt{n}}\langle x, y\rangle\right)^{2t}\right)^{1/2t}.$$

Since z follows an isotropic log-concave distribution, it follows from Lemma 1.15 that $\left(\mathbb{E}z^{2t}\right)^{1/2t} \leq 4t$. For the second term we notice that when x is fixed, the random variable $\frac{1}{\sqrt{n}}\langle x, y\rangle$ follows a one-dimensional log-concave distribution with variance

$\frac{\|x\|_2^2}{n}$. Using Lemma 1.15 again, we have

$$\mathbb{E}\left(\frac{1}{\sqrt{n}}\langle x, y\rangle\right)^{2t} \le \mathbb{E}_{x\sim p}\left(\mathbb{E}_{y\sim q}\left(\frac{1}{\sqrt{n}}\langle x, y\rangle\right)^{2t}\right)$$

$$\le (4t)^{2t}\cdot \mathbb{E}_{x\sim p}\frac{\|x\|_2^{2t}}{n^t} \le (4t)^{4t}.$$

We therefore have

$$\mathbb{E}\left(\frac{1}{\sqrt{n}}\langle x, y\rangle - z\right)^{2t} \le \left(4t + 16t^2\right)^{2t}.$$

Now we bound $\mathbb{P}\left[\left|\frac{1}{\sqrt{n}}\langle x, y\rangle - z\right| \ge R\right]$ as follows. For some constant c_2, $C_R > 0$, whenever $R > C_R$ we have

$$\mathbb{P}\left[\left|\frac{1}{\sqrt{n}}\langle x, y\rangle - z\right| \ge R\right] \le \mathbb{P}\left[\left|\frac{1}{\sqrt{n}}\langle x, y\rangle\right| \ge R/2\right] + \mathbb{P}[|z| \ge R/2]$$

$$\le \mathbb{P}\left[\left|\frac{1}{\sqrt{n}}\langle x, y\rangle\right| \ge R/2\right] + \exp(-c_2 R).$$

Since x follows an isotropic log-concave distribution, we have from Theorem 1.20 that whenever $R > C_R$, there exist constants $c_1, C > 0$ such that

$$\mathbb{P}[\|x\|_2 \ge \sqrt{Cn}] \le \exp(-c_1\sqrt{n}).$$

Whenever $\|x\|_2 < \sqrt{Cn}$ for fixed vector x, the random variable $\frac{1}{\sqrt{n}}\langle x, y\rangle$ follows a one-dimensional log-concave distribution with variance at most C. Therefore when the universal constant C_R is large enough and when $R > C_R$, we have

$$\mathbb{P}\left[\left|\frac{1}{\sqrt{n}}\langle x, y\rangle\right| \ge R/2\right] \le \exp(-c_1\sqrt{n}) + \exp(-c_2 R).$$

Combining everything we have that when $R > C_R$,

$$W_t\left(\frac{1}{\sqrt{n}}\langle x, y\rangle, z\right)^t \le R^{t-s}W_s\left(\frac{1}{\sqrt{n}}\langle x, y\rangle, z\right)^s$$

$$+ (4t + 16t^2)^t \cdot \sqrt{2\left(\exp(-c_2 R) + \exp(-c_1\sqrt{n})\right)}.$$

Optimizing over R, for some constant $c \geq 0$ we have

$$W_t \left(\frac{1}{\sqrt{n}} \langle x, y \rangle, z \right)^t \leq c \cdot W_s \left(\frac{1}{\sqrt{n}} \langle x, y \rangle, z \right)^s \cdot \log^{t-s} \left(\frac{c^t t^{2t}}{W_s \left(\frac{1}{\sqrt{n}} \langle x, y \rangle, z \right)^s} \right)$$

$$+ c^t t^{2t} \exp(-c\sqrt{n}).$$

This finishes the proof of Lemma 1.23.   □

Appendix 2: Missing Proofs in Sect. 1.4.1

In this section, we give proofs of the lemmas in Sect. 1.4.1. Here we repeatedly use the elementary facts that $\mathrm{Tr}(AB) = \mathrm{Tr}(BA)$ and $x^T A y = \mathrm{Tr}\left(A y x^T \right)$.

Lemma 1.53 *For any isotropic log-concave distribution p and symmetric matrices A and B, we have that*

$$T_p(A, B, I) = \sum_i \mathrm{Tr}(A \Delta_i B \Delta_i) \quad and \quad T_p(A, B, I) = \sum_{i,j} A_{ij} \mathrm{Tr}(\Delta_i B \Delta_j),$$

where $\Delta_i = \mathbb{E}_{x \sim p} x x^T x_i$.

Proof Direct calculation shows that

$$T_p(A, B, I) = \mathbb{E}_{x,y \sim p} x^T A y x^T B y x^T y = \sum_i \mathbb{E}_{x,y \sim p} x^T A y x^T B y x_i y_i$$

$$= \sum_i \mathbb{E}_{x,y \sim p} \mathrm{Tr}\left(A x x^T B y y^T x_i y_i \right) = \sum_i \mathrm{Tr}(A \Delta_i B \Delta_i),$$

and

$$T_p(A, B, I) = \mathbb{E}_{x,y \sim p} x^T A y x^T B y x^T y = \sum_{i,j} A_{ij} \mathbb{E}_{x,y \sim p} x_i y_j x^T B y x^T y$$

$$= \sum_{i,j} A_{ij} \mathbb{E}_{x,y \sim p} \mathrm{Tr}\left(x x^T B y y^T x_i y_j \right) = \sum_{i,j} A_{ij} \mathrm{Tr}(\Delta_i B \Delta_j).$$

□

Lemma 1.39 *For any $A_1, A_2, A_3 \succeq 0$, we have that $T(A_1, A_2, A_3) \geq 0$ and for any symmetric matrices B_1, B_2, B_3, we have that*

$$T(B_1, B_2, B_3) \leq T(|B_1|, |B_2|, |B_3|).$$

Proof Fix any isotropic log-concave distribution p. We define $\Delta_i = \mathbb{E}_{x\sim p} x x^T x^T A_3^{1/2} e_i$ which is well defined since $A_3 \succeq 0$. Then, we have that

$$T_p(A_1, A_2, A_3) = \mathbb{E}_{x,y\sim p} x^T A_1 y x^T A_2 y x^T A_3 y = \sum_i \text{Tr}(A_1 \Delta_i A_2 \Delta_i).$$

Since Δ_i is symmetric and $A_1, A_2 \succeq 0$, we have that $A_1^{1/2} \Delta_i A_2 \Delta_i A_1^{1/2} \succeq 0$ and $\text{Tr}(A_1 \Delta_i A_2 \Delta_i) \geq 0$. Therefore, $T(A_1, A_2, A_3) \geq T_p(A_1, A_2, A_3) \geq 0$.

For the second part, we write $B_1 = B_1^{(1)} - B_1^{(2)}$ where $B_1^{(1)} \succeq 0$, $B_1^{(2)} \succeq 0$ and $|B_1| = B_1^{(1)} + B_1^{(2)}$. We define $B_2^{(1)}, B_2^{(2)}, B_3^{(1)}, B_3^{(2)}$ similarly. Note that

$$T(B_1, B_2, B_3) = T\left(B_1^{(1)}, B_2^{(1)}, B_3^{(1)}\right) - T\left(B_1^{(1)}, B_2^{(1)}, B_3^{(2)}\right)$$

$$- T\left(B_1^{(1)}, B_2^{(2)}, B_3^{(1)}\right) + T\left(B_1^{(1)}, B_2^{(2)}, B_3^{(2)}\right)$$

$$- T\left(B_1^{(2)}, B_2^{(1)}, B_3^{(1)}\right) + T\left(B_1^{(2)}, B_2^{(1)}, B_3^{(2)}\right)$$

$$+ T\left(B_1^{(2)}, B_2^{(2)}, B_3^{(1)}\right) - T\left(B_1^{(2)}, B_2^{(2)}, B_3^{(2)}\right).$$

Since $B_j^{(i)} \succeq 0$, the first part of this lemma shows that every term $T\left(B_1^{(i)}, B_2^{(j)}, B_3^{(k)}\right) \geq 0$. Hence, we have that

$$T(B_1, B_2, B_3) \leq T\left(B_1^{(1)}, B_2^{(1)}, B_3^{(1)}\right) + T\left(B_1^{(1)}, B_2^{(1)}, B_3^{(2)}\right)$$

$$+ T\left(B_1^{(1)}, B_2^{(2)}, B_3^{(1)}\right) + T\left(B_1^{(1)}, B_2^{(2)}, B_3^{(2)}\right)$$

$$+ T\left(B_1^{(2)}, B_2^{(1)}, B_3^{(1)}\right) + T\left(B_1^{(2)}, B_2^{(1)}, B_3^{(2)}\right)$$

$$+ T\left(B_1^{(2)}, B_2^{(2)}, B_3^{(1)}\right) + T\left(B_1^{(2)}, B_2^{(2)}, B_3^{(2)}\right)$$

$$= T\left(|B_1|, |B_2|, |B_3|\right).$$

$\square$

Lemma 1.54 *Suppose that $\psi_k \leq \alpha k^\beta$ for all $k \leq n$ for some $0 \leq \beta \leq \frac{1}{2}$ and $\alpha \geq 1$. Given an isotropic log-concave distribution p and a unit vector v, the following two statements hold for $\Delta = \mathbb{E}_{x\sim p} x x^T x^T v$:*

1. For any orthogonal projection matrix P with rank r, we have that

$$\text{Tr}(\Delta P \Delta) \leq O\left(\psi_{\min(2r,n)}^2\right).$$

2. *For any symmetric matrix A, we have that*

$$\mathrm{Tr}(\Delta A \Delta) \leq O\left(\alpha^2 \log n\right) \cdot \left(\mathrm{Tr}\,|A|^{1/(2\beta)}\right)^{2\beta}.$$

Proof We first bound $\mathrm{Tr}(\Delta P \Delta)$. This part of the proof is generalized from a proof by Eldan [10]. Note that $\mathrm{Tr}(\Delta P \Delta) = \mathbb{E}_{x \sim p} x^T P \Delta x x^T v$. Since $\mathbb{E} x^T v = 0$, we have that

$$\mathrm{Tr}(\Delta P \Delta) \leq \sqrt{\mathbb{E}\left(x^T v\right)^2} \sqrt{\mathrm{Var}\left(x^T P \Delta x\right)} \overset{\text{Lem 1.18}}{\leq} O\left(\psi_{\mathrm{rank}(P\Delta + \Delta P)}\right) \cdot \sqrt{\mathrm{Tr}\left(\Delta P \Delta\right)}.$$

This gives $\mathrm{Tr}(\Delta P \Delta) \leq O\left(\psi^2_{\min(2r,n)}\right)$.

Now we bound $\mathrm{Tr}(\Delta A \Delta)$. Since $\mathrm{Tr}(\Delta A \Delta) \leq \mathrm{Tr}(\Delta |A| \Delta)$, we can assume without loss of generality that $A \succeq 0$. We write $A = \sum_i A_i + B$ where each A_i has eigenvalues between $\left(\|A\|_{\mathrm{op}} 2^i / n, \|A\|_{\mathrm{op}} 2^{i+1}/n\right]$ and B has eigenvalues smaller than or equals to $\|A\|_{\mathrm{op}}/n$. Clearly, we only need at most $\lceil \log(n) + 1 \rceil$ many such A_i. Let P_i be the orthogonal projection from $\mathbb{R}^n$ to the span of the range of A_i. Using $\|A_i\|_{\mathrm{op}} P_i \succeq A_i$, we have that

$$\mathrm{Tr}(\Delta A_i \Delta) \leq \|A_i\|_{\mathrm{op}} \mathrm{Tr}(\Delta P_i \Delta)$$

$$\leq O\left(\psi^2_{\min(2\mathrm{rank}(A_i),n)}\right) \cdot \|A_i\|_{\mathrm{op}} \leq O(\alpha^2) \cdot \sum_i \mathrm{rank}(A_i)^{2\beta} \|A_i\|_{\mathrm{op}},$$

where we used the first part of this lemma in the last inequality.

Similarly, we have that

$$\mathrm{Tr}(\Delta B \Delta) \leq O\left(\psi^2_n\right) \cdot \|B\|_{\mathrm{op}} \leq O(n \|B\|_{\mathrm{op}}) \leq O(1) \cdot \|A\|_{\mathrm{op}}.$$

Combining the bounds on $\mathrm{Tr}(\Delta A_i \Delta)$ and $\mathrm{Tr}(\Delta B \Delta)$, we have that

$$\mathrm{Tr}(\Delta A \Delta) \leq O(\alpha^2) \cdot \sum_i \mathrm{rank}(A_i)^{2\beta} \|A_i\|_{\mathrm{op}} + O(1) \cdot \|A\|_{\mathrm{op}}$$

$$\leq O(\alpha^2) \cdot \left(\sum_i \mathrm{rank}(A_i) \|A_i\|_{\mathrm{op}}^{1/(2\beta)}\right)^{2\beta} \log(n)^{1-2\beta}$$

$$\leq O(\alpha^2 \log n) \cdot \left(\mathrm{Tr}\,|A|^{1/(2\beta)}\right)^{2\beta}.$$

$\square$

In the next lemma, we collect tensor inequalities that will be useful for later proofs.

Lemma 1.40 *Suppose that $\psi_k \leq \alpha k^\beta$ for all $k \leq n$ for some fixed $0 \leq \beta \leq \frac{1}{2}$ and $\alpha \geq 1$. For any isotropic log-concave distribution p in $\mathbb{R}^n$ and symmetric matrices A and B, we have that*

1. $T(A, I, I) \leq T(I, I, I) \cdot \|A\|_{\text{op}}$.
2. $T(A, I, I) \leq O(\psi_n^2) \cdot \text{Tr} |A|$.
3. $T(A, B, I) \leq O(\psi_r^2) \cdot \|B\|_{\text{op}} \text{Tr} |A|$ *where* $r = \min(2 \cdot \text{rank}(B), n)$.
4. $T(A, B, I) \leq O(\alpha^2 \log n) \cdot \left(\text{Tr} |B|^{1/(2\beta)}\right)^{2\beta} \text{Tr} |A|$.
5. $T(A, B, I) \leq \left(T\left(|A|^s, I, I\right)\right)^{1/s} \cdot \left(T\left(|B|^t, I, I\right)\right)^{1/t}$, *for any* $s, t \geq 1$ *with* $s^{-1} + t^{-1} = 1$.

Proof Without loss of generality, we can assume A is diagonal by rotating space. In particular, if we want to prove something for $\text{Tr}(A^\alpha \Delta A^\beta \Delta)$ where A, Δ are symmetric matrices, we use the spectral decomposition $A = U\Sigma U^T$ to rewrite this as

$$\text{Tr}\left(U\Sigma^\alpha U^T \Delta U \Sigma^\beta U^T \Delta\right) = \text{Tr}\left(\Sigma^\alpha \left(U^T \Delta U\right) \Sigma^\beta \left(U^T \Delta U\right)\right),$$

which puts us back in the same situation, but with a diagonal matrix A. For all inequalities listed above, it suffices to upper bound T by upper bounding T_p for any isotropic log-concave distribution p.

For inequality 1, we note that

$$T_p(A, I, I) \overset{\text{Lem 1.53}}{=} \sum_i A_{ii} \text{Tr}(\Delta_i^2) \leq \|A\|_{\text{op}} \sum_i \text{Tr}\left(\Delta_i^2\right) \overset{\text{Lem 1.53}}{=} \|A\|_{\text{op}} T(I, I, I),$$

where the last inequality is from the third moment assumption.

For inequality 2, we note that

$$T_p(A, I, I) \overset{\text{Lem 1.53}}{=} \sum_i A_{ii} \text{Tr}(\Delta_i^2) \overset{\text{Lem 1.54}}{\leq} \sum_i |A_{ii}| \cdot O\left(\psi_n^2\right) = O\left(\psi_n^2\right) \cdot \text{Tr} |A|.$$

For inequality 3, we let P be the orthogonal projection from $\mathbb{R}^n$ to the span of the range of B. Then, we have that

$$T_p(A, B, I) \leq T_p(|A|, |B|, I) \qquad\qquad\qquad\qquad \text{(Lem 1.39)}$$

$$= \sum_i |A_{ii}| \text{Tr}(\Delta_i |B| \Delta_i) \qquad\qquad\qquad \text{(Lem 1.53)}$$

$$\overset{①}{\leq} \|B\|_{\text{op}} \sum_i |A_{ii}| \text{Tr}(\Delta_i P \Delta_i)$$

$$\leq O\left(\psi_r^2\right) \cdot \text{Tr} |A| \|B\|_{\text{op}}. \qquad\qquad\qquad \text{(Lem 1.54)}$$

where we used that $|B| \preceq \|B\|_{\text{op}} P$ in ①.

For inequality 4, we note that

$$T_p(A, B, I) \overset{\text{Lem 1.53}}{=} \sum_i A_{ii} \text{Tr}(\Delta_i B \Delta_i) \overset{\text{Lem 1.54}}{\leq} O(\alpha^2 \log n) \cdot \text{Tr}|A| \left(\text{Tr}|B|^{1/(2\beta)} \right)^{2\beta}.$$

For inequality 5, we note that

$$T_p(A, B, I) \leq T_p(|A|, |B|, I) \tag{Lem 1.39}$$

$$= \sum_i \text{Tr}(|A| \Delta_i |B| \Delta_i) \tag{Lem 1.53}$$

$$\leq \sum_i \text{Tr}(|A| |\Delta_i| |B| |\Delta_i|)$$

$$= \sum_i \text{Tr}\left(|\Delta_i|^{1/s} |A| |\Delta_i|^{1/s} |\Delta_i|^{1/t} |B| |\Delta_i|^{1/t} \right)$$

$$\leq \sum_i \left(\text{Tr}\left(\left(|\Delta_i|^{1/s} |A| |\Delta_i|^{1/s} \right)^s \right) \right)^{1/s}$$

$$\cdot \left(\text{Tr}\left(\left(|\Delta_i|^{1/t} |B| |\Delta_i|^{1/t} \right)^t \right) \right)^{1/t} \tag{Lem 1.24}$$

$$\leq \sum_i \left(\text{Tr}\left(|\Delta_i| |A|^s |\Delta_i| \right) \right)^{1/s} \cdot \left(\text{Tr}\left(|\Delta_i| |B|^t |\Delta_i| \right) \right)^{1/t} \tag{Lem 1.25}$$

$$= \sum_i \left(\text{Tr}\left(|A|^s \Delta_i^2 \right) \right)^{1/s} \cdot \left(\text{Tr}\left(|B|^t \Delta_i^2 \right) \right)^{1/t}$$

$$\leq \left(\sum_i \text{Tr}\left(|A|^s \Delta_i^2 \right) \right)^{1/s} \cdot \left(\sum_i \text{Tr}\left(|B|^t \Delta_i^2 \right) \right)^{1/t}$$

$$= \left(T_p\left(|A|^s, I, I \right) \right)^{1/s} \cdot \left(T_p\left(|B|^t, I, I \right) \right)^{1/t}. \tag{Lem 1.53}$$

□

Lemma 1.41 *For any positive semi-definite matrices A, B, C and any $\alpha \in [0, 1]$, then*

$$T\left(B^{1/2} A^\alpha B^{1/2}, B^{1/2} A^{1-\alpha} B^{1/2}, C \right) \leq T\left(B^{1/2} A B^{1/2}, B, C \right).$$

Proof Fix any isotropic log-concave distribution p. Let $\Delta_i = \mathbb{E}_{x \sim p} B^{1/2} x x^T B^{1/2} x^T C^{1/2} e_i$. Then, we have that

$$T_p(B^{1/2}A^{\alpha}B^{1/2}, B^{1/2}A^{1-\alpha}B^{1/2}, C)$$

$$= \mathbb{E}_{x,y\sim p}x^T B^{1/2}A^{\alpha}B^{1/2}yx^T B^{1/2}A^{1-\alpha}B^{1/2}yx^T Cy$$

$$= \sum_i \mathbb{E}\left(\left(y^T B^{1/2}A^{\alpha}B^{1/2}x\right)\left(x^T B^{1/2}A^{1-\alpha}B^{1/2}y\right)x^T C^{1/2}e_i y^T C^{1/2}e_i\right)$$

$$= \sum_i \mathbb{E}\left(\text{Tr}\left(A^{\alpha}B^{1/2}xx^T B^{1/2}A^{1-\alpha}B^{1/2}yy^T B^{1/2}\right)\left(x^T C^{1/2}e_i\right)\left(y^T C^{1/2}e_i\right)\right)$$

$$= \sum_i \text{Tr}(A^{\alpha}\Delta_i A^{1-\alpha}\Delta_i).$$

Using Lemma 1.26, we have that

$$\sum_i \text{Tr}\left(A^{\alpha}\Delta_i A^{1-\alpha}\Delta_i\right) \le \sum_i \text{Tr}\left(A\Delta_i^2\right) = \mathbb{E}_{x,y\sim p}x^T B^{1/2}AB^{1/2}yx^T Byx^T Cy$$

$$= T_p\left(B^{1/2}AB^{1/2}, B, C\right).$$

Taking the supremum over all isotropic log-concave distributions, we get the result.

$\square$

Acknowledgements We thank the anonymous referee for many helpful suggestions.

References

1. Z. Allen-Zhu, Y.T. Lee, L. Orecchia, Using optimization to obtain a width-independent, parallel, simpler, and faster positive SDP solver, in *Proceedings of the Twenty-Seventh Annual ACM-SIAM Symposium on Discrete Algorithms*, pp. 1824–1831 (SIAM, Philadelphia, 2016)
2. M. Anttila, K. Ball, I. Perissinaki, The central limit problem for convex bodies. Trans. Am. Math. Soc. **355**(12), 4723–4735 (2003)
3. S. Artstein-Avidan, A. Giannopoulos, V.D. Milman, *Asymptotic Geometric Analysis, Part I*, vol 202 (AMS, Providence, 2015)
4. S.G Bobkov, A. Koldobsky, On the central limit property of convex bodies, in *Geometric Aspects of Functional Analysis*, pp. 44–52 (Springer, Berlin, 2003)
5. S. Brazitikos, A. Giannopoulos, P. Valettas, B.-H. Vritsiou, *Geometry of Isotropic Convex Bodies*, vol 196 (American Mathematical Society, Providence, 2014)
6. J. Cheeger, *A Lower Bound for the Smallest Eigenvalue of the Laplacian*, pp. 195–199 (Princeton University, Princeton, 1969)
7. B. Cousins, S. Vempala, A cubic algorithm for computing Gaussian volume, in *SODA*, pp. 1215–1228 (2014)
8. K.E. Dambis, On the decomposition of continuous submartingales. Theory Probab. Appl. **10**(3), 401–410 (1965)
9. L.E. Dubins, G. Schwarz, On continuous martingales. Proc. Natl. Acad. Sci. USA **53**(5), 913 (1965)

10. R. Eldan, Thin shell implies spectral gap up to polylog via a stochastic localization scheme. Geom. Funct. Anal. **23**, 532–569 (2013)
11. R. Eldan, D. Mikulincer, A. Zhai, *The CLT in High Dimensions: Quantitative Bounds via Martingale Embedding* (2018). arXiv preprint arXiv:1806.09087
12. B. Fleury, Concentration in a thin Euclidean shell for log-concave measures. J. Funct. Anal. **259**(4), 832–841 (2010)
13. O. Guédon, Concentration phenomena in high dimensional geometry, in *ESAIM: Proceedings*, vol 44, pp. 47–60 (EDP Sciences, Les Ulis, 2014)
14. O. Guedon, E. Milman, Interpolating thin-shell and sharp large-deviation estimates for isotropic log-concave measures. Geom. Funct. Anal. **21**(5), 1043–1068 (2011)
15. K. Itô, 109. stochastic integral. Proc. Imp. Acad. **20**(8), 519–524 (1944)
16. B. Klartag, A central limit theorem for convex sets. Invent. Math. **168**, 91–131 (2007)
17. B. Klartag, Power-law estimates for the central limit theorem for convex sets. J. Funct. Anal. **245**, 284–310 (2007)
18. Y.T. Lee, S.S. Vempala, *Eldan's Stochastic Localization and the KLS Hyperplane Conjecture: Isoperimetry, Concentration and Mixing* (2016). arXiv:1612.01507 (updated Jan 2019)
19. Y.T. Lee, S.S. Vempala, Eldan's stochastic localization and the kls hyperplane conjecture: an improved lower bound for expansion, in *2017 IEEE 58th Annual Symposium on Foundations of Computer Science (FOCS)*, pp. 998–1007 (IEEE, Piscataway , 2017)
20. Y.T. Lee, S.S. Vempala, The kannan-lovász-simonovits conjecture. CoRR, abs/1807.03465 (2018)
21. Y.T. Lee, S.S. Vempala, Stochastic localization+ stieltjes barrier= tight bound for log-sobolev, in *Proceedings of the 50th Annual ACM SIGACT Symposium on Theory of Computing*, pp. 1122–1129 (ACM, New York, 2018)
22. E. Lieb, W. Thirring, Inequalities for the moments of the eigenvalues of the Schrödinger equation and their relation to Sobolev inequalities, in *Studies in Mathematical Physics: Essays in honor of Valentine Bargman* (ed.) E. Lieb, B. Simon, A.S. Wightman, pp. 269–303 (1976)
23. L. Lovász, S. Vempala, The geometry of logconcave functions and sampling algorithms. Random Struct. Algoritm. **30**(3), 307–358 (2007)
24. V.G. Maz'ja, Classes of domains and imbedding theorems for function spaces. Dokl. Acad. Nauk SSSR (Engl. transl. Soviet Math. Dokl., 1 (1961) 882–885) **3**, 527–530 (1960)
25. E.S. Meckes, M.W. Meckes, On the equivalence of modes of convergence for log-concave measures, in *Geometric Aspects of Functional Analysis*, pp. 385–394 (Springer, Berlin, 2014)
26. E. Milman, On the role of convexity in isoperimetry, spectral gap and concentration. Invent. Math. **177**(1), 1–43 (2009)
27. G. Paouris, Concentration of mass on convex bodies. Geom. Funct. Anal. **16**, 1021–1049 (2006)
28. A. Zhai, A high-dimensional CLT in w_2 distance with near optimal convergence rate. Probab. Theory Relat. Fields **170**(3–4), 821–845 (2018)

Chapter 2
The Lower Bound for Koldobsky's Slicing Inequality via Random Rounding

Bo'az Klartag and Galyna V. Livshyts

Abstract We study the lower bound for Koldobsky's slicing inequality. We show that there exists a measure μ and a symmetric convex body $K \subseteq \mathbb{R}^n$, such that for all $\xi \in \mathbb{S}^{n-1}$ and all $t \in \mathbb{R}$,

$$\mu^+(K \cap (\xi^\perp + t\xi)) \leq \frac{c}{\sqrt{n}} \mu(K) |K|^{-\frac{1}{n}}.$$

Our bound is optimal, up to the value of the universal constant. It improves slightly upon the results of the first named author and Koldobsky, which included a doubly-logarithmic error. The proof is based on an efficient way of discretizing the unit sphere.

2.1 Introduction

We shall work in the Euclidean n-dimensional space $\mathbb{R}^n$. The unit ball shall be denoted by B_2^n and the unit sphere by $\mathbb{S}^{n-1}$. The Lebesgue volume of a measurable set $A \subset \mathbb{R}^n$ is denoted by $|A|$. Throughout the paper, c, C, C' etc stand for positive absolute constants whose value may change from line to line.

Given a measure μ with a continuous density f on $\mathbb{R}^n$ and a set $A \subseteq \mathbb{R}^n$ of Hausdorff dimension $n - 1$, we write

$$\mu^+(A) = \int_A f(x)dx,$$

B. Klartag (✉)
School of Mathematical Sciences, Tel Aviv University, Tel Aviv, Israel

Department of Mathematics, Weizmann Institute of Science, Rehovot, Israel
e-mail: boaz.klartag@weizmann.ac.il

G. V. Livshyts
Department of Mathematics, Georgia Institute of Technology, Atlanta, GA, USA

© Springer Nature Switzerland AG 2020
B. Klartag, E. Milman (eds.), *Geometric Aspects of Functional Analysis*,
Lecture Notes in Mathematics 2266,
https://doi.org/10.1007/978-3-030-46762-3_2

where the integration is with respect to the $(n-1)$-dimensional Hausdorff measure.

For a measure μ on $\mathbb{R}^n$ with a continuous density and for an origin symmetric convex body K in $\mathbb{R}^n$ (i.e., $K = -K$), define the quantity

$$S_{\mu,K} = \inf_{\xi \in \mathbb{S}^{n-1}} \frac{\mu(K)}{|K|^{\frac{1}{n}} \mu^+(K \cap \xi^\perp)},$$

where $\xi^\perp = \{x \in \mathbb{R}^n, \langle x, \xi \rangle = 0\}$ is the hyperplane orthogonal to ξ. We let

$$S_n = \sup_\mu \sup_{K \subset \mathbb{R}^n} S_{\mu,K},$$

where the suprema run over measures μ with a continuous density f in $\mathbb{R}^n$ and all origin-symmetric convex bodies $K \subseteq \mathbb{R}^n$

Koldobsky in a series of papers [12–14] investigated the question *how large can S_n be?* The discrete version of this question was studied by Alexander, Henk, Zvavitch [1] and Regev [19]. In [12], where the question has first arisen, Koldobsky gave upper and lower bounds on $S(\mu, K)$, that are independent of the dimension in the case when K is an intersection body. In [13], he established the general bound $S_n \leq \sqrt{n}$. In [14], he has shown that $S_{\mu,K}$ is bounded from above by an absolute constant in the case when K is an unconditional convex body (invariant under coordinate reflections). Further, Koldobsky and Pajor [15] have shown that $S_{\mu,K} \leq C\sqrt{p}$ when K is the unit ball of an n-dimensional section of L_p.

In the case when μ is the Lebesgue measure, it was conjectured by Bourgain [5, 6] that $S_{\mu,K} \leq C$, for an arbitrary origin-symmetric convex body K. The best currently known bound in this case is $S_{\mu,K} \leq Cn^{\frac{1}{4}}$, established by the first named author [10], slightly improving upon Bourgain's estimate from [7]. However, it was shown by the first named author and Koldobsky [11] that $S_n \geq \frac{c\sqrt{n}}{\sqrt{\log\log n}}$. Moreover, it was shown there that for every n there exists a measure μ with continuous density and a symmetric convex body $K \subseteq \mathbb{R}^n$ such that for all $\xi \in \mathbb{S}^{n-1}$ and for all $t \geq 0$,

$$\mu^+(K \cap (\xi^\perp + t\xi)) \leq C \frac{\sqrt{\log\log n}}{\sqrt{n}} \mu(K)|K|^{-\frac{1}{n}}, \tag{2.1}$$

where $C > 0$ is some absolute constant. Here $A + x = \{y + x; y \in A\}$ for a set $A \subseteq \mathbb{R}^n$ and a vector $x \in \mathbb{R}^m$. In this note we improve the bound (2.1), and obtain:

Theorem 2.1.1 *For every n there exists a measure μ and a convex symmetric body $L \subseteq \mathbb{R}^n$ such that for all $\xi \in \mathbb{S}^{n-1}$ and for all $t \geq 0$,*

$$\mu^+(L \cap (\xi^\perp + t\xi)) \leq \frac{C}{\sqrt{n}} \mu(L)|L|^{-\frac{1}{n}}, \tag{2.2}$$

where $C > 0$ is a universal constant.

In [4], the first named author, Bobkov and Koldobsky explored the connections of (2.1) and the maximal "distance" of convex bodies to subspaces of L_p. Write L_p^n for the collection of origin-symmetric convex bodies in $\mathbb{R}^n$ that are linear images of unit balls of n-dimensional subspaces of the Banach space L_p. The *outer volume ratio* of a symmetric convex body K in $\mathbb{R}^n$ to the subspaces of L_p is defined as

$$d_{ovr}(K, L_p^n) := \inf_{D \in L_p^n : K \subset D} \left(\frac{|D|}{|K|} \right)^{\frac{1}{n}}.$$

John's theorem, and the fact that l_2^n embeds in L_p, entails that $d_{ovr}(K, L_p^n) \leq \sqrt{n}$, for any symmetric convex body K. Combined with the consideration from [4], Theorem 2.1.1 implies a doubly-logarithmic improvement of a result of [4]:

Corollary 2.1.2 *There exists an absolute constant $c > 0$ and an origin-symmetric convex body L in $\mathbb{R}^n$ such that for any $p \geq 1$,*

$$d_{ovr}(L, L_p^n) \geq c \frac{\sqrt{n}}{\sqrt{p}}.$$

The construction of μ and K is randomized, and follows the idea from [11]. The question boils down to estimating the supremum of a certain random function. The method of the proof is based on an efficient way of discretizing the unit sphere. We consider, for every point in $\mathbb{S}^{n-1}$, a "rounding" to a point in a scaled integer lattice, chosen at random, see Raghavan and Thompson [17]. This construction was recently used in [2] for efficiently computing sketches of high-dimensional data. It is somewhat reminiscent of the method used in discrepancy theory, called jittered sampling. For instance, using this method, Beck [3] has obtained strong bounds for the L_2-discrepancy.

In Sect. 2.2 we describe the net construction. In Sect. 2.3 we derive the key estimate for our random function. In Sect. 2.4 we conclude the proof of Theorem 2.1.1. In Sect. 2.5 we briefly outline some further applications, in particular in relation to random matrices; this discussion in detail shall appear in a separate paper.

We use the notation $\log^{(k)}(\cdot)$ for the logarithm iterated k times, and $\log^* n$ for the smallest positive integer m such that $\log^{(m)} n \leq 1$. Denote $\|x\|_p = \left(\sum_i |x_i|^p \right)^{1/p}$ for $x \in \mathbb{R}^n$, and also $\|x\|_\infty = \max_i |x_i|$ and $|x| = \|x\|_2 = \sqrt{\langle x, x \rangle}$. Write $B_p^n = \{x \in \mathbb{R}^n ; \|x\|_p \leq 1\}$. We also write $A + B = \{x + y ; x \in A, y \in B\}$ for the Minkowski sum.

2.2 The Random Rounding and the Net Construction

We fix a dimension n and a parameter $\rho \in (0, 1/2)$. We define $\mathcal{F}_\rho$ as the set of all vectors of Euclidean norm between $1 - 2\rho$ and $1 + \rho$ in which every coordinate is an integer multiple of $\rho/\sqrt{n}$. That is,

$$\mathcal{F}_\rho = \left((1+\rho)B_2^n \setminus (1-2\rho)B_2^n\right) \cap \frac{\rho}{\sqrt{n}}\mathbb{Z}^n.$$

Lemma 2.2.1 *The set $\mathcal{F}_\rho$ satisfies $\#\mathcal{F}_\rho \leq \left(\frac{C}{\rho}\right)^n$, where C is a universal constant. Moreover, let $\xi \in \mathbb{S}^{n-1}$, and suppose that $\eta \in (\rho/\sqrt{n})\mathbb{Z}^n$ satisfies $\|\xi - \eta\|_\infty \leq \rho/\sqrt{n}$. Then $\eta \in \mathcal{F}_\rho$.*

Proof Any $x \in \mathcal{F}_\rho$ satisfies $\|x\|_1 \leq \sqrt{n}|x| \leq 2\sqrt{n}$. Hence all vectors in the scaled set $(\sqrt{n}/\rho) \cdot \mathcal{F}_\rho$ have integer coordinates whose absolute values sum to a number which is at most $2n/\rho$. Recall that the number of vectors $x \in \mathbb{R}^n$ with non-negative, integer coordinates and $\|x\|_1 \leq R$ equals

$$\binom{R+n}{n} \leq \left(e\frac{R+n}{n}\right)^n$$

where R is a non-negative integer. Consequently,

$$\#\mathcal{F}_\rho \leq 2^n \cdot \left(e\frac{2\rho^{-1}n+n}{n}\right)^n \leq \left(\frac{C}{\rho}\right)^n.$$

We move on to the "Moreover" part. We have $|\xi - \eta| \leq \sqrt{n}\|\xi - \eta\|_\infty \leq \rho$. Therefore $1 - 2\rho < |\eta| \leq \rho$ and consequently $\eta \in ((1+\rho)B_2^n \setminus (1-2\rho)B_2^n) \cap \frac{\rho}{\sqrt{n}}\mathbb{Z}^n = \mathcal{F}_\rho$. $\qquad\square$

Definition 2.2.2 For $\xi \in \mathbb{S}^{n-1}$ consider a random vector $\eta^\xi \in (\rho/\sqrt{n})\mathbb{Z}^n$ with independent coordinates such that $\|\xi - \eta^\xi\|_\infty \leq \rho/\sqrt{n}$ with probability one and $\mathbb{E}\eta^\xi = \xi$. Namely, for $i = 1, \ldots, n$, writing $\xi_i = \frac{\rho}{\sqrt{n}}(k_i + p_i)$ for an integer k_i and $p_i \in [0, 1)$,

$$\eta_i^\xi = \begin{cases} \frac{\rho}{\sqrt{n}}k_i, & \text{with probability } 1 - p_i \\ \frac{\rho}{\sqrt{n}}(k_i + 1), & \text{with probability } p_i. \end{cases}$$

For any $\xi \in \mathbb{S}^{n-1}$, the random vector η^ξ belongs to $\mathcal{F}_\rho$ with probability one, according to Lemma 2.2.1. The random vector $\eta^\xi - \xi$ is a centered random vector with independent coordinates, all belonging to the interval $[-\rho/\sqrt{n}, \rho/\sqrt{n}]$. We shall make use of Hoeffding's inequality for bounded random variables (see, e.g., Theorem 2.2.6 and Theorem 2.6.2 in Vershynin [20]).

Lemma 2.2.3 (Hoeffding's Inequality) *Let $X_1, \ldots, X_n$ be independent random variables taking values in $[m_i, M_i]$, $i = 1, \ldots, n$. Then for any $\beta > 0$,*

$$P\left(\left|\sum_{i=1}^{n} X_i - \mathbb{E}X_i\right| \geq \beta\right) \leq 2e^{-\frac{c\beta^2}{\sum_{i=1}^{n}(M_i - m_i)^2}},$$

where $c > 0$ is an absolute constant.

The next Lemma follows immediately from Hoeffding's inequality with $X_i = (\eta_i^\xi - \xi_i)\theta_i$ and $[m_i, M_i] = [-\frac{\rho}{\sqrt{n}}\theta_i, \frac{\rho}{\sqrt{n}}\theta_i]$:

Lemma 2.2.4 *For any $\xi \in \mathbb{S}^{n-1}$, $\beta > 0$ and $\theta \in \mathbb{R}^n$,*

$$P(|\langle \eta^\xi - \xi, \theta \rangle| \geq \beta) \leq 2\exp\left(-\frac{cn\beta^2}{|\theta|^2\rho^2}\right).$$

Here $c > 0$ is an absolute constant.

2.3 The Key Estimate

Let N be a positive integer, and consider independent random vectors $\theta_1, \ldots, \theta_N$ uniformly distributed on the unit sphere $\mathbb{S}^{n-1}$. Unless specified otherwise, the expectation and the probability shall be considered with respect to their distribution.

For $r > 0$, abbreviate

$$\varphi(r) = e^{-\frac{r^2}{2}}.$$

The main result of this section is the following Proposition.

Proposition 2.3.1 *There exist absolute constants $C_1, \ldots, C_5 > 0$ with the following property. Let $n \geq 5$, consider $r \in [C_2\sqrt{n}, n]$ and suppose that $N \geq n$ satisfies $N \in [C_1 n \log \frac{Nr}{n\sqrt{n}}, n^{10}]$. Then with probability at least $1 - e^{-5n}$, for all $\xi \in \mathbb{S}^{n-1}$, and for all $t \in \mathbb{R}$,*

$$\frac{1}{N}\sum_{k=1}^{N} \varphi(r\langle \xi, \theta_k \rangle + t) \leq C_3\sqrt{\frac{n\log\frac{Nr}{n\sqrt{n}}}{N}} + \left(1 + \frac{C_4\sqrt{n}}{r}\right)\frac{\sqrt{n}}{r}\varphi\left(\frac{q\sqrt{n}}{r}t\right),$$

where $q \geq 1 - \frac{C_5\sqrt{n}}{r}$.

We shall require a few Lemmas, before we proceed with the proof of Proposition 2.3.1.

2.3.1 Asymptotic Estimates

For a fixed vector $\eta \in \mathbb{R}^n$ and $t \in \mathbb{R}$, denote

$$F(\eta, t) = \frac{1}{N} \sum_{k=1}^{N} \varphi(r \langle \eta, \theta_k \rangle + t). \qquad (2.3)$$

Observe that $F(\eta, t) \leq 1$ with probability one. First, we shall show a sharpening of [11, Lemma 3.2].

Lemma 2.3.2 *Let $n \geq 1$. Let θ be a random vector uniformly distributed on $\mathbb{S}^{n+2}$. For any $r > 0$, for any $t \in \mathbb{R}$, for any fixed $\eta \in \mathbb{R}^{n+3}$, one has*

$$\mathbb{E}\varphi(r \langle \theta, \eta \rangle + t) \leq \left(1 + \frac{c(\log n)^2}{n}\right) \frac{\sqrt{n}}{\sqrt{n + r^2 |\eta|^2}} \varphi\left(\frac{t\sqrt{n}}{\sqrt{n + r^2 |\eta|^2}}\right).$$

Here $c > 0$ is an absolute constant.

Proof Observe that the formulation of the Lemma allows to assume, without loss of generality, that $|\eta| = 1$: indeed, in the case $\eta = 0$ the statement is straight-forward, and otherwise it follows from the case $|\eta| = 1$ by scaling. The random variable $\langle \theta, \eta \rangle$ is distributed on $[-1, 1]$ according to the density

$$\frac{(1 - s^2)^{\frac{n}{2}}}{\int_{-1}^{1} (1 - s^2)^{\frac{n}{2}} ds}.$$

Recall that for any $x \in [0, 1]$,

$$\log(1 - x) = -x - \frac{x^2}{2} + O(x^3), \qquad (2.4)$$

and hence there is an absolute constant $C > 0$ such that for any $x \in [0, \frac{2 \log n}{n}]$,

$$\log(1 - x) \geq -x - \frac{C(\log n)^2}{n^2}. \qquad (2.5)$$

Applying (2.5) with $x = s^2$, we estimate

$$\int_{-1}^{1} (1 - s^2)^{\frac{n}{2}} ds \geq \int_{-\sqrt{\frac{2 \log n}{n}}}^{\sqrt{\frac{2 \log n}{n}}} (1 - s^2)^{\frac{n}{2}} ds$$

$$\geq \int_{-\sqrt{\frac{2 \log n}{n}}}^{\sqrt{\frac{2 \log n}{n}}} e^{-\frac{ns^2}{2} - \frac{C(\log n)^2}{2n}} ds$$

$$\geq \left(1 - \frac{c'(\log n)^2}{n}\right) \int_{-\sqrt{\frac{2\log n}{n}}}^{\sqrt{\frac{2\log n}{n}}} e^{-\frac{ns^2}{2}} ds$$

$$= \frac{1}{\sqrt{n}} \left(1 - \frac{c'(\log n)^2}{n}\right) \int_{-\sqrt{2\log n}}^{\sqrt{2\log n}} e^{-\frac{s^2}{2}} ds. \qquad (2.6)$$

Recall that for any $a > 0$, one has

$$\int_a^\infty e^{-\frac{y^2}{2}} dy \leq \frac{1}{a} e^{-\frac{a^2}{2}}, \qquad (2.7)$$

and therefore

$$\int_{-\sqrt{2\log n}}^{\sqrt{2\log n}} e^{-\frac{s^2}{2}} ds \geq \sqrt{2\pi} - \frac{\sqrt{2}}{n\sqrt{\log n}}. \qquad (2.8)$$

By (2.8) and (2.6), we conclude that there exists an absolute constant $\tilde{c} > 0$ such that

$$\int_{-1}^1 (1 - s^2)^{\frac{n}{2}} ds \geq \frac{\sqrt{2\pi}}{\sqrt{n}} \left(1 - \frac{\tilde{c}(\log n)^2}{n}\right). \qquad (2.9)$$

We remark that the second order term estimate is of course not sharp, yet it is more than sufficient for our purposes.

Next, using the inequality $1 - x \leq e^{-x}$ for $x = s^2$, we estimate from above

$$\int_{-1}^1 (1 - s^2)^{\frac{n}{2}} e^{-\frac{(rs+t)^2}{2}} ds \leq \int_{-\infty}^\infty e^{-\frac{ns^2+(rs+t)^2}{2}} ds. \qquad (2.10)$$

It remains to observe that

$$ns^2 + (rs + t)^2 = \left(\sqrt{n + r^2} s + \frac{tr}{\sqrt{n + r^2}}\right)^2 + \frac{nt^2}{n + r^2},$$

and to conclude, by (2.10), that

$$\int_{-1}^1 (1 - s^2)^{\frac{n}{2}} e^{-\frac{(rs+t)^2}{2}} ds \leq \sqrt{2\pi} \frac{1}{\sqrt{n + r^2}} \varphi\left(\frac{\sqrt{n}t}{\sqrt{n + r^2}}\right). \qquad (2.11)$$

From (2.9) and (2.11) we note, for every unit vector η :

$$\mathbb{E}\varphi(r\langle\theta, \eta\rangle + t) \leq \left(1 + \frac{c(\log n)^2}{n}\right) \frac{\sqrt{n}}{\sqrt{n + r^2}} \varphi\left(\frac{t\sqrt{n}}{\sqrt{n + r^2}}\right). \qquad (2.12)$$

$\square$

As an immediate corollary of Lemma 2.3.2 and Hoeffding's inequality, we get:

Lemma 2.3.3 *Let* $N \geq n \geq 4$, $r \geq \sqrt{n}$ *and* $\rho \in (0, \frac{1}{3})$. *There exist absolute constants* $c, C, C' > 0$ *such that for all* $\eta \in (1 + \rho)B_2^n \setminus (1 - 2\rho)B_2^n$ *and* $t \in \mathbb{R}$, $\beta > 0$,

$$
P\left(F(\eta, t) > \beta + (1 + c(\rho + \frac{(\log n)^2}{n} + \frac{n}{r^2}))\frac{\sqrt{n}}{r}\varphi\left(\frac{qt\sqrt{n}}{r}\right)\right) \leq e^{-C\beta^2 N},
$$

where $q \geq 1 - C'(\rho + \frac{n}{r^2})$.

Proof In view of Lemma 2.2.3 (Hoeffding's inequality), it suffices to show that under the assumptions of the Lemma,

$$
\mathbb{E}\varphi(r\langle\theta, \eta\rangle + t) \leq (1 + c(\rho + \frac{(\log n)^2}{n} + \frac{n}{r^2}))\frac{\sqrt{n}}{r}\varphi\left(\frac{qt\sqrt{n}}{r}\right). \tag{2.13}
$$

Indeed, by Lemma 2.3.2, for some $c_1 > 0$,

$$
\mathbb{E}\varphi(r\langle\theta, \eta\rangle + t) \leq \left(1 + \frac{c_1(\log n)^2}{n}\right)\frac{\sqrt{n}}{\sqrt{n + r^2|\eta|^2}}\varphi\left(\frac{t\sqrt{n}}{\sqrt{n + r^2|\eta|^2}}\right).
$$

It remains to observe, that since $r \geq \sqrt{n}$,

$$
\frac{|t|\sqrt{n}}{\sqrt{n + r^2|\eta|^2}} \geq \frac{q|t|\sqrt{n}}{r},
$$

where $q = 1 + O(\rho + \frac{n}{r^2})$, and

$$
\left(1 + \frac{c_1(\log n)^2}{n}\right)\frac{\sqrt{n}}{\sqrt{n + r^2|\eta|^2}} \leq \left(1 + c(\rho + \frac{(\log n)^2}{n} + \frac{n}{r^2})\right)\frac{\sqrt{n}}{r},
$$

with an appropriate constant $c > 0$. $\square$

2.3.2 Union Bound

Given $\rho > 0$, recall the notation $\mathcal{F}_\rho$ for the net from Lemma 2.2.1. Our next Lemma is a combination of the union bound with Lemma 2.3.3.

Lemma 2.3.4 (Union Bound) *There exist absolute constants* $C_1, C_2, C' > 0$ *such that the following holds. Let* $\rho \in (0, \frac{1}{3})$. *Let* $N \in [C_1 n \log \frac{1}{\rho}, n^{10}]$ *be an integer. Fix*

$r \in [C_2\sqrt{n}, n]$. Then with probability at least $1 - e^{-5n}$, for every $\eta \in \mathcal{F}_\rho$, and for every $t \in \mathbb{R}$,

$$F(\eta, t) \le C_6\sqrt{\frac{n}{N}\log\frac{1}{\rho}} + \left(1 + C_7(\rho + \frac{n}{r^2} + \frac{(\log n)^2}{n} + \frac{1}{r})\right)\frac{\sqrt{n}}{r}\varphi\left(\frac{q\sqrt{n}t}{r}\right),$$

for large enough absolute constants $C_6, C_7 > 0$, which depend only on C_1 and C_2, and for $q \ge 1 - C'(\rho + \frac{n}{r^2})$.

Proof Let

$$\alpha = C_6\sqrt{\frac{n}{N}\log\frac{1}{\rho}} + \left(1 + C_7(\rho + \frac{n}{r^2} + \frac{(\log n)^2}{n} + \frac{1}{r})\right)\frac{\sqrt{n}}{r}\varphi\left(\frac{q\sqrt{n}t}{r}\right),$$

where $q \ge 1 - C'(\rho + n/r^2)$ and the constants shall be appropriately chosen later. Note that

$$\alpha \ge C_6\sqrt{\frac{n}{N}\log\frac{1}{\rho}} \ge n^{-4.5} \cdot C_6\sqrt{\log 2}, \tag{2.14}$$

since $\rho \le \frac{1}{2}$ and $N \le n^{10}$.

Observe also that for any pair of vectors $\theta \in \mathbb{S}^{n-1}$, $\eta \in \mathcal{F}_\rho \subset 2B_2^n$ and for any $t \ge 3r$, we have

$$|r\langle\eta, \theta\rangle + t| \ge r,$$

and hence

$$e^{-\frac{1}{2}(r\langle\eta,\theta\rangle+t)^2} \le e^{-\frac{r^2}{2}}. \tag{2.15}$$

In view of (2.14), (2.15), and the fact that $r \ge \sqrt{n}$, we have, for $t \ge 3r$:

$$F(\eta, t) \le e^{-\frac{r^2}{2}} \le e^{-\frac{n}{2}} \le n^{-4.5}C_6\sqrt{\log 2} \le \alpha,$$

where the inequality follows as long as C_6 is chosen to be larger than $1 + o(1)$. This implies the statement of the Lemma in the range $t \ge 3r$.

Next, suppose $t \in [0, 3r]$. Let $\epsilon = \frac{1}{r^2}$. Consider an ϵ-net $\mathcal{N} = \{t_1, \dots, t_m\}$ on the interval $[0, 3r]$ with $t_j = \epsilon \cdot j$. Note that

$$\#\mathcal{N} \le [3r^3] + 1 \le 4r^3, \tag{2.16}$$

since $r \ge \sqrt{n} \ge 1$.

For any $A \in \mathbb{R}$, for any $\epsilon > 0$, and for any $t_1, t_2 \in \mathbb{R}$ such that $|t_1 - t_2| \le \epsilon$, we have

$$|A + t_2|^2 \le |A + t_1|^2 + 2\epsilon|A + t_1| + \epsilon^2,$$

and hence

$$\varphi(A + t_1) \le \varphi(A + t_2)e^{|A+t_1|\epsilon + \frac{\epsilon^2}{2}}. \tag{2.17}$$

Observe that for all $t \in [0, 3r]$, for an arbitrary $\eta \in \mathcal{F}_\rho \subset 2B_2^n$, and any $\theta \in \mathbb{S}^{n-1}$, we have

$$|r\langle \eta, \theta \rangle + t| \le 5r,$$

and hence

$$e^{|r\langle \eta,\theta \rangle + t|\epsilon + \frac{\epsilon^2}{2}} \le e^{5r\epsilon + \frac{\epsilon^2}{2}} = e^{\frac{5}{r} + \frac{1}{2r^2}} \le 1 + \frac{C'}{r}, \tag{2.18}$$

for an absolute constant C'.

By (2.17) and (2.18), for each $t \subset [0, 3r]$ there exists $\tau \in \mathcal{N}$, such that

$$F(\eta, t) \le \left(1 + \frac{C'}{r}\right) F(\eta, \tau).$$

Therefore, by the union bound,

$$P\left(\exists t \in [0, 3r], \ \exists \eta \in \mathcal{F}_\rho : \ F(\eta, t) > \alpha\right)$$

$$\le P\left(\exists \tau \in \mathcal{N}, \ \exists \eta \in \mathcal{F}_\rho : \ F(\eta, \tau) > \frac{\alpha}{1 + \frac{C'}{r}}\right)$$

$$\le \#\mathcal{N} \cdot \#\mathcal{F}_\rho \cdot P\left(F(\eta, \tau) > \frac{\alpha}{1 + \frac{C'}{r}}\right). \tag{2.19}$$

By Lemma 2.2.1 and (2.16),

$$\#\mathcal{N} \cdot \#\mathcal{F}_\rho \le 4r^3 \left(\frac{C}{\rho}\right)^n \le \left(\frac{\tilde{C}}{\rho}\right)^n. \tag{2.20}$$

We used above that $r \le n$.

Let

$$\beta := \left(1 + \frac{C'}{r}\right)^{-1} C_6 \sqrt{\frac{n}{N} \log \frac{1}{\rho}}.$$

Provided that C_6 and C_7 are chosen large enough, we have:

$$\frac{\alpha}{1 + \frac{C'}{r}} \geq \beta + (1 + c(\rho + \frac{(\log n)^2}{n} + \frac{n}{r^2}))\frac{\sqrt{n}}{r}\varphi\left(\frac{qt\sqrt{n}}{r}\right), \tag{2.21}$$

and

$$C\beta^2 N = C(1 + \frac{C'}{r})^{-2}C_6^2 n \log \frac{1}{\rho} \geq 5n + n \log \frac{\tilde{C}}{\rho}, \tag{2.22}$$

where c and C are the constants from Lemma 2.3.3 and $\tilde{C}$ is the constant from (2.20).

By Lemma 2.3.3, (2.21) and (2.22), we have

$$P\left(F(\eta, t) > \frac{\alpha}{1 + \frac{C'}{r}}\right) \leq e^{-C\beta^2 N} \leq e^{-5n - n \log \frac{\tilde{C}}{\rho}}. \tag{2.23}$$

By (2.19), (2.20) and (2.23), we conclude that the desired event holds with probability at least

$$1 - \left(\frac{\tilde{C}}{\rho}\right)^n e^{-5n - n \log \frac{\tilde{C}}{\rho}} = 1 - e^{-5n}.$$

This finishes the proof.   □

2.3.3 An Application of Random Rounding and Conclusion of the Proof of the Proposition 2.3.1

We begin by formulating a general fact about sub-Gaussian random variables, which complements the estimate from Lemma 2.3.2.

Lemma 2.3.5 *Let* $M \geq 10$. *Let* Y *be a sub-Gaussian random variable with constant* $\frac{1}{M}$: *that is, suppose for any* $s > 0$,

$$P(|Y| > s) \leq e^{-M^2 s^2}. \tag{2.24}$$

Then there exists an absolute constant $C > 0$, *such that for any* $a \in \mathbb{R}$,

$$\mathbb{E}\varphi(Y + a) \geq \varphi(a) - \frac{C}{M}.$$

Here the expectation is taken with respect to Y.

Proof Since the condition (2.24) applies for both Y and $-Y$, and since φ is an even function, we may assume, without loss of generality, that $a \geq 0$ (alternatively, we may replace a with $|a|$ in the calculations below).

We begin by writing

$$
\mathbb{E}\varphi(Y+a) = \int_0^1 P(\varphi(Y+a) > \lambda)d\lambda = \int_0^\infty se^{-\frac{s^2}{2}} P(|Y+a| < s)ds
$$

$$
\geq \int_a^\infty se^{-\frac{s^2}{2}} \left(1 - P(|Y+a| \geq s)\right) ds
$$

$$
= e^{-\frac{a^2}{2}} - \int_a^\infty se^{-\frac{s^2}{2}} P(|Y+a| \geq s)ds. \tag{2.25}
$$

Note that for $s \geq a \geq 0$, we have

$$
P(|Y+a| \geq s) = P(Y \geq s-a) + P(-Y \geq s+a) \leq 2P(|Y| \geq s-a). \tag{2.26}
$$

By (2.24) and (2.26), we estimate

$$
\int_a^\infty se^{-\frac{s^2}{2}} P(|Y+a| \geq s)ds \leq 2 \int_a^\infty se^{-\frac{s^2}{2}} e^{-M^2(s-a)^2}ds
$$

$$
= 2 \int_0^\infty (t+a)e^{-\frac{(t+a)^2}{2}} e^{-M^2 t^2} dt. \tag{2.27}
$$

Recall that

$$
(t+a)e^{-\frac{(t+a)^2}{2}} \leq \frac{1}{\sqrt{e}}, \tag{2.28}
$$

and that

$$
\int_0^\infty e^{-M^2 t^2} dt = \frac{\sqrt{\pi}}{2M}. \tag{2.29}
$$

By (2.25), (2.27), (2.28) and (2.29), letting $C = \frac{\sqrt{\pi}}{\sqrt{e}}$, we have

$$
\mathbb{E}\varphi(Y+a) \geq \varphi(a) - \frac{C}{M}, \tag{2.30}
$$

yielding the conclusion.

$\square$

Next, we shall demonstrate the following corollary of Lemmas 2.2.4 and 2.3.5.

Corollary 2.3.6 *There exist absolute constants $C, c > 0$ such that for any $M, r > 0$ and $\rho \in (0, \frac{c\sqrt{n}}{rM}]$, and for any $\xi \in \mathbb{S}^{n-1}$,*

$$F(\xi, t) \leq \mathbb{E}_\eta F(\eta^\xi, t) + \frac{C}{M},$$

with function F defined in (2.3) and η^ξ defined in Definition 2.2, and the expectation taken with respect to η^ξ.

Proof By Lemma 2.2.4, for any fixed $\theta \in \mathbb{S}^{n-1}$, for an absolute constant $c > 0$, the random variable $r\langle \eta^\xi - \xi, \theta \rangle$ is sub-Gaussian with constant $\frac{r\rho}{c\sqrt{n}} \leq \frac{1}{M}$. Therefore, applying Lemma 2.3.5 N times with $Y = r\langle \eta^\xi - \xi, \theta_k \rangle$ and $a = r\langle \xi, \theta_k \rangle + t$, we get

$$\mathbb{E}_\eta \frac{1}{N} \sum_{k=1}^{N} \varphi(r\langle \eta^\xi, \theta_k \rangle + t) \geq \frac{1}{N} \sum_{k=1}^{N} \varphi(r\langle \xi, \theta_k \rangle + t) - \frac{1}{N} \sum_{k=1}^{N} \frac{C}{M}$$

$$= \frac{1}{N} \sum_{k=1}^{N} \varphi(r\langle \xi, \theta_k \rangle + t) - \frac{C}{M},$$

finishing the proof. □

We are ready to prove Proposition 2.3.1.

Proof of the Proposition 2.3.1 Let $\rho = \frac{n\sqrt{n}}{Nr}$. By Corollary 2.3.6, applied with $M = c\frac{N}{n}$, we have, for every $\xi \in \mathbb{S}^{n-1}$,

$$\frac{1}{N} \sum_{k=1}^{N} \varphi(r\langle \xi, \theta_k \rangle + t) \leq \mathbb{E}_\eta \frac{1}{N} \sum_{k=1}^{N} \varphi\left(r\langle \eta^\xi, \theta_k \rangle + t\right) + \frac{C'n}{N}$$

$$\leq \max_{\eta \in \mathcal{F}_\rho} \frac{1}{N} \sum_{k=1}^{N} \varphi\left(r\langle \eta, \theta_k \rangle + t\right) + \frac{C'n}{N}. \tag{2.31}$$

By Lemma 2.3.4 and with our choice of ρ, with probability $1 - e^{-5n}$, (2.31) is bounded from above by

$$C_6 \sqrt{\frac{n}{N} \log \frac{Nr}{n\sqrt{n}}} + \left(1 + C_7\left(\frac{n\sqrt{n}}{Nr} + \frac{n}{r^2} + \frac{(\log n)^2}{n} + \frac{1}{r}\right)\right) \frac{\sqrt{n}}{r} \varphi\left(\frac{q\sqrt{n}t}{r}\right) + \frac{C'n}{N},$$

where $q = 1 - C'(\rho + \frac{n}{r^2}) \geq 1 - C_5(\frac{\sqrt{n}}{r})$, in view of our choice of ρ. It remains to note, in view of the fact that $N \geq nC_1 \log 2$ and $r \geq C_2\sqrt{n}$, that for an appropriate absolute constant $C_3 > 0$, one has

$$C_6 \sqrt{\frac{n}{N} \log \frac{Nr}{n\sqrt{n}}} + \frac{C'n}{N} \leq C_3 \sqrt{\frac{n}{N} \log \frac{Nr}{n\sqrt{n}}},$$

and for an appropriate absolute constant $C_4 > 0$,

$$C_7 \left(\frac{n\sqrt{n}}{Nr} + \frac{n}{r^2} + \frac{(\log n)^2}{n} + \frac{1}{r} \right) \leq C_4 \frac{\sqrt{n}}{r}.$$

The proposition follows. $\qquad\qquad\qquad\qquad\qquad\qquad\qquad\qquad\qquad\qquad$ $\square$

2.4 Proof of Theorem 2.1.1

Let m be the largest positive integer such that $\log^{(m)} n \geq C_0$, for a sufficiently large absolute constant $C_0 > 0$ to be determined shortly. Note that, hence,

$$\log^{(m)} n \leq C_0', \qquad\qquad\qquad\qquad (2.32)$$

for some absolute constant C_0'.

Consider, for $k = 1, \ldots, m$,

$$N_1 = n^{10}, \quad N_2 = n(\log n)^5, \ldots, \quad N_k = n \left(\log^{(k-1)} n \right)^5, \ldots$$

Let also

$$R_1 = \frac{n}{\log n}, \ldots, \quad R_k = \frac{n}{\log^{(k)} n}, \ldots$$

Consider independent unit random vectors $\theta_{kj} \in \mathbb{S}^{n-1}$, where $k = 1, \ldots, m$ and $j = 1, \ldots, N_k$. Following [11], consider the convex body

$$K = conv\{\pm R_k \theta_{kj}, \pm n e_i\},$$

and the probability measures

$$\mu_k = \frac{1}{N_k} \sum_{j=1}^{N_k} \delta_{R_k \theta_{kj}}, \qquad\qquad \mu_{-k} = \frac{1}{N_k} \sum_{j=1}^{N_k} \delta_{-R_k \theta_{kj}},$$

where δ stands for the Dirac measure. We now set

$$\mu = \gamma_n * \frac{\mu_1 * \mu_2 * \ldots * \mu_m + \mu_{-1} * \mu_{-2} * \ldots * \mu_{-m}}{2}.$$

Here γ_n stands for the standard Gaussian measure on $\mathbb{R}^n$. We shall show that there exists a configuration of θ_{kj}, such that μ and $L = 4K$ satisfy the conclusion of the theorem.

Step 1 Firstly, we estimate the volume of the body $L = 4K$ from above, following the method of [11]. Note that for all $k = 1, \ldots, m$ we have $\varphi\left(\frac{5n}{R_k}\right) \leq c$, for some absolute constant $c \in (0, 1)$, and hence there exists an absolute constant $\hat{C} > 0$ such that

$$\log\left[1 - \varphi\left(\frac{5n}{R_k}\right)\right] \geq -\hat{C}\varphi\left(\frac{5n}{R_k}\right), \tag{2.33}$$

for all $k = 1, \ldots, m$.

By Khatri-Sidak lemma (see, e.g. [9] for a simple proof), applied together with the Blaschke-Santalo inequality, and in view of (2.33), we have

$$|4K|^{-1} \geq c_1^n |5nK^o| \geq c_2^n \gamma_n(5nK^o) \geq c^n \prod_{k=1}^{m}\left(1 - \varphi\left(\frac{5n}{R_k}\right)\right)^{N_k}$$

$$\geq c^n \exp\left(-\hat{C}\sum_{k=1}^{m} N_k e^{-\frac{25n^2}{2R_k^2}}\right). \tag{2.34}$$

Plugging the values of N_k and R_k, and using $\frac{25}{2} > 7$, we get

$$\sum_{k=1}^{m} N_k e^{-\frac{25n^2}{2R_k^2}} \leq n^{10} e^{-7(\log n)^2} + n\sum_{k=2}^{m}(\log^{(k-1)} n)^5 e^{-7(\log^{(k)} n)^2} \leq c'n, \tag{2.35}$$

since the sum converges faster than exponentially.

By (2.34) and (2.35), we conclude that

$$|4K| \leq c_0^n, \tag{2.36}$$

for some absolute constant $c_0 > 0$.

Step 2 Next, we estimate the sections from above. Note that (see [11] for details),

$$\mu^+(\xi^\perp + t\xi) = \frac{A + B}{2} \tag{2.37}$$

where

$$A = \frac{1}{\sqrt{2\pi}} \frac{1}{N_1 \dots N_m} \sum_{j_1=1}^{N_1} \sum_{j_2=1}^{N_2} \dots \sum_{j_m=1}^{N_m} \varphi(t + R_1\langle\xi, \theta_{1j_1}\rangle + \dots + R_m\langle\xi, \theta_{mj_m}\rangle)$$

(2.38)

and

$$B = \frac{1}{\sqrt{2\pi}} \frac{1}{N_1 \dots N_m} \sum_{j_1=1}^{N_1} \sum_{j_2=1}^{N_2} \dots \sum_{j_m=1}^{N_m} \varphi(-t + R_1\langle\xi, \theta_{1j_1}\rangle + \dots + R_m\langle\xi, \theta_{mj_m}\rangle)$$

(2.39)

For $r \geq C_2\sqrt{n}$ we set $q(r) = 1 - C_5\sqrt{n}/r$ where C_5 is the constant coming from Proposition 2.3.1. We define $r_1, r_2, \dots, r_m \in [C_2\sqrt{n}, n]$ such that

$$r_1 := R_1$$

and for $k \geq 1$,

$$r_{k+1} := \frac{q(r_k)\sqrt{n}R_{k+1}}{R_k} = \left(\prod_{j=1}^{k} \frac{q(r_j)\sqrt{n}}{r_j}\right) \cdot R_{k+1}.$$

Denote

$$\alpha_k := \prod_{j=1}^{k-1}\left[\left(1 + \frac{C_4\sqrt{n}}{r_j}\right) \frac{1}{q(r_j)}\right] \leq \prod_{j=1}^{k-1}\left[\left(1 + \frac{\hat{C}\sqrt{n}}{r_j}\right)\right].$$

(2.40)

The reason for the definition of α_k, is the inequality

$$\prod_{j=1}^{k-1}\left[\left(1 + \frac{C_4\sqrt{n}}{r_j}\right) \frac{\sqrt{n}}{r_j}\right] \leq C\frac{\alpha_k\sqrt{n}}{R_{k-1}},$$

which we will use below in a repeated application of Proposition 2.3.1. Observe that there exists an absolute constant $\tilde{C} > 0$ such that for every $k = 1, \dots, m$, we have

$$\alpha_k \leq (1 + \hat{C}\frac{\sqrt{n}}{R_1}) \prod_{j=2}^{k-1}(1 + \check{C}\frac{\log^{(j)} n}{\log^{(j-1)} n}) \leq Ce^{\bar{C}\sum_{j=2}^{k-1} \frac{\log^{(j)} n}{\log^{(j-1)} n}} \leq \tilde{C},$$

(2.41)

since the sum converges faster than exponentially.

Provided that $C_0 > 0$ is selected large enough, we have that for each k, the pair $N = N_k$ and $r = r_k$ satisfies the assumptions of Proposition 2.3.1. Applying Proposition 2.3.1 consecutively m times with $N = N_k$ and $r = r_k$ for $k = 1, \ldots, m$, we get that with probability at least $1 - me^{-5n} = 1 - o(1)$, for every $\xi \in \mathbb{S}^{n-1}$ and for every $t \in \mathbb{R}$, the term A from (2.38) is bounded from above by a constant multiple of

$$\sqrt{\frac{n \log \frac{N_1 r_1}{n\sqrt{n}}}{N_1}} + \frac{\alpha_2 \sqrt{n}}{R_1} \sqrt{\frac{n \log \frac{N_2 r_2}{n\sqrt{n}}}{N_2}} + \ldots + \frac{\alpha_m \sqrt{n}}{R_{m-1}} \sqrt{\frac{n \log \frac{N_m r_m}{n\sqrt{n}}}{N_m}} + \frac{\alpha_{m+1} \sqrt{n}}{R_m}$$

$$\leq \frac{c'}{\sqrt{n}} + \frac{c''}{\sqrt{n}} \sum_{k=1}^{m} \alpha_k \frac{1}{\log^{(k)} n} + \frac{\alpha_m \log^{(m)} n}{\sqrt{n}} \leq \frac{C}{\sqrt{n}},$$

for an appropriate constant $C > 0$, where we used (2.41) to bound α_k, and (2.32) to bound $\log^{(m)} n$.

The same bound applies also to the term B from (2.39). We conclude, in view of (2.37) that with high probability, for all $\xi \in \mathbb{S}^{n-1}$ and for all $t \in \mathbb{R}$,

$$\mu(\xi^{\perp} + t\xi) \leq \frac{C}{\sqrt{n}}. \tag{2.42}$$

Step 3 Recall that μ is an average of translates of the Gaussian measure, centered at the vertices of K. As was shown in [11, Lemma 3.8], using the fact that $\sqrt{n} B_2^n \subset K$, and since $4K = 2K + 2K$ contains $2\sqrt{n} B_2^n + 2K$, one has

$$\mu(4K) \geq \gamma_n(2\sqrt{n} B_2^n) \geq \frac{1}{2}, \tag{2.43}$$

where, e.g. Markov's inequality is used in the last passage.

Combining (2.36), (2.42) and (2.43), we arrive to the conclusion of the theorem, with $L = 4K$.   □

2.5 Further Applications

2.5.1 Comparison via the Hilbert-Schmidt Norm for Arbitrary Matrices

As another consequence of the Lemma 2.2.4, we have:

Lemma 2.5.1 (Comparison via the Hilbert-Schmidt Norm) *Let* $\rho \in (0, \frac{1}{2})$. *There exists a collection of points* $\mathcal{N} \subset 2B_2^n \setminus \frac{1}{2} B_2^n$ *with* $\#\mathcal{N} \leq (\frac{C}{\rho})^n$ *such that* *for any matrix $A : \mathbb{R}^n \to \mathbb{R}^N$, for every $\xi \in \mathbb{S}^{n-1}$ there exists an $\eta \in \mathcal{N}$ satisfying*

$$|A\eta|^2 \leq C_1 |A\xi|^2 + C_2 \frac{\rho^2}{n} ||A||_{HS}^2. \tag{2.44}$$

Here C, C_1, C_2 are absolute constants.

Proof Recall that $|Ax|^2 = \sum_{i=1}^{N} \langle X_i, x \rangle^2$, where X_i are the rows of A. In order to prove the Lemma, it suffices to show, for every vector $g \in \mathbb{R}^n$, that

$$\mathbb{E}_\eta \langle \eta^\xi, g \rangle^2 \leq C_1 \langle \xi, g \rangle^2 + C_2 \frac{\rho^2}{n} |g|^2; \tag{2.45}$$

the Lemma shall follow by applying (2.45) to the rows of A and summing up.

We shall show (2.45). Using the inequality $a^2 = (a - b + b)^2 \leq 2(a-b)^2 + 2b^2$, we see

$$|\langle \eta^\xi, g \rangle|^2 \leq 2|\langle \eta^\xi, g \rangle - \langle \xi, g \rangle|^2 + 2|\langle \xi, g \rangle|^2,$$

and hence

$$\mathbb{E}_\eta |\langle \eta^\xi, g \rangle|^2 \leq 2\mathbb{E}_\eta |\langle \eta^\xi, g \rangle - \langle \xi, g \rangle|^2 + 2|\langle \xi, g \rangle|^2. \tag{2.46}$$

By Lemma 2.2.4, $|\langle \eta^\xi, g \rangle - \langle \xi, g \rangle|$ is sub-Gaussian with constant $c' \frac{\rho |g|}{\sqrt{n}}$, and hence

$$\mathbb{E}_\eta |\langle \eta^\xi, g \rangle - \langle \xi, g \rangle|^2 \leq 2 \int_0^\infty t e^{-\frac{cnt^2}{\rho^2 |g|^2}} dt \leq C \frac{\rho^2 |g|^2}{n}, \tag{2.47}$$

for some absolute constant $C > 0$; (2.46) and (2.47) entail (2.45). $\qquad\square$

A fact similar to Lemma 2.5.1 was recently shown and used by Lytova and Tikhomirov [16].

Lemma 2.5.1 shows that there exists a net of cardinality C^n, such that for any random matrix $A : \mathbb{R}^n \to \mathbb{R}^N$ whose entries have bounded second moments, with probability at least

$$1 - P(||A||_{HS}^2 \geq 10\mathbb{E}||A||_{HS}^2) \geq \frac{9}{10}$$

one has (2.44), with $\mathbb{E}||A||_{HS}^2$ in place of $||A||_{HS}^2$. However, such probability estimate is unsatisfactory when studying small ball estimates for the smallest singular values of random matrices. In the soon-to-follow paper, we significantly strengthen Lemma 2.5.1: we employ the idea of Rebrova and Tikhomirov [18], and in place of the covering by cubes, we consider a covering by parallelepipeds of sufficiently large volume. This leads us to consider the following *refinement of the Hilbert-Schmidt norm*: with $\kappa > 1$, for an $N \times n$ matrix A, define

$$\mathcal{B}_\kappa(A) = \min_{\alpha_i \in [0,1], \, \prod_{i=1}^n \alpha_i \geq \kappa^{-n}} \sum_{i=1}^n \alpha_i^2 |Ae_i|^2.$$

$\mathcal{B}_\kappa$ acts as an averaging on the columns of A. In a separate paper we shall show that there exists a net $\mathcal{N} \subset 2B_2^n \setminus \frac{1}{2}B_2^n$, of cardinality $\left(\frac{C}{\rho}\right)^n$, such that for all $N \times n$ matrices A, for every $\xi \in \mathbb{S}^{n-1}$ there exists an $\eta \in \mathcal{N}$ satisfying

$$|A\eta|^2 \leq C_1 |A\xi|^2 + \frac{\rho^2}{n} \mathcal{B}_{10}(A). \tag{2.48}$$

The proof shall be a combination of the argument similar to the proof of Lemma 2.5.1 along with the construction of a net on the family of admissible nets. The bound on the cardinality of that net shall follow, in fact, again from Lemma 2.2.1.

The advantage of (2.48) over (2.44) is the strong large deviation properties of $\mathcal{B}_{10}(A)$. For example, we shall show an elementary fact that for any random matrix A with independent columns and $\mathbb{E}\|A\|_{HS}^2 < \infty$,

$$P(\mathcal{B}_{10}(A) \geq 2\mathbb{E}\|A\|_{HS}^2) \leq e^{-cn}. \tag{2.49}$$

The detailed proofs of the mentioned facts, and applications to sharp estimates for the small ball probability of the smallest singular value of heavy-tailed matrices shall be outlined in a separate paper.

2.5.2 Covering Spheres with Strips

For $\theta \in \mathbb{S}^{n-1}$, $\tau \in \mathbb{R}$ and $\alpha > 0$, consider a strip

$$S(\theta, \alpha, \tau) := \{\xi \in \mathbb{S}^{n-1} : |\langle \xi, \theta \rangle + \tau| \leq \alpha\}.$$

Observe that

$$\sum_{k=1}^N \mathbf{1}_{S(\theta_k, \frac{1}{r}, \frac{t}{r})}(\xi) \leq C \sum_{k=1}^N \varphi(r \langle \xi, \theta_k \rangle + t).$$

Therefore, Proposition 2.3.1 implies

Proposition 2.5.2 *For any N and for any $\alpha \leq \frac{c}{\sqrt{n}}$ with $N \in [cn \log \frac{N}{\alpha n^{3/2}}, n^{10}]$ there exists a collection of points $\theta_1, \ldots, \theta_N \in \mathbb{S}^{n-1}$ such that every strip of width 2α contains no more than*

$$\tilde{C}\left[\sqrt{Nn \log \frac{N}{\alpha n^{3/2}}} + N\sqrt{n}\alpha\right]$$

points in this collection.

We note that in view of the point-strip duality, bounding $\sum_{k=1}^{N} 1_{S(\theta_k, \frac{1}{r}, \frac{t}{r})}(\xi)$ yields estimates of the form stated in Proposition 2.5.2.

The direct consideration of the characteristic functions in place of the Gaussian functions gives exactly the same bound as an application of Proposition 2.3.1.

In [8], Frankl, Nagy and Naszodi conjecture that for every collection of N points on $\mathbb{S}^2$ there exists a strip of width $\frac{2}{N}$ containing at least $f(N)$ points, where $f(N) \to \infty$ as $N \to \infty$. Proposition 2.5.2 generalizes Theorem 4.2 by Frankl, Nagy, Naszodi [8] from the two-dimensional case to an arbitrary dimension, with good dimensional constant, although it does not shed any light on the dependence on N.

Acknowledgements The second named author is supported in part by the NSF CAREER DMS-1753260. The work was partially supported by the National Science Foundation under Grant No. DMS-1440140 while the authors were in residence at the Mathematical Sciences Research Institute in Berkeley, California, during the Fall 2017 semester. The authors are grateful to Alexander Koldobsky for fruitful discussions and helpful comments. The authors are thankful to the anonymous referee for valuable suggestions.

References

1. M. Alexander, M. Henk, A. Zvavitch, A discrete version of Koldobsky's slicing inequality. Israel J. Math. **222**(1), 261–278 (2017)
2. N. Alon, B. Klartag, Optimal compression of approximate inner products and dimension reduction, in *Symposium on Foundations of Computer Science (FOCS 2017)*, pp. 639–650
3. J. Beck, Irregularities of distribution, I. Acta Math. **159**(1–2), 1–49 (1987)
4. S. Bobkov, B. Klartag, A. Koldobsly, Estimates for moments of general measures on convex bodies. Proc. Amer. Math. Soc. **146**, 4879–4888 (2018)
5. J. Bourgain, On high-dimensional maximal functions associated to convex bodies. Amer. J. Math. **108**, 1467–1476 (1986)
6. J. Bourgain, Geometry of Banach spaces and harmonic analysis, in *Proceedings of the International Congress of Mathematicians*, vol. 1, 2 (Berkeley, 1986) (American Mathematical Society, Providence, 1987), pp. 871–878
7. J. Bourgain, On the distribution of polynomials on high-dimensional convex sets, in Geometric Aspects of Functional Analysis. Lecture Notes in Mathematics, vol. 1469 (Springer, Berlin, 1991), pp. 127–137
8. N. Frankl, A. Nagy, M. Naszodi, Coverings: variations on a result of Rogers and on the Epsilon-net theorem of Haussler and Welzl. Discret. Math. **341**(3), 863–874 (2018)
9. A. Giannopoulos, On some vector balancing problems. Studia Math. **122**, 225–234 (1997)

10. B. Klartag, On convex perturbations with a bounded isotropic constant. Geom. Funct. Anal. (GAFA) **16**, 1274–1290 (2006)
11. B. Klartag, A. Koldobsky, An example related to the slicing inequality for general measures. J. Funct. Anal. **274**(7), 2089–2112 (2018)
12. A. Koldobsky, A hyperplane inequality for measures of convex bodies in $\mathbb{R}^n$, $n \leq 4$. Discrete Comput. Geom. **47**, 538–547 (2012)
13. A. Koldobsky, A $\sqrt{n}$ estimate for measures of hyperplane sections of convex bodies. Adv. Math. **254**, 33–40 (2014)
14. A. Koldobsky, Slicing inequalities for measures of convex bodies. Adv. Math. **283**, 473–488 (2015)
15. A. Koldobsky, A. Pajor, A remark on measures of sections of L_p -balls, in Geometric Aspects of Functional Analysis. Lecture Notes in Mathematics, vol. 2169 (Springer, Cham, 2017), pp. 213–220
16. A. Lytova, K. Tikhomirov, On delocalization of eigenvectors of random non-Hermitian matrices. Preprint (2018)
17. P. Raghavan, C.D. Thompson, Randomized rounding: a technique for provably good algorithms and algorithmic proofs. Combinatorica **7**(4), 365–374 (1987)
18. E. Rebrova, K. Tikhomirov, Coverings of random ellipsoids, and invertibility of matrices with i.i.d. heavy-tailed entries. Israel J. Math. **227**(2), 507–544 (2018)
19. O. Regev, A note on Koldobsky's lattice slicing inequality, preprint https://arxiv.org/abs/1608.04945
20. R. Vershynin, *High-Dimensional Probability: An Introduction with Applications in Data Science* (Cambridge University Press, Cambridge, 2018)

Chapter 3
Two-Sided Estimates for Order Statistics of Log-Concave Random Vectors

Rafał Latała and Marta Strzelecka

Abstract We establish two-sided bounds for expectations of order statistics (k-th maxima) of moduli of coordinates of centered log-concave random vectors with uncorrelated coordinates. Our bounds are exact up to multiplicative universal constants in the unconditional case for all k and in the isotropic case for $k \leq n - cn^{5/6}$. We also derive two-sided estimates for expectations of sums of k largest moduli of coordinates for some classes of random vectors.

3.1 Introduction and Main Results

For a vector $x \in \mathbb{R}^n$ let $k\text{-}\max x_i$ (or $k\text{-}\min x_i$) denote its k-th *maximum* (respectively its k-th *minimum*), i.e. its k-th maximal (respectively k-th minimal) coordinate. For a random vector $X = (X_1, \ldots, X_n)$, $k\text{-}\min X_i$ is also called the k-th order statistic of X.

Let $X = (X_1, \ldots, X_n)$ be a random vector with finite first moment. In this note we try to estimate $\mathbb{E}k\text{-}\max_i |X_i|$ and

$$
\mathbb{E} \max_{|I|=k} \sum_{i \in I} |X_i| = \mathbb{E} \sum_{l=1}^{k} l\text{-}\max_i |X_i|.
$$

Order statistics play an important role in various statistical applications and there is an extensive literature on this subject (cf. [2, 5] and references therein).

We put special emphasis on the case of log-concave vectors, i.e. random vectors X satisfying the property $\mathbb{P}(X \in \lambda K + (1 - \lambda)L) \geq \mathbb{P}(X \in K)^{\lambda}\mathbb{P}(X \in L)^{1-\lambda}$ for any $\lambda \in [0, 1]$ and any nonempty compact sets K and L. By the result of Borell [3] a vector X with full dimensional support is log-concave if and only if it has

R. Latała (✉) · M. Strzelecka
Institute of Mathematics, University of Warsaw, Warsaw, Poland
e-mail: rlatala@mimuw.edu.pl; martast@mimuw.edu.pl

© Springer Nature Switzerland AG 2020
B. Klartag, E. Milman (eds.), *Geometric Aspects of Functional Analysis*,
Lecture Notes in Mathematics 2266,
https://doi.org/10.1007/978-3-030-46762-3_3

a log-concave density, i.e. the density of a form $e^{-h(x)}$ where h is convex with values in $(-\infty, \infty]$. A typical example of a log-concave vector is a vector uniformly distributed over a convex body. In recent years the study of log-concave vectors attracted attention of many researchers, cf. monographs [1, 4].

To bound the sum of k largest coordinates of X we define

$$t(k, X) := \inf\left\{t > 0: \frac{1}{t}\sum_{i=1}^{n}\mathbb{E}|X_i|\mathbf{1}_{\{|X_i|\geq t\}} \leq k\right\}. \tag{3.1}$$

and start with an easy upper bound.

Proposition 3.1 *For any random vector X with finite first moment we have*

$$\mathbb{E}\max_{|I|=k}\sum_{i\in I}|X_i| \leq 2kt(k, X). \tag{3.2}$$

Proof For any $t > 0$ we have

$$\max_{|I|=k}\sum_{i\in I}|X_i| \leq tk + \sum_{i=1}^{n}|X_i|\mathbf{1}_{\{|X_i|\geq t\}}.$$

$\square$

It turns out that this bound may be reversed for vectors with independent coordinates or, more generally, vectors satisfying the following condition

$$\mathbb{P}(|X_i| \geq s, |X_j| \geq t) \leq \alpha\mathbb{P}(|X_i| \geq s)\mathbb{P}(|X_j| \geq t) \quad \text{for all } i \neq j \text{ and all } s, t > 0. \tag{3.3}$$

If $\alpha = 1$ this means that moduli of coordinates of X are negatively correlated.

Theorem 3.2 *Suppose that a random vector X satisfies condition (3.3) with some $\alpha \geq 1$. Then there exists a constant $c(\alpha) > 0$ which depends only on α such that for any $1 \leq k \leq n$,*

$$c(\alpha)kt(k, X) \leq \mathbb{E}\max_{|I|=k}\sum_{i\in I}|X_i| \leq 2kt(k, X).$$

We may take $c(\alpha) = (288(5 + 4\alpha)(1 + 2\alpha))^{-1}$.

In the case of i.i.d. coordinates two-sided bounds for $\mathbb{E}\max_{|I|=k}\sum_{i\in I}|a_iX_i|$ in terms of an Orlicz norm (related to the distribution of X_i) of a vector $(a_i)_{i\leq n}$ where known before, see [7].

Log-concave vectors with diagonal covariance matrices behave in many aspects like vectors with independent coordinates. This is true also in our case.

Theorem 3.3 *Let X be a log-concave random vector with uncorrelated coordinates (i.e. $\mathrm{Cov}(X_i, X_j) = 0$ for $i \neq j$). Then for any $1 \leq k \leq n$,*

$$ckt(k, X) \leq \mathbb{E} \max_{|I|=k} \sum_{i \in I} |X_i| \leq 2kt(k, X).$$

In the above statement and in the sequel c and C denote positive universal constants.

The next two examples show that the lower bound cannot hold if $n \gg k$ and only marginal distributions of X_i are log-concave or the coordinates of X are highly correlated.

Example 3.1 Let $X = (\varepsilon_1 g, \varepsilon_2 g, \ldots, \varepsilon_n g)$, where $\varepsilon_1, \ldots, \varepsilon_n, g$ are independent, $\mathbb{P}(\varepsilon_i = \pm 1) = 1/2$ and g has the normal $\mathcal{N}(0, 1)$ distribution. Then $\mathrm{Cov} X = \mathrm{Id}$ and it is not hard to check that $\mathbb{E} \max_{|I|=k} \sum_{i \in I} |X_i| = k\sqrt{2/\pi}$ and $t(k, X) \sim \ln^{1/2}(n/k)$ if $k \leq n/2$.

Example 3.2 Let $X = (g, \ldots, g)$, where $g \sim \mathcal{N}(0, 1)$. Then, as in the previous example, $\mathbb{E} \max_{|I|=k} \sum_{i \in I} |X_i| = k\sqrt{2/\pi}$ and $t(k, X) \sim \ln^{1/2}(n/k)$.

Question 3.1 Let $X' = (X'_1, X'_2, \ldots, X'_n)$ be a decoupled version of X, i.e. X'_i are independent and X'_i has the same distribution as X_i. Due to Theorem 3.2 (applied to X'), the assertion of Theorem 3.3 may be stated equivalently as

$$\mathbb{E} \max_{|I|=k} \sum_{i \in I} |X_i| \sim \mathbb{E} \max_{|I|=k} \sum_{i \in I} |X'_i|.$$

Is the more general fact true that for any symmetric norm and any log-concave vector X with uncorrelated coordinates

$$\mathbb{E} \|X\| \sim \mathbb{E} \|X'\|?$$

Maybe such an estimate holds at least in the case of unconditional log-concave vectors?

We turn our attention to bounding k-maxima of $|X_i|$. This was investigated in [8] (under some strong assumptions on the function $t \mapsto \mathbb{P}(|X_i| \geq t)$) and in the weighted i.i.d. setting in [7, 9, 15]. We will give different bounds valid for log-concave vectors, in which we do not have to assume independence, nor any special conditions on the growth of the distribution function of the coordinates of X. To this end we need to define another quantity:

$$t^*(p, X) := \inf \left\{ t > 0 : \sum_{i=1}^n \mathbb{P}(|X_i| \geq t) \leq p \right\} \quad \text{for } 0 < p < n.$$

Theorem 3.4 *Let X be a mean zero log-concave n-dimensional random vector with uncorrelated coordinates and $1 \leq k \leq n$. Then*

$$\mathbb{E}k\text{-}\max_{i \leq n}|X_i| \geq \frac{1}{2}\text{Med}\left(k\text{-}\max_{i \leq n}|X_i|\right) \geq ct^*\left(k - \frac{1}{2}, X\right).$$

Moreover, if X is additionally unconditional then

$$\mathbb{E}k\text{-}\max_{i \leq n}|X_i| \leq Ct^*\left(k - \frac{1}{2}, X\right).$$

The next theorem provides an upper bound in the general log-concave case.

Theorem 3.5 *Let X be a mean zero log-concave n-dimensional random vector with uncorrelated coordinates and $1 \leq k \leq n$. Then*

$$\mathbb{P}\left(k\text{-}\max_{i \leq n}|X_i| \geq Ct^*\left(k - \frac{1}{2}, X\right)\right) \leq 1 - c \tag{3.4}$$

and

$$\mathbb{E}k\text{-}\max_{i \leq n}|X_i| \leq Ct^*\left(k - \frac{1}{2}k^{5/6}, X\right). \tag{3.5}$$

In the isotropic case (i.e. $\mathbb{E}X_i = 0$, $\text{Cov}X = \text{Id}$) one may show that $t^*(k/2, X) \sim t^*(k, X) \sim t(k, X)$ for $k \leq n/2$ and $t^*(p, X) \sim \frac{n-p}{n}$ for $p \geq n/4$ (see Lemma 3.24 below). In particular $t^*(n-k+1-(n-k+1)^{5/6}/2, X) \sim k/n+n^{-1/6}$ for $k \leq n/2$. This together with the two previous theorems implies the following corollary.

Corollary 3.6 *Let X be an isotropic log-concave n-dimensional random vector and $1 \leq k \leq n/2$. Then*

$$\mathbb{E}k\text{-max}_{i \leq n}|X_i| \sim t^*(k, X) \sim t(k, X)$$

and

$$c\frac{k}{n} \leq \mathbb{E}k\text{-min}_{i \leq n}|X_i| = \mathbb{E}(n - k + 1)\text{-max}_{i \leq n}|X_i| \leq C\left(\frac{k}{n} + n^{-1/6}\right).$$

If X is additionally unconditional then

$$\mathbb{E}k\text{-min}_{i \leq n}|X_i| = \mathbb{E}(n - k + 1)\text{-max}_{i \leq n}|X_i| \sim \frac{k}{n}.$$

Question 3.2 Does the second part of Theorem 3.4 hold without the unconditionality assumptions? In particular, is it true that in the isotropic log-concave case $\mathbb{E}k\text{-}\min_{i \leq n}|X_i| \sim k/n$ for $1 \leq k \leq n/2$?

Notation Throughout this paper by letters C, c we denote universal positive constants and by $C(\alpha), c(\alpha)$ constants depending only on the parameter α. The values of constants $C, c, C(\alpha), c(\alpha)$ may differ at each occurrence. If we need to fix a value of constant, we use letters $C_0, C_1, \ldots$ or $c_0, c_1, \ldots$. We write $f \sim g$ if $cf \leq g \leq Cg$. For a random variable Z we denote $\|Z\|_p = (\mathbb{E}|Z|^p)^{1/p}$. Recall that a random vector X is called isotropic, if $\mathbb{E}X = 0$ and $\mathrm{Cov}X = \mathrm{Id}$.

This note is organised as follows. In Sect. 3.2 we provide a lower bound for the sum of k largest coordinates, which involves the Poincaré constant of a vector. In Sect. 3.3 we use this result to obtain Theorem 3.3. In Sect. 3.4 we prove Theorem 3.2 and provide its application to comparison of weak and strong moments. In Sect. 3.5 we prove the first part of Theorem 3.4 and in Sect. 3.6 we prove the second part of Theorems 3.4, 3.5, and Lemma 3.24.

3.2 Exponential Concentration

A probability measure μ on $\mathbb{R}^n$ satisfies *exponential concentration with constant* $\alpha > 0$ if for any Borel set A with $\mu(A) \geq 1/2$,

$$1 - \mu(A + uB_2^n) \leq e^{-u/\alpha} \quad \text{for all } u > 0.$$

We say that a random n-dimensional vector satisfies exponential concentration if its distribution has such a property.

It is well known that exponential concentration is implied by the Poincaré inequality

$$\mathrm{Var}_\mu f \leq \beta \int |\nabla f|^2 d\mu \quad \text{for all bounded smooth functions } f \colon \mathbb{R}^n \mapsto \mathbb{R}$$

and $\alpha \leq 3\sqrt{\beta}$ (cf. [12, Corollary 3.2]).

Obviously, the constant in the exponential concentration is not linearly invariant. Typically one assumes that the vector is isotropic. For our purposes a more natural normalization will be that all coordinates have L_1-norm equal to 1.

The next proposition states that bound (3.2) may be reversed under the assumption that X satisfies the exponential concentration.

Proposition 3.7 *Assume that $Y = (Y_1, \ldots, Y_n)$ satisfies the exponential concentration with constant $\alpha > 0$ and $\mathbb{E}|Y_i| \geq 1$ for all i. Then for any sequence $a = (a_i)_{i=1}^n$ of real numbers and $X_i := a_i Y_i$ we have*

$$\mathbb{E} \max_{|I|=k} \sum_{i \in I} |X_i| \geq \left(8 + 64\frac{\alpha}{\sqrt{k}}\right)^{-1} kt(k, X),$$

where $t(k, X)$ is given by (3.1).

We begin the proof with a few simple observations.

Lemma 3.8 *For any real numbers $z_1, \ldots, z_n$ and $1 \leq k \leq n$ we have*

$$\max_{|I|=k} \sum_{i \in I} |z_i| = \int_0^\infty \min\left\{k, \sum_{i=1}^n \mathbf{1}_{\{|z_i| \geq s\}}\right\} ds.$$

Proof Without loss of generality we may assume that $z_1 \geq z_2 \geq \ldots \geq z_n \geq 0$. Then

$$\int_0^\infty \min\left\{k, \sum_{i=1}^n \mathbf{1}_{\{|z_i| \geq s\}}\right\} ds = \sum_{l=1}^{k-1} \int_{z_{l+1}}^{z_l} l \, ds + \int_0^{z_k} k \, ds = \sum_{l=1}^{k-1} l(z_l - z_{l+1}) + k z_k$$

$$= z_1 + \ldots + z_k = \max_{|I|=k} \sum_{i \in I} |z_i|.$$

$\square$

Fix a sequence $(X_i)_{i \leq n}$ and define for $s \geq 0$,

$$N(s) := \sum_{i=1}^n \mathbf{1}_{\{|X_i| \geq s\}}. \tag{3.6}$$

Corollary 3.9 *For any $k = 1, \ldots, n$,*

$$\mathbb{E} \max_{|I|=k} \sum_{i \in I} |X_i| = \int_0^\infty \sum_{l=1}^k \mathbb{P}(N(s) \geq l) ds,$$

and for any $t > 0$,

$$\mathbb{E} \sum_{i=1}^n |X_i| \mathbf{1}_{\{|X_i| \geq t\}} = t \mathbb{E} N(t) + \int_t^\infty \sum_{l=1}^\infty \mathbb{P}(N(s) \geq l) ds.$$

In particular

$$\mathbb{E} \sum_{i=1}^n |X_i| \mathbf{1}_{\{|X_i| \geq t\}} \leq \mathbb{E} \max_{|I|=k} \sum_{i \in I} |X_i| + \sum_{l=k+1}^\infty \left(t \mathbb{P}(N(t) \geq l) + \int_t^\infty \mathbb{P}(N(s) \geq l) ds \right).$$

Proof We have

$$\int_0^\infty \sum_{l=1}^k \mathbb{P}(N(s) \geq l) ds = \int_0^\infty \mathbb{E} \min\{k, N(s)\} ds = \mathbb{E} \int_0^\infty \min\{k, N(s)\} ds$$

$$= \mathbb{E} \max_{|I|=k} \sum_{i \in I} |X_i|,$$

where the last equality follows by Lemma 3.8.

Moreover,

$$t\mathbb{E}N(t) + \int_t^\infty \sum_{l=1}^\infty \mathbb{P}(N(s) \geq l)ds = t\mathbb{E}N(t) + \int_t^\infty \mathbb{E}N(s)ds$$

$$= \mathbb{E}\sum_{i=1}^n \left(t\mathbf{1}_{\{|X_i|\geq t\}} + \int_t^\infty \mathbf{1}_{\{|X_i|\geq s\}}ds \right)$$

$$= \mathbb{E}\sum_{i=1}^n |X_i|\mathbf{1}_{\{|X_i|\geq t\}}.$$

The last part of the assertion easily follows, since

$$t\mathbb{E}N(t) = t\sum_{l=1}^n \mathbb{P}(N(t) \geq l) \leq \int_0^t \sum_{l=1}^k \mathbb{P}(N(s) \geq l)ds + \sum_{l=k+1}^\infty t\mathbb{P}(N(t) \geq l).$$

□

Proof of Proposition 3.7 To shorten the notation put $t_k := t(k, X)$. Without loss of generality we may assume that $a_1 \geq a_2 \geq \ldots \geq a_n \geq 0$ and $a_{\lceil k/4 \rceil} = 1$. Observe first that

$$\mathbb{E}\max_{|I|=k} \sum_{i\in I} |X_i| \geq \sum_{i=1}^{\lceil k/4 \rceil} a_i \mathbb{E}|Y_i| \geq k/4,$$

so we may assume that $t_k \geq 16\alpha/\sqrt{k}$.

Let μ be the law of Y and

$$A := \left\{ y \in \mathbb{R}^n: \sum_{i=1}^n \mathbf{1}_{\{|a_i y_i|\geq \frac{1}{2} t_k\}} < \frac{k}{2} \right\}.$$

We have

$$\mathbb{E}\max_{|I|=k} \sum_{i\in I} |X_i| \geq \frac{k}{4} t_k \mathbb{P}\left(\sum_{i=1}^k \mathbf{1}_{\{|a_i Y_i|\geq \frac{1}{2} t_k\}} \geq \frac{k}{2} \right) = \frac{k}{4} t_k (1 - \mu(A)),$$

so we may assume that $\mu(A) \geq 1/2$.

Observe that if $y \in A$ and $\sum_{i=1}^n \mathbf{1}_{\{|a_i z_i|\geq s\}} \geq l > k$ for some $s \geq t_k$ then

$$\sum_{i=1}^n (z_i - y_i)^2 \geq \sum_{i=\lceil k/4 \rceil}^n (a_i z_i - a_i y_i)^2 \geq (l - 3k/4)(s - t_k/2)^2 > \frac{ls^2}{16}.$$

Thus we have

$$\mathbb{P}(N(s) \geq l) \leq 1 - \mu\left(A + \frac{s\sqrt{l}}{4} B_2^n\right) \leq e^{-\frac{s\sqrt{l}}{4\alpha}} \quad \text{for } l > k, \ s \geq t_k.$$

Therefore

$$\int_{t_k}^{\infty} \mathbb{P}(N(s) \geq l) ds \leq \int_{t_k}^{\infty} e^{-\frac{s\sqrt{l}}{4\alpha}} ds = \frac{4\alpha}{\sqrt{l}} e^{-\frac{t_k\sqrt{l}}{4\alpha}} \quad \text{for } l > k,$$

and

$$\sum_{l=k+1}^{\infty} \left(t_k \mathbb{P}(N(t_k) \geq l) + \int_{t_k}^{\infty} \mathbb{P}(N(s) \geq l) ds \right) \leq \sum_{l=k+1}^{\infty} \left(t_k + \frac{4\alpha}{\sqrt{l}} \right) e^{-\frac{t_k\sqrt{l}}{4\alpha}}$$

$$\leq \left(t_k + \frac{4\alpha}{\sqrt{k+1}} \right) \int_k^{\infty} e^{-\frac{t_k\sqrt{u}}{4\alpha}} du \leq \left(t_k + \frac{4\alpha}{\sqrt{k+1}} \right) e^{-\frac{t_k\sqrt{k}}{4\sqrt{2}\alpha}} \int_k^{\infty} e^{-\frac{t_k\sqrt{u-k}}{4\sqrt{2}\alpha}} du$$

$$= \left(t_k + \frac{4\alpha}{\sqrt{k+1}} \right) \frac{64\alpha^2}{t_k^2} e^{-\frac{t_k\sqrt{k}}{4\sqrt{2}\alpha}} \leq \left(t_k + \frac{1}{4} t_k \right) \frac{k}{4} \leq \frac{1}{2} k t_k,$$

where to get the next-to-last inequality we used the fact that $t_k \geq 16\alpha/\sqrt{k}$.
Hence Corollary 3.9 and the definition of t_k yields

$$k t_k \leq \mathbb{E} \sum_{i=1}^{n} |X_i| \mathbf{1}_{\{|X_i| \geq t_k\}}$$

$$\leq \mathbb{E} \max_{|I|=k} \sum_{i \in I} |X_i| + \sum_{l=k+1}^{\infty} \left(t_k \mathbb{P}(N(t_k) \geq l) + \int_{t_k}^{\infty} \mathbb{P}(N(s) \geq l) ds \right)$$

$$\leq \mathbb{E} \max_{|I|=k} \sum_{i \in I} |X_i| + \frac{1}{2} k t_k,$$

so $\mathbb{E} \max_{|I|=k} \sum_{i \in I} |X_i| \geq \frac{1}{2} k t_k$. $\square$

We finish this section with a simple fact that will be used in the sequel.

Lemma 3.10 *Suppose that a measure μ satisfies exponential concentration with constant α. Then for any $c \in (0, 1)$ and any Borel set A with $\mu(A) > c$ we have*

$$1 - \mu(A + u B_2^n) \leq \exp\left(-\left(\frac{u}{\alpha} + \ln c \right)_+ \right) \quad \text{for } u \geq 0.$$

Proof Let $D := \mathbb{R}^n \setminus (A + r B_2^n)$. Observe that $D + r B_2^n$ has an empty intersection with A so if $\mu(D) \geq 1/2$ then

$$c < \mu(A) \leq 1 - \mu(D + r B_2^n) \leq e^{-r/\alpha},$$

and $r < \alpha \ln(1/c)$. Hence $\mu(A + \alpha \ln(1/c) B_2^n) \geq 1/2$, therefore for $s \geq 0$,

$$1 - \mu(A + (s + \alpha \ln(1/c)) B_2^n) = 1 - \mu((A + \alpha \ln(1/c) B_2^n) + s B_2^n) \leq e^{-s/\alpha},$$

and the assertion easily follows. □

3.3 Sums of Largest Coordinates of Log-Concave Vectors

We will use the regular growth of moments of norms of log-concave vectors multiple times. By [4, Theorem 2.4.6], if $f : \mathbb{R}^n \to \mathbb{R}$ is a seminorm, and X is log-concave, then

$$(\mathbb{E} f(X)^p)^{1/p} \leq C_1 \frac{p}{q} (\mathbb{E} f(X)^q)^{1/q} \quad \text{for } p \geq q \geq 1, \tag{3.7}$$

where C_1 is a universal constant.

We will also apply a few times the functional version of the Grünbaum inequality (see [14, Lemma 5.4]) which states that

$$\mathbb{P}(Z \geq 0) \geq \frac{1}{e} \quad \text{for any mean-zero log-concave random variable Z.} \tag{3.8}$$

Let us start with a few technical lemmas. The first one will be used to reduce proofs of Theorem 3.3 and lower bound in Theorem 3.4 to the symmetric case.

Lemma 3.11 *Let X be a log-concave n-dimensional vector and X' be an independent copy of X. Then for any $1 \leq k \leq n$,*

$$\mathbb{E} \max_{|I|=k} \sum_{i \in I} |X_i - X_i'| \leq 2\mathbb{E} \max_{|I|=k} \sum_{i \in I} |X_i|,$$

$$t(k, X) \leq et(k, X - X') + \frac{2}{k} \max_{|I|=k} \sum_{i \in I} \mathbb{E}|X_i|, \tag{3.9}$$

and

$$t^*(2k, X - X') \leq 2t^*(k, X). \tag{3.10}$$

Proof The first estimate follows by the easy bound

$$\mathbb{E} \max_{|I|=k} \sum_{i \in I} |X_i - X_i'| \leq \mathbb{E} \max_{|I|=k} \sum_{i \in I} |X_i| + \mathbb{E} \max_{|I|=k} \sum_{i \in I} |X_i'| = 2\mathbb{E} \max_{|I|=k} \sum_{i \in I} |X_i|.$$

To get the second bound we may and will assume that $\mathbb{E}|X_1| \geq \mathbb{E}|X_2| \geq \ldots \geq \mathbb{E}|X_n|$. Let us define $Y := X - \mathbb{E}X$, $Y' := X' - \mathbb{E}X$ and $M := \frac{1}{k}\sum_{i=1}^{k}\mathbb{E}|X_i| \geq \max_{i \geq k}\mathbb{E}|X_i|$. Obviously

$$\sum_{i=1}^{k}\mathbb{E}|X_i|\mathbf{1}_{\{|X_i|\geq t\}} \leq kM \quad \text{for } t \geq 0. \tag{3.11}$$

We have $\mathbb{E}Y_i = 0$, thus $\mathbb{P}(Y_i \leq 0) \geq 1/e$ by (3.8). Hence

$$\mathbb{E}Y_i\mathbf{1}_{\{Y_i>t\}} \leq e\mathbb{E}Y_i\mathbf{1}_{\{Y_i>t,Y_i'\leq 0\}} \leq e\mathbb{E}|Y_i - Y_i'|\mathbf{1}_{\{Y_i-Y_i'>t\}} = e\mathbb{E}|X_i - X_i'|\mathbf{1}_{\{X_i-X_i'>t\}}$$

for $t \geq 0$. In the same way we show that

$$\mathbb{E}|Y_i|\mathbf{1}_{\{Y_i<-t\}} \leq e\mathbb{E}|Y_i|\mathbf{1}_{\{Y_i<-t,Y_i'\geq 0\}} \leq e\mathbb{E}|X_i - X_i'|\mathbf{1}_{\{X_i'-X_i>t\}}$$

Therefore

$$\mathbb{E}|Y_i|\mathbf{1}_{\{|Y_i|>t\}} \leq e\mathbb{E}|X_i - X_i'|\mathbf{1}_{\{|X_i-X_i'|>t\}}.$$

We have

$$\sum_{i=k+1}^{n}\mathbb{E}|X_i|\mathbf{1}_{\{|X_i|>et(k,X-X')+M\}} \leq \sum_{i=k+1}^{n}\mathbb{E}|X_i|\mathbf{1}_{\{|Y_i|>et(k,X-X')\}}$$

$$\leq \sum_{i=k+1}^{n}\mathbb{E}|Y_i|\mathbf{1}_{\{|Y_i|>t(k,X-X')\}} + \sum_{i=k+1}^{n}|\mathbb{E}X_i|\mathbb{P}(|Y_i| > et(k,X-X'))$$

$$\leq e\sum_{i=1}^{n}\mathbb{E}|X_i - X_i'|\mathbf{1}_{\{|X_i-X_i'|>t(k,X-X')\}} + M\sum_{i=1}^{n}\mathbb{P}(|Y_i| > et(k,X-X'))$$

$$\leq ekt(k,X-X') + M\sum_{i=1}^{n}\big(et(k,X-X')\big)^{-1}\mathbb{E}|Y_i|\mathbf{1}_{\{|Y_i|>et(k,X-X')\}}$$

$$\leq ekt(k,X-X') + Mt(k,X-X')^{-1}\sum_{i=1}^{n}\mathbb{E}|X_i - X_i'|\mathbf{1}_{\{|X_i-X_i'|>t(k,X-X')\}}$$

$$\leq ekt(k,X-X') + kM.$$

Together with (3.11) we get

$$\sum_{i=1}^{n}\mathbb{E}|X_i|\mathbf{1}_{\{|X_i|>et(k,X-X')+M\}} \leq k(et(k,X-X') + 2M)$$

and (3.9) easily follows.

In order to prove (3.10), note that for $u > 0$,

$$\mathbb{P}(|X_i - X_i'| \geq 2u) \leq \mathbb{P}\big(\max\{|X_i|, |X_i'|\} \geq u\big) \leq 2\mathbb{P}(|X_i| \geq u),$$

thus the last part of the assertion follows by the definition of parameters t^*. □

Lemma 3.12 *Suppose that V is a real symmetric log-concave random variable. Then for any $t > 0$ and $\lambda \in (0, 1]$,*

$$\mathbb{E}|V|\mathbf{1}_{\{|V|\geq t\}} \leq \frac{4}{\lambda}\mathbb{P}(|V| \geq t)^{1-\lambda}\mathbb{E}|V|\mathbf{1}_{\{|V|\geq \lambda t\}}.$$

Moreover, if $\mathbb{P}(|V| \geq t) \leq 1/4$, then $\mathbb{E}|V|\mathbf{1}_{\{|V|\geq t\}} \leq 4t\mathbb{P}(|V| \geq t)$.

Proof Without loss of generality we may assume that $\mathbb{P}(|V| \geq t) \leq 1/4$ (otherwise the first estimate is trivial).

Observe that $\mathbb{P}(|V| \geq s) = \exp(-N(s))$ where $N : [0, \infty) \to [0, \infty]$ is convex and $N(0) = 0$. In particular

$$\mathbb{P}(|V| \geq \gamma t) \leq \mathbb{P}(|V| \geq t)^{\gamma} \quad \text{for } \gamma > 1$$

and

$$\mathbb{P}(|V| \geq \gamma t) \geq \mathbb{P}(|V| \geq t)^{\gamma} \quad \text{for } \gamma \in [0, 1].$$

We have

$$\mathbb{E}|V|\mathbf{1}_{\{|V|\geq t\}} \leq \sum_{k=0}^{\infty} 2^{k+1} t \mathbb{P}(|V| \geq 2^k t) \leq 2t \sum_{k=0}^{\infty} 2^k \mathbb{P}(|V| \geq t)^{2^k}$$

$$\leq 2t\mathbb{P}(|V| \geq t) \sum_{k=0}^{\infty} 2^k 4^{1-2^k} \leq 4t\mathbb{P}(|V| \geq t).$$

This implies the second part of the lemma.

To conclude the proof of the first bound it is enough to observe that

$$\mathbb{E}|V|\mathbf{1}_{\{|V|\geq \lambda t\}} \geq \lambda t \mathbb{P}(|V| \geq \lambda t) \geq \lambda t \mathbb{P}(|V| \geq t)^{\lambda}.$$

□

Proof of Theorem 3.3 By Proposition 3.1 it is enough to show the lower bound. By Lemma 3.11 we may assume that X is symmetric. We may also obviously assume that $\|X_i\|_2^2 = \mathbb{E}X_i^2 > 0$ for all i.

Let $Z = (Z_1, \ldots, Z_n)$, where $Z_i = X_i/\|X_i\|_2$. Then Z is log-concave, isotropic and, by (3.7), $\mathbb{E}|Z_i| \geq 1/(2C_1)$ for all i. Set $Y := 2C_1 Z$. Then $X_i = a_i Y_i$ and $\mathbb{E}|Y_i| \geq 1$. Moreover, since any m-dimensional projection of Z is a log-concave, isotropic m-dimensional vector, we know by the result of Lee and Vempala [13], that it satisfies the exponential concentration with a constants $Cm^{1/4}$. (In fact an easy modification of the proof below shows that for our purposes it would be enough to have exponential concentration with a constant Cm^γ for some $\gamma < 1/2$, so one may also use Eldan's result [6] which gives such estimates for any $\gamma > 1/3$). So any m-dimensional projection of Y satisfies exponential concentration with constant $C_2 m^{1/4}$.

Let us fix k and set $t := t(k, X)$, then (since X_i has no atoms)

$$\sum_{i=1}^{n} \mathbb{E}|X_i|\mathbf{1}_{\{|X_i|\geq t\}} = kt. \tag{3.12}$$

For $l = 1, 2, \ldots$ define

$$I_l := \{i \in [n]: \ \beta^{l-1} \geq \mathbb{P}(|X_i| \geq t) \geq \beta^l\},$$

where $\beta = 2^{-8}$. By (3.12) there exists l such that

$$\sum_{i \in I_l} \mathbb{E}|X_i|\mathbf{1}_{\{|X_i|\geq t\}} \geq kt2^{-l}.$$

Let us consider three cases.

(1) $l = 1$ and $|I_1| \leq k$. Then

$$\mathbb{E} \max_{|I|=k} \sum_{i \in I} |X_i| \geq \sum_{i \in I_1} \mathbb{E}|X_i|\mathbf{1}_{\{|X_i|\geq t\}} \geq \frac{1}{2}kt.$$

(2) $l = 1$ and $|I_1| > k$. Choose $J \subset I_1$ of cardinality k. Then

$$\mathbb{E} \max_{|I|=k} \sum_{i \in I} |X_i| \geq \sum_{i \in J} \mathbb{E}|X_i| \geq \sum_{i \in J} t\mathbb{P}(|X_i| \geq t) \geq \beta kt.$$

(3) $l > 1$. By Lemma 3.12 (applied with $\lambda = 1/8$) we have

$$\sum_{i \in I_l} \mathbb{E}|X_i|\mathbf{1}_{\{|X_i|\geq t/8\}} \geq \frac{1}{32}\beta^{-7(l-1)/8} \sum_{i \in I_l} \mathbb{E}|X_i|\mathbf{1}_{\{|X_i|\geq t\}} \geq \frac{1}{32}\beta^{-7(l-1)/8}2^{-l}kt. \tag{3.13}$$

Moreover for $i \in I_l$, $\mathbb{P}(|X_i| \geq t) \leq \beta^{l-1} \leq 1/4$, so the second part of Lemma 3.12 yields

$$4t|I_l|\beta^{l-1} \geq \sum_{i \in I_l} \mathbb{E}|X_i|\mathbf{1}_{\{|X_i|\geq t\}} \geq kt2^{-l}$$

and $|I_l| \geq \beta^{1-l}2^{-l-2}k = 2^{7l-10}k \geq k$.

Set $k' := \beta^{-7l/8}2^{-l}k = 2^{6l}k$. If $k' \geq |I_l|$ then, using (3.13), we estimate

$$\mathbb{E}\max_{|I|=k}\sum_{i\in I}|X_i| \geq \frac{k}{|I_l|}\sum_{i\in I_l}\mathbb{E}|X_i| \geq \beta^{7l/8}2^l\sum_{i\in I_l}\mathbb{E}|X_i|\mathbf{1}_{\{|X_i|\geq t/8\}} \geq \frac{1}{32}\beta^{7/8}kt$$

$$= 2^{-12}kt.$$

Otherwise set $X' = (X_i)_{i\in I_l}$ and $Y' = (Y_i)_{i\in I_l}$. By (3.12) we have

$$kt \geq \sum_{i\in I_l}\mathbb{E}|X_i|\mathbf{1}_{\{|X_i|\geq t\}} \geq |I_l|t\beta^l,$$

so $|I_l| \leq k\beta^{-l}$ and Y' satisfies exponential concentration with constant $\alpha' = C_2k^{1/4}\beta^{-l/4}$. Estimate (3.13) yields

$$\sum_{i\in I_l}\mathbb{E}|X_i|\mathbf{1}_{\{|X_i|\geq 2^{-12}t\}} \geq \sum_{i\in I_l}\mathbb{E}|X_i|\mathbf{1}_{\{|X_i|\geq t/8\}} \geq 2^{-12}k't,$$

so $t(k', X') \geq 2^{-12}t$. Moreover, by Proposition 3.7 we have (since $k' \leq |I_l|$)

$$\mathbb{E}\max_{I\subset I_l,|I|=k'}\sum_{i\in I}|X_i| \geq \frac{1}{8+64\alpha'/\sqrt{k'}}k't(k', X').$$

To conclude observe that

$$\frac{\alpha'}{\sqrt{k'}} = C_2 2^{-l}k^{-1/4} \leq \frac{C_2}{4}$$

and since $k' \geq k$,

$$\mathbb{E}\max_{|I|=k}\sum_{i\in I}|X_i| \geq \frac{k}{k'}\mathbb{E}\max_{I\subset I_l,|I|=k'}\sum_{i\in I}|X_i| \geq \frac{1}{8+16C_2}2^{-12}tk.$$

$\square$

3.4 Vectors Satisfying Condition (3.3)

Proof of Theorem 3.2 By Proposition 3.1 we need to show only the lower bound. Assume first that variables X_i have no atoms and $k \geq 4(1+\alpha)$.

Let $t_k = t(k, X)$. Then $\mathbb{E}\sum_{i=1}^n|X_i|\mathbf{1}_{\{|X_i|\geq t_k\}} = kt_k$. Note, that (3.3) implies that for all $i \neq j$ we have

$$\mathbb{E}|X_iX_j|\mathbf{1}_{\{|X_i|\geq t_k,|X_j|\geq t_k\}} \leq \alpha\mathbb{E}|X_i|\mathbf{1}_{\{|X_i|\geq t_k\}}\mathbb{E}|X_j|\mathbf{1}_{\{|X_j|\geq t_k\}}. \tag{3.14}$$

We may assume that $\mathbb{E}\max_{|I|=k}\sum_{i\in I}|X_i| \leq \frac{1}{6}kt_k$, because otherwise the lower bound holds trivially.

Let us define

$$Y := \sum_{i=1}^{n}|X_i|\mathbf{1}_{\{kt_k\geq|X_i|\geq t_k\}} \quad \text{and} \quad A := (\mathbb{E}Y^2)^{1/2}.$$

Since

$$\mathbb{E}\max_{|I|=k}\sum_{i\in I}|X_i| \geq \mathbb{E}\left[\frac{1}{2}kt_k\mathbf{1}_{\{Y\geq kt_k/2\}}\right] = \frac{1}{2}kt_k\mathbb{P}\left(Y \geq \frac{kt_k}{2}\right),$$

it suffices to bound below the probability that $Y \geq kt_k/2$ by a constant depending only on α.

We have

$$A^2 = \mathbb{E}Y^2 \leq \sum_{i=1}^{n}\mathbb{E}X_i^2\mathbf{1}_{\{kt_k\geq|X_i|\geq t_k\}} + \sum_{i\neq j}\mathbb{E}|X_iX_j|\mathbf{1}_{\{|X_i|\geq t_k,|X_j|\geq t_k\}}$$

$$\overset{(3.14)}{\leq} kt_k\mathbb{E}Y + \alpha\sum_{i\neq j}\mathbb{E}|X_i|\mathbf{1}_{\{|X_i|\geq t_k\}}\mathbb{E}|X_j|\mathbf{1}_{\{|X_j|\geq t_k\}}$$

$$\leq kt_kA + \alpha\left(\sum_{i=1}^{n}\mathbb{E}|X_i|\mathbf{1}_{\{|X_i|\geq t_k\}}\right)^2 \leq \frac{1}{2}(k^2t_k^2 + A^2) + \alpha k^2t_k^2.$$

Therefore $A^2 \leq (1 + 2\alpha)k^2t_k^2$ and for any $l \geq k/2$ we have

$$\mathbb{E}Y\mathbf{1}_{\{Y\geq kt_k/2\}} \leq lt_k\mathbb{P}(Y \geq kt_k/2) + \frac{1}{lt_k}\mathbb{E}Y^2$$

$$\leq lt_k\mathbb{P}(Y \geq kt_k/2) + (1 + 2\alpha)k^2l^{-1}t_k. \qquad (3.15)$$

By Corollary 3.9 we have (recall definition (3.6))

$$\sum_{i=1}^{n}\mathbb{E}|X_i|\mathbf{1}_{\{|X_i|\geq kt_k\}} \leq \mathbb{E}\max_{|I|=k}\sum_{i\in I}|X_i|$$

$$+ \sum_{l=k+1}^{\infty}\left(kt_k\mathbb{P}(N(kt_k) \geq l) + \int_{kt_k}^{\infty}\mathbb{P}(N(s) \geq l)ds\right)$$

$$\leq \frac{1}{6}kt_k + \sum_{l=k+1}^{\infty}\left(kt_k\mathbb{E}N(kt_k)^2l^{-2} + \int_{kt_k}^{\infty}\mathbb{E}N(s)^2l^{-2}ds\right)$$

$$\leq \frac{1}{6}kt_k + \frac{1}{k}\left(kt_k\mathbb{E}N(kt_k)^2 + \int_{kt_k}^{\infty}\mathbb{E}N(s)^2ds\right). \qquad (3.16)$$

Assumption (3.3) implies that

$$\mathbb{E}N(s)^2 = \sum_{i=1}^{n} \mathbb{P}(|X_i| \geq s) + \sum_{i \neq j} \mathbb{P}(|X_i| \geq s, |X_j| \geq s)$$

$$\leq \sum_{i=1}^{n} \mathbb{P}(|X_i| \geq s) + \alpha \left(\sum_{i=1}^{n} \mathbb{P}(|X_i| \geq s) \right)^2.$$

Moreover for $s \geq kt_k$ we have

$$\sum_{i=1}^{n} \mathbb{P}(|X_i| \geq s) \leq \frac{1}{s} \sum_{i=1}^{n} \mathbb{E}|X_i| \mathbf{1}_{\{|X_i| \geq s\}} \leq \frac{kt_k}{s} \leq 1,$$

so

$$\mathbb{E}N(s)^2 \leq (1 + \alpha) \sum_{i=1}^{n} \mathbb{P}(|X_i| \geq s) \quad \text{for } s \geq kt_k.$$

Thus

$$kt_k \mathbb{E}N(kt_k)^2 \leq kt_k(1 + \alpha) \sum_{i=1}^{n} \mathbb{P}(|X_i| \geq kt_k) \leq (1 + \alpha) \sum_{i=1}^{n} \mathbb{E}|X_i| \mathbf{1}_{\{|X_i| \geq kt_k\}},$$

and

$$\int_{kt_k}^{\infty} \mathbb{E}N(s)^2 ds \leq (1 + \alpha) \sum_{i=1}^{n} \int_{kt_k}^{\infty} \mathbb{P}(|X_i| \geq s) ds \leq (1 + \alpha) \sum_{i=1}^{n} \mathbb{E}|X_i| \mathbf{1}_{\{|X_i| \geq kt_k\}}.$$

This together with (3.16) and the assumption that $k \geq 4(1 + \alpha)$ implies

$$\sum_{i=1}^{n} \mathbb{E}|X_i| \mathbf{1}_{\{|X_i| \geq kt_k\}} \leq \frac{1}{3} kt_k$$

and

$$\mathbb{E}Y = \sum_{i=1}^{n} \mathbb{E}|X_i| \mathbf{1}_{\{|X_i| \geq t_k\}} - \sum_{i=1}^{n} \mathbb{E}|X_i| \mathbf{1}_{\{|X_i| \geq kt_k\}} \geq \frac{2}{3} kt_k.$$

Therefore

$$\mathbb{E}Y \mathbf{1}_{\{Y \geq kt_k/2\}} \geq \mathbb{E}Y - \frac{1}{2} kt_k \geq \frac{1}{6} kt_k.$$

This applied to (3.15) with $l = (12 + 24\alpha)k$ gives us $\mathbb{P}(Y \geq kt_k/2) \geq (144 + 288\alpha)^{-1}$ and in consequence

$$\mathbb{E} \max_{|I|=k} \sum_{i \in I} |X_i| \geq \frac{1}{288(1 + 2\alpha)} kt(k, X).$$

Since $k \mapsto kt(k, X)$ is non-decreasing, in the case $k \leq \lceil 4(1 + \alpha) \rceil =: k_0$ we have

$$\mathbb{E} \max_{|I|=k} |X_i| \geq \frac{k}{k_0} \mathbb{E} \max_{|I|=k_0} |X_i| \geq \frac{k}{5 + 4\alpha} \cdot \frac{1}{288(1 + 2\alpha)} k_0 t(k_0, X)$$

$$\geq \frac{1}{288(5 + 4\alpha)(1 + 2\alpha)} kt(k, X).$$

The last step is to loose the assumption that X_i has no atoms. Note that both assumption (3.3) and the lower bound depend only on $(|X_i|)_{i=1}^n$, so we may assume that X_i are nonnegative almost surely. Consider $X^\varepsilon := (X_i + \varepsilon Y_i)_{i=1}^n$, where $Y_1, \ldots, Y_n$ are i.i.d. nonnegative r.v's with $\mathbb{E}Y_i < \infty$ and a density g, independent of X. Then for every $s, t > 0$ we have (observe that (3.3) holds also for $s < 0$ or $t < 0$).

$$\mathbb{P}(X_i^\varepsilon \geq s, X_j^\varepsilon \geq t)$$

$$= \int_0^\infty \int_0^\infty \mathbb{P}(X_i + \varepsilon y_i \geq s, \ X_j + \varepsilon y_j \geq t) g(y_i) g(y_j) dy_i dy_j$$

$$\overset{(3.3)}{\leq} \alpha \int_0^\infty \int_0^\infty \mathbb{P}(X_i \geq s - \varepsilon y_i) \mathbb{P}(X_j \geq t - \varepsilon y_j) g(y_i) g(y_j) dy_i dy_j$$

$$= \alpha \mathbb{P}(X_i^\varepsilon \geq s) \mathbb{P}(X_j^\varepsilon \geq t).$$

Thus X^ε satisfies assumption (3.3) and has the density function for every $\varepsilon > 0$. Therefore for all natural k we have

$$\mathbb{E} \max_{|I|=k} \sum_{i=1}^n X_i^\varepsilon \geq c(\alpha) kt(k, X^\varepsilon) \geq c(\alpha) kt(k, X).$$

Clearly, $\mathbb{E} \max_{|I|=k} \sum_{i=1}^n X_i^\varepsilon \to \mathbb{E} \max_{|I|=k} \sum_{i=1}^n X_i$ as $\varepsilon \to 0$, so the lower bound holds in the case of arbitrary X satisfying (3.3). $\qquad\square$

We may use Theorem 3.2 to obtain a comparison of weak and strong moments for the supremum norm:

Corollary 3.13 *Let X be an n-dimensional centered random vector satisfying condition (3.3). Assume that*

$$\|X_i\|_{2p} \leq \beta \|X_i\|_p \qquad \textit{for every } p \geq 2 \textit{ and } i = 1, \ldots, n. \tag{3.17}$$

Then the following comparison of weak and strong moments for the supremum norm holds: for all $a \in \mathbb{R}^n$ and all $p \geq 1$,

$$\left(\mathbb{E} \max_{i \leq n} |a_i X_i|^p\right)^{1/p} \leq C(\alpha, \beta) \left[\mathbb{E} \max_{i \leq n} |a_i X_i| + \max_{i \leq n} \left(\mathbb{E} |a_i X_i|^p\right)^{1/p}\right],$$

where $C(\alpha, \beta)$ is a constant depending only on α and β.

Proof Let $X' = (X'_i)_{i \leq n}$ be a decoupled version of X. For any $p > 0$ a random vector $(|a_i X_i|^p)_{i \leq n}$ satisfies condition (3.3), so by Theorem 3.2

$$\left(\mathbb{E} \max_{i \leq n} |a_i X_i|^p\right)^{1/p} \sim \left(\mathbb{E} \max_{i \leq n} |a_i X'_i|^p\right)^{1/p}$$

for all $p > 0$, up to a constant depending only on α. The coordinates of X' are independent and satisfy condition (3.17), so due to [11, Theorem 1.1] the comparison of weak and strong moments of X' holds, i.e. for $p \geq 1$,

$$\left(\mathbb{E} \max_{i \leq n} |a_i X'_i|^p\right)^{1/p} \leq C(\beta) \left[\mathbb{E} \max_{i \leq n} |a_i X'_i| + \max_{i \leq n} \left(\mathbb{E} |a_i X'_i|^p\right)^{1/p}\right],$$

where $C(\beta)$ depends only on β. These two observations yield the assertion. $\quad\square$

3.5   Lower Estimates for Order Statistics

The next lemma shows the relation between $t(k, X)$ and $t^*(k, X)$ for log-concave vectors X.

Lemma 3.14 *Let X be a symmetric log-concave random vector in $\mathbb{R}^n$. For any $1 \leq k \leq n$ we have*

$$\frac{1}{3} \left(t^*(k, X) + \frac{1}{k} \max_{|I|=k} \sum_{i \in I} \mathbb{E}|X_i|\right) \leq t(k, X) \leq 4 \left(t^*(k, X) + \frac{1}{k} \max_{|I|=k} \sum_{i \in I} \mathbb{E}|X_i|\right).$$

Proof Let $t_k := t(k, X)$ and $t^*_k := t^*(k, X)$. We may assume that any X_i is not identically equal to 0. Then $\sum_{i=1}^n \mathbb{P}(|X_i| \geq t^*_k) = k$ and $\sum_{i=1}^n \mathbb{E}|X_i| \mathbf{1}_{\{|X_i| \geq t_k\}} = k t_k$. Obviously $t^*_k \leq t_k$. Also for any $|I| = k$ we have

$$\sum_{i \in I} \mathbb{E}|X_i| \leq \sum_{i \in I} \left(t_k + \mathbb{E}|X_i| \mathbf{1}_{\{|X_i| \geq t_k\}}\right) \leq |I| t_k + k t_k = 2k t_k.$$

To prove the upper bound set

$$I_1 := \{i \in [n]: \mathbb{P}(|X_i| \geq t^*_k) \geq 1/4\}.$$

We have

$$k \geq \sum_{i \in |I_1|} \mathbb{P}(|X_i| \geq t_k^*) \geq \frac{1}{4}|I_1|,$$

so $|I_1| \leq 4k$. Hence

$$\sum_{i \in I_1} \mathbb{E}|X_i| \mathbf{1}_{\{|X_i| \geq t_k^*\}} \leq \sum_{i \in I_1} \mathbb{E}|X_i| \leq 4 \max_{|I|=k} \sum_{i \in I} \mathbb{E}|X_i|.$$

Moreover by the second part of Lemma 3.12 we get

$$\mathbb{E}|X_i| \mathbf{1}_{\{|X_i| \geq t_k^*\}} \leq 4t_k^* \mathbb{P}(|X_i| \geq t_k^*) \quad \text{for } i \notin I_1,$$

so

$$\sum_{i \notin I_1} \mathbb{E}|X_i| \mathbf{1}_{\{|X_i| \geq t_k^*\}} \leq 4t_k^* \sum_{i=1}^{n} \mathbb{P}(|X_i| \geq t_k^*) \leq 4kt_k^*.$$

Hence if $s = 4t_k^* + \frac{4}{k} \max_{|I|=k} \sum_{i \in I} \mathbb{E}|X_i|$ then

$$\sum_{i=1}^{n} \mathbb{E}|X_i| \mathbf{1}_{\{|X_i| \geq s\}} \leq \sum_{i=1}^{n} \mathbb{E}|X_i| \mathbf{1}_{\{|X_i| \geq t_k^*\}} \leq 4 \max_{|I|=k} \sum_{i \in I} \mathbb{E}|X_i| + 4kt_k^* = ks,$$

that is $t_k \leq s$. $\qquad\square$

To derive bounds for order statistics we will also need a few facts about log-concave vectors.

Lemma 3.15 *Assume that Z is an isotropic one- or two-dimensional log-concave random vector with a density g. Then $g(t) \leq C$ for all t. If Z is one-dimensional, then also $g(t) \geq c$ for all $|t| \leq t_0$, where $t_0 > 0$ is an absolute constant.*

Proof We will use a classical result (see [4, Theorem 2.2.2, Proposition 3.3.1, Proposition 3.3.2, and Proposition 2.5.9]): $\|g\|_{\sup} \sim g(0) \sim 1$ (note that here we use the assumption that Z is isotropic, in particular that $\mathbb{E}Z = 0$, and that the dimension of Z is 1 or 2). This implies the upper bound on g.

In order to get the lower bound in the one-dimensional case, it suffices to prove that $g(u) \geq c$ for $|u| = \varepsilon \mathbb{E}|Z| \geq (2C_1)^{-1}\varepsilon$, where $1/4 > \varepsilon > 0$ is fixed and its value will be chosen later (then by the log-concavity we get $g(u)^s g(0)^{1-s} \leq g(su)$ for all $s \in (0, 1)$). Since $-Z$ is again isotropic we may assume that $u \geq 0$.

If $g(u) \geq g(0)/e$, then we are done. Otherwise by log-concavity of g we get

$$\mathbb{P}(Z \geq u) = \int_u^{\infty} g(s)ds \leq \int_u^{\infty} g(u)^{s/u} g(0)^{-s/u+1} ds \leq g(0) \int_u^{\infty} e^{-s/u} ds \leq C_0 u \leq C_0 \varepsilon.$$

On the other hand, Z has mean zero, so $\mathbb{E}|Z| = 2\mathbb{E}Z_+$ and by the Paley–Zygmund inequality and (3.7) we have

$$\mathbb{P}(Z \geq u) = \mathbb{P}(Z_+ \geq 2\varepsilon\mathbb{E}Z_+) \geq (1 - 2\varepsilon)^2 \frac{(\mathbb{E}Z_+)^2}{\mathbb{E}Z_+^2} \geq \frac{1}{16}\frac{(\mathbb{E}|Z|)^2}{\mathbb{E}Z^2} \geq c_0.$$

For $\varepsilon < c_0/C_0$ we get a contradiction. $\qquad\square$

Lemma 3.16 *Let Y be a mean zero log-concave random variable and let $\mathbb{P}(|Y| \geq t) \leq p$ for some $p > 0$. Then*

$$\mathbb{P}\left(|Y| \geq \frac{t}{2}\right) \geq \frac{1}{\sqrt{ep}}\mathbb{P}(|Y| \geq t).$$

Proof By the Grünbaum inequality (3.8) we have $\mathbb{P}(Y \geq 0) \geq 1/e$, hence

$$\mathbb{P}\left(Y \geq \frac{t}{2}\right) \geq \sqrt{\mathbb{P}(Y \geq t)\mathbb{P}(Y \geq 0)} \geq \frac{1}{\sqrt{e}}\sqrt{\mathbb{P}(Y \geq t)} \geq \frac{1}{\sqrt{ep}}\mathbb{P}(Y \geq t).$$

Since $-Y$ satisfies the same assumptions as Y we also have

$$\mathbb{P}\left(-Y \geq \frac{t}{2}\right) \geq \frac{1}{\sqrt{ep}}\mathbb{P}(-Y \geq t).$$

$\qquad\square$

Lemma 3.17 *Let Y be a mean zero log-concave random variable and let $\mathbb{P}(|Y| \geq t) \geq p$ for some $p > 0$. Then there exists a universal constant C such that*

$$\mathbb{P}(|Y| \leq \lambda t) \leq \frac{C\lambda}{\sqrt{p}}\mathbb{P}(|Y| \leq t) \quad \text{for } \lambda \in [0, 1].$$

Proof Without loss of generality we may assume that $\mathbb{E}Y^2 = 1$. Then by Chebyshev's inequality $t \leq p^{-1/2}$. Let g be the density of Y. By Lemma 3.15 we know that $\|g\|_\infty \leq C$ and $g(t) \geq c$ on $[-t_0, t_0]$, where c, C and $t_0 \in (0, 1)$ are universal constants. Thus

$$\mathbb{P}(|Y| \leq t) \geq \mathbb{P}(|Y| \leq t_0\sqrt{p}t) \geq 2ct_0\sqrt{p}t,$$

and

$$\mathbb{P}(|Y| \leq \lambda t) \leq 2\|g\|_\infty \lambda t \leq 2C\lambda t \leq \frac{C\lambda}{ct_0\sqrt{p}}\mathbb{P}(|Y| \leq t).$$

$\qquad\square$

Now we are ready to give a proof of the lower bound in Theorem 3.4. The next proposition is a key part of it.

Proposition 3.18 *Let X be a mean zero log-concave n-dimensional random vector with uncorrelated coordinates and let $\alpha > 1/4$. Suppose that*

$$\mathbb{P}\big(|X_i| \geq t^*(\alpha, X)\big) \leq \frac{1}{C_3} \quad \text{for all } i.$$

Then

$$\mathbb{P}\left(\lfloor 4\alpha \rfloor\text{-}\max_i |X_i| \geq \frac{1}{C_4} t^*(\alpha, X)\right) \geq \frac{3}{4}.$$

Proof Let $t^* = t^*(\alpha, X)$, $k := \lfloor 4\alpha \rfloor$ and $L = \lfloor \frac{\sqrt{C_3}}{4\sqrt{e}} \rfloor$. We will choose C_3 in such a way that L is large, in particular we may assume that $L \geq 2$. Observe also that $\alpha = \sum_{i=1}^{n} \mathbb{P}(|X_i| \geq t^*(\alpha, X)) \leq nC_3^{-1}$, thus $Lk \leq C_3^{1/2}e^{-1/2}\alpha \leq e^{-1/2}C_3^{-1/2}n \leq n$ if $C_3 \geq 1 > \frac{1}{e}$. Hence

$$k\text{-}\max_i |X_i| \geq \frac{1}{k(L-1)} \sum_{l=k+1}^{Lk} l\text{-}\max_i |X_i|$$

$$= \frac{1}{k(L-1)} \left(\max_{|I|=Lk} \sum_{i \in I} |X_i| - \max_{|I|=k} \sum_{i \in I} |X_i| \right). \quad (3.18)$$

Lemma 3.16 and the definition of $t^*(\alpha, X)$ yield

$$\sum_{i=1}^{n} \mathbb{P}\left(|X_i| \geq \frac{1}{2} t^*\right) \geq \frac{\sqrt{C_3}}{\sqrt{e}} \alpha \geq Lk.$$

This yields $t(Lk, X) \geq t^*(Lk, X) \geq \frac{t^*}{2}$ and by Theorem 3.3 we have

$$\mathbb{E} \max_{|I|=Lk} \sum_{i \in I} |X_i| \geq c_1 Lk \frac{t^*}{2}.$$

Since for any norm $\mathbb{P}(\|X\| \leq t\mathbb{E}\|X\|) \leq Ct$ for $t > 0$ (see [10, Corollary 1]) we have

$$\mathbb{P}\left(\max_{|I|=Lk} \sum_{i \in I} |X_i| \geq c_2 Lkt^* \right) \geq \frac{7}{8}. \quad (3.19)$$

Let X' be an independent copy of X. By the Paley-Zygmund inequality and (3.7), $\mathbb{P}(|X_i| \geq \frac{1}{2}\mathbb{E}|X_i|) \geq \frac{(\mathbb{E}|X_i|)^2}{4\mathbb{E}|X_i|^2} > \frac{1}{C_3}$ if $C_3 > 16C_1^2$, so $\frac{1}{2}\mathbb{E}|X_i| \leq t^*$. Moreover it is easy to verify that $k = \lfloor 4\alpha \rfloor > \alpha$ for $\alpha > 1/4$, thus $t^*(k, X) \leq t^*(\alpha, X) = t^*$. Hence Proposition 3.1, Lemma 3.14, and inequality (3.10) yield

$$\mathbb{E} \max_{|I|=k} \sum_{i \in I} |X_i| = \mathbb{E} \max_{|I|=k} \sum_{i \in I} |X_i - \mathbb{E} X_i'| \leq \mathbb{E} \max_{|I|=k} \sum_{i \in I} |X_i - X_i'|$$

$$\leq \mathbb{E} \max_{|I|=2k} \sum_{i \in I} |X_i - X_i'|$$

$$\leq 4kt(2k, X - X') \leq 16k\big(t^*(2k, X - X') + \max_i \mathbb{E}|X_i - X_i'|\big)$$

$$\leq 16k\big(2t^*(k, X) + 2\max_i \mathbb{E}|X_i|\big) \leq 96kt^*.$$

Therefore

$$\mathbb{P}\left(\max_{|I|=k} \sum_{i \in I} |X_i| \geq 800kt^*\right) \leq \frac{1}{8}. \tag{3.20}$$

Estimates (3.18)–(3.20) yield

$$\mathbb{P}\left(k\text{-}\max_i |X_i| \geq \frac{1}{L-1}(c_2L - 800)t^*\right) \geq \frac{3}{4},$$

so it is enough to choose C_3 in such a way that $L \geq 1600/c_2$. $\qquad\square$

Proof of the First Part of Theorem 3.4 Let $t^* = t^*(k - 1/2, X)$ and C_3 be as in Proposition 3.18. It is enough to consider the case when $t^* > 0$, then $\mathbb{P}(|X_i| = t^*) = 0$ for all i and $\sum_{i=1}^n \mathbb{P}(|X_i| \geq t^*) = k - 1/2$. Define

$$I_1 := \left\{i \leq n\colon \ \mathbb{P}(|X_i| \geq t^*) \leq \frac{1}{C_3}\right\}, \quad \alpha := \sum_{i \in I_1} \mathbb{P}(|X_i| \geq t^*),$$

$$I_2 := \left\{i \leq n\colon \ \mathbb{P}(|X_i| \geq t^*) > \frac{1}{C_3}\right\}, \quad \beta := \sum_{i \in I_2} \mathbb{P}(|X_i| \geq t^*).$$

If $\beta = 0$ then $\alpha = k - 1/2$, $|I_1| = \{1, \ldots, n\}$, and the assertion immediately follows by Proposition 3.18 since $4\alpha \geq k$.

Otherwise define

$$\tilde{N}(t) := \sum_{i \in I_2} \mathbf{1}_{\{|X_i| \leq t\}}.$$

We have by Lemma 3.17 applied with $p = 1/C_3$

$$\mathbb{E}\tilde{N}(\lambda t^*) = \sum_{i \in I_2} \mathbb{P}(|X_i| \leq \lambda t^*) \leq C_5\lambda \sum_{i \in I_2} \mathbb{P}(|X_i| \leq t^*) = C_5\lambda(|I_2| - \beta).$$

Thus

$$\mathbb{P}\left(\lceil\beta\rceil\text{-}\max_{i\in I_2}|X_i|\leq \lambda t^*\right)=\mathbb{P}(\tilde{N}(\lambda t^*)\geq |I_2|+1-\lceil\beta\rceil)$$

$$\leq \frac{1}{|I_2|+1-\lceil\beta\rceil}\mathbb{E}\tilde{N}(\lambda t^*)\leq C_5\lambda.$$

Therefore

$$\mathbb{P}\left(\lceil\beta\rceil\text{-}\max_{i\in I_2}|X_i|\geq \frac{1}{4C_5}t^*\right)\geq \frac{3}{4}.$$

If $\alpha < 1/2$ then $\lceil\beta\rceil = k$ and the assertion easily follows. Otherwise Proposition 3.18 yields

$$\mathbb{P}\left(\lfloor 4\alpha\rfloor\text{-}\max_{i\in I_1}|X_i|\geq \frac{1}{C_4}t^*\right)\geq \frac{3}{4}.$$

Observe that for $\alpha\geq 1/2$ we have $\lfloor 4\alpha\rfloor+\lceil\beta\rceil\geq 4\alpha-1+\beta\geq\alpha+1/2+\beta=k$, so

$$\mathbb{P}\left(k\text{-}\max_i|X_i|\geq\min\left\{\frac{t^*}{C_4},\frac{t^*}{4C_5}\right\}\right)$$

$$\geq\mathbb{P}\left(\lfloor 4\alpha\rfloor\text{-}\max_{i\in I_1}|X_i|\geq\frac{1}{C_4}t^*,\lceil\beta\rceil\text{-}\max_{i\in I_2}|X_i|\geq\frac{1}{4C_5}t^*\right)\geq\frac{1}{2}.$$

□

Remark 3.19 A modification of the proof above shows that under the assumptions of Theorem 3.4 for any $p < 1$ there exists $c(p) > 0$ such that

$$\mathbb{P}\left(k\text{-}\max_{i\leq n}|X_i|\geq c(p)t^*(k-1/2,X)\right)\geq p.$$

3.6   Upper Estimates for Order Statistics

We will need a few more facts concerning log-concave vectors.

Lemma 3.20 *Suppose that X is a mean zero log-concave random vector with uncorrelated coordinates. Then for any $i \neq j$ and $s > 0$,*

$$\mathbb{P}(|X_i|\leq s,|X_j|\leq s)\leq C_6\mathbb{P}(|X_i|\leq s)\mathbb{P}(|X_j|\leq s).$$

Proof Let C_7, c_3 and t_0 be the constants from Lemma 3.15. If $s > t_0\|X_i\|_2$ then, by Lemma 3.15, $\mathbb{P}(|X_i|\leq s)\geq 2c_3t_0$ and the assertion is obvious (with any $C_6\geq (2c_3t_0)^{-1}$). Thus we will assume that $s\leq t_0\min\{\|X_i\|_2,\|X_j\|_2\}$.

Let $\tilde{X}_i = X_i / \|X_i\|_2$ and let g_{ij} be the density of $(\tilde{X}_i, \tilde{X}_j)$. By Lemma 3.15 we know that $\|g_{i,j}\|_\infty \leq C_7$, so

$$\mathbb{P}(|X_i| \leq s, |X_j| \leq s) = \mathbb{P}(|\tilde{X}_i| \leq s/\|X_i\|_2, |\tilde{X}_j| \leq s/\|X_j\|_2) \leq C_7 \frac{s^2}{\|X_i\|_2 \|X_j\|_2}.$$

On the other hand the second part of Lemma 3.15 yields

$$\mathbb{P}(|X_i| \leq s)\mathbb{P}(|X_j| \leq s) \geq \frac{4c_3^2 s^2}{\|X_i\|_2 \|X_j\|_2}.$$

$\square$

Lemma 3.21 *Let Y be a log-concave random variable. Then*

$$\mathbb{P}(|Y| \geq ut) \leq \mathbb{P}(|Y| \geq t)^{(u-1)/2} \quad for\ u \geq 1, t \geq 0.$$

Proof We may assume that Y is non-degenerate (otherwise the statement is obvious), in particular Y has no atoms. Log-concavity of Y yields

$$\mathbb{P}(Y \geq t) \geq \mathbb{P}(Y \geq -t)^{\frac{u-1}{u+1}} \mathbb{P}(Y \geq ut)^{\frac{2}{u+1}}.$$

Hence

$$\mathbb{P}(Y \geq ut) \leq \left(\frac{\mathbb{P}(Y \geq t)}{\mathbb{P}(Y \geq -t)} \right)^{\frac{u+1}{2}} \mathbb{P}(Y \geq -t) = \left(1 - \frac{\mathbb{P}(|Y| \leq t)}{\mathbb{P}(Y \geq -t)} \right)^{\frac{u+1}{2}} \mathbb{P}(Y \geq -t)$$

$$\leq (1 - \mathbb{P}(|Y| \leq t))^{\frac{u+1}{2}} \mathbb{P}(Y \geq -t) = \mathbb{P}(|Y| \geq t)^{\frac{u+1}{2}} \mathbb{P}(Y \geq -t).$$

Since $-Y$ satisfies the same assumptions as Y, we also have

$$\mathbb{P}(Y \leq -ut) \leq \mathbb{P}(|Y| \geq t)^{\frac{u+1}{2}} \mathbb{P}(Y \leq t).$$

Adding both estimates we get

$$\mathbb{P}(|Y| \geq ut) \leq \mathbb{P}(|Y| \geq t)^{\frac{u+1}{2}}(1 + \mathbb{P}(|Y| \leq t)) = \mathbb{P}(|Y| \geq t)^{\frac{u-1}{2}}(1 - \mathbb{P}(|Y| \leq t)^2).$$

$\square$

Lemma 3.22 *Suppose that Y is a log-concave random variable and $\mathbb{P}(|Y| \leq t) \leq \frac{1}{10}$. Then $\mathbb{P}(|Y| \leq 21t) \geq 5\mathbb{P}(|Y| \leq t)$.*

Proof Let $\mathbb{P}(|Y| \leq t) = p$ then by Lemma 3.21

$$\mathbb{P}(|Y| \leq 21t) = 1 - \mathbb{P}(|Y| > 21t) \geq 1 - \mathbb{P}(|Y| > t)^{10} = 1 - (1-p)^{10}$$

$$\geq 10p - 45p^2 \geq 5p.$$

$\square$

Let us now prove (3.4) and see how it implies the second part of Theorem 3.4. Then we give a proof of (3.5).

Proof of (3.4) Fix k and set $t^* := t^*(k - 1/2, X)$. Then $\sum_{i=1}^n \mathbb{P}(|X_i| \geq t^*) = k - 1/2$. Define

$$I_1 := \left\{ i \leq n \colon \mathbb{P}(|X_i| \geq t^*) \leq \frac{9}{10} \right\}, \quad \alpha := \sum_{i \in I_1} \mathbb{P}(|X_i| \geq t^*), \tag{3.21}$$

$$I_2 := \left\{ i \leq n \colon \mathbb{P}(|X_i| \geq t^*) > \frac{9}{10} \right\}, \quad \beta := \sum_{i \in I_2} \mathbb{P}(|X_i| \geq t^*). \tag{3.22}$$

Observe that for $u > 3$ and $1 \leq l \leq |I_1|$ we have by Lemma 3.21

$$\mathbb{P}(l\text{-}\max_{i \in I_1} |X_i| \geq ut^*) \leq \mathbb{E} \frac{1}{l} \sum_{i \in I_1} \mathbf{1}_{\{|X_i| \geq ut^*\}} = \frac{1}{l} \sum_{i \in I_1} \mathbb{P}(|X_i| \geq ut^*) \tag{3.23}$$

$$\leq \frac{1}{l} \sum_{i \in I_1} \mathbb{P}(|X_i| \geq t^*)^{(u-1)/2} \leq \frac{\alpha}{l} \left(\frac{9}{10} \right)^{(u-3)/2}.$$

Consider two cases.

Case 1 $\beta > |I_2| - 1/2$. Then $|I_2| < \beta + 1/2 \leq k$, so $k - |I_2| \geq 1$ and

$$\alpha = k - \frac{1}{2} - \beta \leq k - |I_2|.$$

Therefore by (3.23)

$$\mathbb{P}\left(k\text{-}\max |X_i| \geq 5t^* \right) \leq \mathbb{P}\left((k - |I_2|)\text{-}\max_{i \in I_1} |X_i| \geq 5t^* \right) \leq \frac{9}{10}.$$

Case 2 $\beta \leq |I_2| - 1/2$. Observe that for any disjoint sets J_1, J_2 and integers l, m such that $l \leq |J_1|$, $m \leq |J_2|$ we have

$$(l + m - 1)\text{-}\max_{i \in J_1 \cup J_2} |x_i| \leq \max \left\{ l\text{-}\max_{i \in J_1} |x_i|, m\text{-}\max_{i \in J_2} |x_i| \right\}$$

$$\leq l\text{-}\max_{i \in J_1} |x_i| + m\text{-}\max_{i \in J_2} |x_i|. \tag{3.24}$$

Since

$$\lceil \alpha \rceil + \lceil \beta \rceil \leq \alpha + \beta + 2 < k + 2$$

we have $\lceil \alpha \rceil + \lceil \beta \rceil \leq k + 1$ and, by (3.24),

$$k\text{-}\max_i |X_i| \leq \lceil \alpha \rceil\text{-}\max_{i \in I_1} |X_i| + \lceil \beta \rceil\text{-}\max_{i \in I_2} |X_i|.$$

Estimate (3.23) yields

$$\mathbb{P}\left(\lceil\alpha\rceil\text{-}\max_{i\in I_1}|X_i|\geq ut^*\right)\leq\left(\frac{9}{10}\right)^{(u-3)/2}\qquad\text{for }u\geq 3.$$

To estimate $\lceil\beta\rceil\text{-}\max_{i\in I_2}|X_i| = (|I_2|+1-\lceil\beta\rceil)\text{-}\min_{i\in I_2}|X_i|$ observe that by Lemma 3.22, the definition of I_2 and assumptions on β,

$$\sum_{i\in I_2}\mathbb{P}(|X_i|\leq 21t^*)\geq 5\sum_{i\in I_2}\mathbb{P}(|X_i|\leq t^*) = 5(|I_2|-\beta)\geq 2(|I_2|+1-\lceil\beta\rceil).$$

Set $l := (|I_2|+1-\lceil\beta\rceil)$ and

$$\tilde{N}(t) := \sum_{i\in I_2}\mathbf{1}_{\{|X_i|\leq t\}}.$$

Note that we know already that $\mathbb{E}\tilde{N}(21t^*)\geq 2l$. Thus the Paley-Zygmund inequality implies

$$\mathbb{P}\left(\lceil\beta\rceil\text{-}\max_{i\in I_2}|X_i|\leq 21t^*\right) = \mathbb{P}\left(l\text{-}\min_{i\in I_2}|X_i|\leq 21t^*\right)\geq\mathbb{P}(\tilde{N}(21t^*)\geq l)$$

$$\geq\mathbb{P}\left(\tilde{N}(21t^*)\geq\frac{1}{2}\mathbb{E}\tilde{N}(21t^*)\right)\geq\frac{1}{4}\frac{(\mathbb{E}\tilde{N}(21t^*))^2}{\mathbb{E}\tilde{N}(21t^*)^2}.$$

However Lemma 3.20 yields

$$\mathbb{E}\tilde{N}(21t^*)^2\leq\mathbb{E}\tilde{N}(21t^*)+C_6(\mathbb{E}\tilde{N}(21t^*)))^2\leq(C_6+1)(\mathbb{E}\tilde{N}(21t^*))^2.$$

Therefore

$$\mathbb{P}\left(k\text{-}\max_i|X_i| > (21+u)t^*\right)\leq\mathbb{P}\left(\lceil\alpha\rceil\text{-}\max_{i\in I_1}|X_i|\geq ut^*\right)$$

$$+\mathbb{P}\left(\lceil\beta\rceil\text{-}\max_{i\in I_2}|X_i| > 21t^*\right)$$

$$\leq\left(\frac{9}{10}\right)^{(u-3)/2}+1-\frac{1}{4(C_6+1)}\leq 1-\frac{1}{5(C_6+1)}$$

for sufficiently large u. $\qquad\qquad\square$

The unconditionality assumption plays a crucial role in the proof of the next lemma, which allows to derive the second part of Theorem 3.4 from estimate (3.4).

Lemma 3.23 *Let X be an unconditional log-concave n-dimensional random vector. Then for any $1 \leq k \leq n$,*

$$\mathbb{P}\left(k\text{-}\max_{i \leq n} |X_i| \geq ut\right) \leq \mathbb{P}\left(k\text{-}\max_{i \leq n} |X_i| \geq t\right)^u \quad for\ u > 1, t > 0.$$

Proof Let ν be the law of $(|X_1|, \ldots, |X_n|)$. Then ν is log-concave on $\mathbb{R}_n^+$. Define for $t > 0$,

$$A_t := \left\{x \in \mathbb{R}_n^+ : k\text{-}\max_{i \leq n} |x_i| \geq t\right\}.$$

It is easy to check that $\frac{1}{u} A_{ut} + (1 - \frac{1}{u})\mathbb{R}_+^n \subset A_t$, hence

$$\mathbb{P}\left(k\text{-}\max_{i \leq n} |X_i| \geq t\right) = \nu(A_t) \geq \nu(A_{ut})^{1/u} \nu(\mathbb{R}_+^n)^{1-1/u} = \mathbb{P}\left(k\text{-}\max_{i \leq n} |X_i| \geq ut\right)^{1/u}.$$

□

Proof of the Second Part of Theorem 3.4 Estimate (3.4) together with Lemma 3.23 yields

$$\mathbb{P}\left(k\text{-}\max_{i \leq n} |X_i| \geq Cut^*(k - 1/2, X)\right) \leq (1 - c)^u \quad for\ u \geq 1,$$

and the assertion follows by integration by parts.   □

Proof of (3.5) Define I_1, I_2, α and β by (3.21) and (3.22), where this time $t^* = t^*(k - k^{5/6}/2, X)$. Estimate (3.23) is still valid so integration by parts yields

$$\mathbb{E}l\text{-}\max_{i \in I_1} |X_i| \leq \left(3 + 20\frac{\alpha}{l}\right)t^*.$$

Set

$$k_\beta := \left\lceil \beta + \frac{1}{2} k^{5/6} \right\rceil.$$

Observe that

$$\lceil \alpha \rceil + k_\beta < \alpha + \beta + \frac{1}{2} k^{5/6} + 2 = k + 2.$$

Hence $\lceil \alpha \rceil + k_\beta \leq k + 1$.

If $k_\beta > |I_2|$, then $k - |I_2| \geq \lceil \alpha \rceil + k_\beta - 1 - |I_2| \geq \lceil \alpha \rceil$, so

$$\mathbb{E}k\text{-}\max_i |X_i| \leq \mathbb{E}(k - |I_2|)\text{-}\max_{i \in I_1} |X_i| \leq \mathbb{E}\lceil \alpha \rceil\text{-}\max_{i \in I_1} |X_i| \leq 23t^*.$$

Therefore it suffices to consider case $k_\beta \leq |I_2|$ only.

Since $\lceil \alpha \rceil + k_\beta - 1 \le k$ and $k_\beta \le |I_2|$, we have by (3.24),

$$\mathbb{E} k\text{-}\max_i |X_i| \le \mathbb{E}\lceil \alpha \rceil\text{-}\max_{i \in I_1} |X_i| + \mathbb{E} k_\beta\text{-}\max_{i \in I_2} |X_i| \le 23t^* + \mathbb{E} k_\beta\text{-}\max_{i \in I_2} |X_i|.$$

Since $\beta \le k - \frac{1}{2}k^{5/6}$ and $x \to x - \frac{1}{2}x^{5/6}$ is increasing for $x \ge 1/2$ we have

$$\beta \le \beta + \frac{1}{2}k^{5/6} - \frac{1}{2}\left(\beta + \frac{1}{2}k^{5/6}\right)^{5/6} \le k_\beta - \frac{1}{2}k_\beta^{5/6}.$$

Therefore, considering $(X_i)_{i \in I_2}$ instead of X and k_β instead of k it is enough to show the following claim:

Let $s > 0$, $n \ge k$ and let X be an n-dimensional log-concave vector with uncorrelated coordinates. Suppose that

$$\sum_{i \le n} \mathbb{P}(|X_i| \ge s) \le k - \frac{1}{2}k^{5/6} \quad \text{and} \quad \min_{i \le n} \mathbb{P}(|X_i| \ge s) \ge 9/10$$

then

$$\mathbb{E} k\text{-}\max_{i \le n} |X_i| \le C_8 s.$$

We will show the claim by induction on k. For $k = 1$ the statement is obvious (since the assumptions are contradictory). Suppose now that $k \ge 2$ and the assertion holds for $k - 1$.

Case 1 $\mathbb{P}(|X_{i_0}| \ge s) \ge 1 - \frac{5}{12}k^{-1/6}$ for some $1 \le i_0 \le n$. Then

$$\sum_{i \ne i_0} \mathbb{P}(|X_i| \ge s) \le k - \frac{1}{2}k^{5/6} - \left(1 - \frac{5}{12}k^{-1/6}\right) \le k - 1 - \frac{1}{2}(k-1)^{5/6},$$

where to get the last inequality we used that $x^{5/6}$ is concave on $\mathbb{R}_+$, so $(1-t)^{5/6} \le 1 - \frac{5}{6}t$ for $t = 1/k$. Therefore by the induction assumption applied to $(X_i)_{i \ne i_0}$,

$$\mathbb{E} k\text{-}\max_i |X_i| \le \mathbb{E}(k-1)\text{-}\max_{i \ne i_0} |X_i| \le C_8 s.$$

Case 2 $\mathbb{P}(|X_i| \le s) \ge \frac{5}{12}k^{-1/6}$ for all i. Applying Lemma 3.15 we get

$$\frac{5}{12}k^{-1/6} \le \mathbb{P}\left(\frac{|X_i|}{\|X_i\|_2} \le \frac{s}{\|X_i\|_2}\right) \le C\frac{s}{\|X_i\|_2},$$

so $\max_i \|X_i\|_2 \le Ck^{1/6}s$. Moreover $n \le \frac{10}{9}k$. Therefore by the result of Lee and Vempala [13] X satisfies the exponential concentration with $\alpha \le C_9 k^{5/12}s$.

Let $l = \lceil k - \frac{1}{2}(k^{5/6} - 1) \rceil$ then $s \geq t_*(l - 1/2, X)$ and $k - l + 1 \geq \frac{1}{2}(k^{5/6} - 1) \geq \frac{1}{9}k^{5/6}$. Let

$$A := \left\{ x \in \mathbb{R}^n : l\text{-}\max_i |x_i| \leq C_{10}s \right\}.$$

By (3.4) (applied with l instead of k) we have $\mathbb{P}(X \in A) \geq c_4$. Observe that

$$k\text{-}\max_i |x_i| \geq C_{10}s + u \Rightarrow \text{dist}(x, A) \geq \sqrt{k - l + 1}u \geq \frac{1}{3}k^{5/12}u.$$

Therefore by Lemma 3.10 we get

$$\mathbb{P}\left(k\text{-}\max_i |X_i| \geq C_{10}s + 3C_9us \right) \leq \exp\left(-(u + \ln c_4)_+\right).$$

Integration by parts yields

$$\mathbb{E}k\text{-}\max_i |X_i| \leq (C_{10} + 3C_9(1 - \ln c_4))\, s$$

and the induction step is shown in this case provided that $C_8 \geq C_{10} + 3C_9(1 - \ln c_4)$. $\qquad\square$

To obtain Corollary 3.6 we used the following lemma.

Lemma 3.24 *Assume that X is a symmetric isotropic log-concave vector in $\mathbb{R}^n$. Then*

$$t^*(p, X) \sim \frac{n - p}{n} \quad for\ n > p \geq n/4. \tag{3.25}$$

and

$$t^*(k/2, X) \sim t^*(k, X) \sim t(k, X) \quad for\ k \leq n/2. \tag{3.26}$$

Proof Observe that

$$\sum_{i=1}^n \mathbb{P}(|X_i| \leq t^*(p, X)) = n - p.$$

Thus Lemma 3.15 implies that for $p \geq c_5n$ (with $c_5 \in (\frac{1}{2}, 1)$) we have $t^*(p, X) \sim \frac{n-p}{n}$. Moreover, by the Markov inequality

$$\sum_{i=1}^n \mathbb{P}(|X_i| \geq 4) \leq \frac{n}{16},$$

so $t^*(n/4, X) \leq 4$. Since $p \mapsto t^*(p, X)$ is non-increasing, we know that $t^*(p, X) \sim 1$ for $n/4 \leq p \leq c_5 n$.

Now we will prove (3.26). We have

$$t^*(k, X) \leq t^*(k/2, X) \leq t(k/2, X) \leq 2t(k, X),$$

so it suffices to show that $t^*(k, X) \geq ct(k, X)$. To this end we fix $k \leq n/2$. By (3.25) we know that $t := C_{11} t^*(k, X) \geq C_{11} t^*(n/2, X) \geq e$, so the isotropicity of X and Markov's inequality yield $\mathbb{P}(|X_i| \geq t) \leq e^{-2}$ for all i. We may also assume that $t \geq t^*(k, X)$. Integration by parts and Lemma 3.21 yield

$$\mathbb{E}|X_i| \mathbf{1}_{\{|X_i| \geq t\}} \leq 3t\mathbb{P}(|X_i| \geq t) + t \int_0^\infty \mathbb{P}(|X_i| \geq (s+3)t)ds$$

$$\leq 3t\mathbb{P}(|X_i| \geq t) + t \int_0^\infty \mathbb{P}(|X_i| \geq t)e^{-s}ds \leq 4t\mathbb{P}(|X_i| \geq t).$$

Therefore

$$\sum_{i=1}^n \mathbb{E}|X_i| \mathbf{1}_{\{|X_i| \geq t\}} \leq 4t \sum_{i=1}^n \mathbb{P}(|X_i| \geq t) \leq 4t \sum_{i=1}^n \mathbb{P}(|X_i| \geq t^*(k, X)) \leq 4kt,$$

so $t(k, X) \leq 4C_{11} t^*(k, X)$. □

Acknowledgements The research of RL was supported by the National Science Centre, Poland grant 2015/18/A/ST1/00553 and of MS by the National Science Centre, Poland grants 2015/19/N/ST1/02661 and 2018/28/T/ST1/00001.

References

1. S. Artstein-Avidan, A. Giannopoulos, V.D. Milman, *Asymptotic Geometric Analysis. Part I* (American Mathematical Society, Providence, 2015)
2. N. Balakrishnan, A.C. Cohen, *Order Statistics and Inference* (Academic Press, New York, 1991)
3. C. Borell, Convex measures on locally convex spaces. Ark. Math. **12**, 239–252 (1974)
4. S. Brazitikos, A. Giannopoulos, P. Valettas, B.H. Vritsiou, *Geometry of Isotropic Convex Bodies* (American Mathematical Society, Providence, 2014)
5. H.A. David, H.N. Nagaraja, *Order Statistics*, 3rd edn. (Wiley-Interscience, Hoboken, 2003)
6. R. Eldan, Thin shell implies spectral gap up to polylog via a stochastic localization scheme. Geom. Funct. Anal. **23**, 532–569 (2013)
7. Y. Gordon, A. Litvak, C. Schütt, E. Werner, Orlicz norms of sequences of random variables. Ann. Probab. **30**(4), 1833–1853 (2002)
8. Y. Gordon, A. Litvak, C. Schütt, E. Werner, On the minimum of several random variables. Proc. Amer. Math. Soc. **134**(12), 3665–3675 (2006)
9. Y. Gordon, A. Litvak, C. Schütt, E. Werner, Uniform estimates for order statistics and Orlicz functions. Positivity **16**(1), 1–28 (2012)

10. R. Latała, On the equivalence between geometric and arithmetic means for log-concave measures, in *Convex Geometric Analysis* (Berkeley, 1996). Mathematical Sciences Research Institute Publications, vol. 34 (Cambridge University Press, Cambridge, 1999), pp. 123–127

11. R. Latała, M. Strzelecka, Comparison of weak and strong moments for vectors with independent coordinates. Mathematika **64**(1), 211–229 (2018)

12. M. Ledoux, *The Concentration of Measure Phenomenon* (American Mathematical Society, Providence, 2001)

13. Y.T. Lee, S. Vempala, Eldan's stochastic localization and the KLS hyperplane conjecture: an improved lower bound for expansion, in *58th Annual IEEE Symposium on Foundations of Computer Science – FOCS 2017* (IEEE Computer Society, Los Alamitos, 2017), pp. 998–1007

14. L. Lovász, S. Vempala, The geometry of logconcave functions and sampling algorithms. Random Struct. Algoritm. **30**(3), 307–358 (2007)

15. J. Prochno, S. Riemer, On the maximum of random variables on product spaces. Houston J. Math. **39**(4), 1301–1311 (2013)

Chapter 4
Further Investigations of Rényi Entropy Power Inequalities and an Entropic Characterization of s-Concave Densities

Jiange Li, Arnaud Marsiglietti, and James Melbourne

Abstract We investigate the role of convexity in Rényi entropy power inequalities. After proving that a general Rényi entropy power inequality in the style of Bobkov and Chistyakov (IEEE Trans Inform Theory 61(2):708–714, 2015) fails when the Rényi parameter $r \in (0, 1)$, we show that random vectors with s-concave densities do satisfy such a Rényi entropy power inequality. Along the way, we establish the convergence in the Central Limit Theorem for Rényi entropies of order $r \in (0, 1)$ for log-concave densities and for compactly supported, spherically symmetric and unimodal densities, complementing a celebrated result of Barron (Ann Probab 14:336–342, 1986). Additionally, we give an entropic characterization of the class of s-concave densities, which extends a classical result of Cover and Zhang (IEEE Trans Inform Theory 40(4):1244–1246, 1994).

4.1   Introduction

Let X be a random vector in $\mathbb{R}^d$. Suppose that X has the density f with respect to the Lebesgue measure. For $r \in (0, 1) \cup (1, \infty)$, the Rényi entropy of order r (or simply, r-Rényi entropy) is defined as

$$h_r(X) = \frac{1}{1 - r} \log \int_{\mathbb{R}^d} f(x)^r dx. \tag{4.1}$$

J. Li
Einstein Institute of Mathematics, Hebrew University of Jerusalem, Jerusalem, Israel
e-mail: jiange.li@mail.huji.ac.il

A. Marsiglietti (✉)
Department of Mathematics, University of Florida, Gainesville, FL, USA
e-mail: a.marsiglietti@ufl.edu

J. Melbourne
Electrical and Computer Engineering, University of Minnesota, Minneapolis, MN, USA
e-mail: melbo013@umn.edu

© Springer Nature Switzerland AG 2020
B. Klartag, E. Milman (eds.), *Geometric Aspects of Functional Analysis*,
Lecture Notes in Mathematics 2266,
https://doi.org/10.1007/978-3-030-46762-3_4

For $r \in \{0, 1, \infty\}$, the r-Rényi entropy can be extended continuously such that the RHS of (4.1) is $\log |\text{supp}(f)|$ for $r = 0$; $- \int_{\mathbb{R}^d} f(x) \log f(x) dx$ for $r = 1$; and $- \log \|f\|_\infty$ for $r = \infty$. The case $r = 1$ corresponds to the classical Shannon differential entropy. Here, we denote by $|\text{supp}(f)|$ the Lebesgue measure of the support of f, and $\|f\|_\infty$ represents the essential supremum of f. The r-Rényi entropy power is defined by

$$N_r(X) = e^{2h_r(X)/d}.$$

In the following, we drop the subscript when $r = 1$.

The classical Entropy Power Inequality (henceforth, EPI) of Shannon [39] and Stam [41], states that the entropy power $N(X)$ is super-additive on the sum of independent random vectors. There has been recent success in obtaining extensions of the EPI from the Shannon differential entropy to r-Rényi entropy. In [7, 8], Bobkov and Chistyakov showed that, at the expense of an absolute constant $c > 0$, the following Rényi EPI of order $r \in [1, \infty]$ holds

$$N_r(X_1 + \cdots + X_n) \geq c \sum_{i=1}^{n} N_r(X_i). \tag{4.2}$$

Ram and Sason soon after gave a sharpened constant depending on the number of summands [36]. Madiman, Melbourne, and Xu sharpened constants in the $r = \infty$ case by identifying extremizers in [31, 32]. Savaré and Toscani [38] showed that a modified Rényi entropy power is concave along the solution of a nonlinear heat equation, which generalizes Costa's concavity of entropy power [19]. Bobkov and Marsiglietti [10] proved the following variant of Rényi EPI

$$N_r(X + Y)^\alpha \geq N_r(X)^\alpha + N_r(Y)^\alpha \tag{4.3}$$

for $r > 1$ and some exponent α only depending on r. It is clear that (4.3) holds for more than two summands. Improvement of the exponent α was given by Li [27].

One of our goals is to establish analogues of (4.2) and (4.3) when the Rényi parameter $r \in (0, 1)$. Both (4.2) and (4.3) can be derived from Young's convolution inequality in conjunction with the entropic comparison inequality $h_{r_1}(X) \geq h_{r_2}(X)$ for any $0 \leq r_1 \leq r_2$. The latter fact is an immediate consequence of Jensen's inequality. When the Rényi parameter $r \in (0, 1)$, analogues of (4.2) and (4.3) require a converse of the entropic comparison inequality aforementioned. This technical issue prevents a general Rényi EPI of order $r \in (0, 1)$ for generic random vectors. Our first result shows that a general Rényi EPI of the form (4.2) indeed fails for all $r \in (0, 1)$.

Theorem 4.1 *For any $r \in (0, 1)$ and $\varepsilon > 0$, there exist independent random vectors $X_1, \cdots, X_n$ in $\mathbb{R}^d$, for some $d \geq 1$ and $n \geq 2$, such that*

$$N_r(X_1 + \cdots + X_n) < \varepsilon \sum_{i=1}^{n} N_r(X_i). \qquad (4.4)$$

We have an explicit construction of such random vectors. They are essentially truncations of some spherically symmetric random vectors with finite covariance matrices and infinite Rényi entropies of order $r \in (0, 1)$. The key point is the convergence along the Central Limit Theorem (henceforth, CLT) for Rényi entropies of order $r \in (0, 1)$; that is, the r-Rényi entropy of their normalized sum converges to the r-Rényi entropy of a Gaussian. This implies that, after appropriate normalization, the LHS of (4.4) is finite, but the RHS of (4.4) can be as large as possible. The entropic CLT has been studied for a long time. A celebrated result of Barron [3] shows the convergence in the CLT for Shannon differential entropy (see [26] for a multidimensional setting). The recent work of Bobkov and Marsiglietti [11] studies the convergence in the CLT for Rényi entropy of order $r > 1$ for real-valued random variables (see also [12] for convergence in Rényi divergence, which is not equivalent to convergence in Rényi entropy unless $r = 1$). In Sect. 4.2, we establish the analogue of [11, Theorem 1.1] in higher dimensions and we prove convergence along the CLT for Rényi entropies of order $r \in (0, 1)$ for a large class of densities.

As mentioned above, the reverse entropic comparison inequality prevents Rényi EPIs of order $r \in (0, 1)$ for generic random vectors. However, a large class of random vectors with the so-called s-concave densities do satisfy such a reverse entropic comparison inequality. Our next results show that Rényi EPI of order $r \in (0, 1)$ holds for such densities. This extends the earlier work of Marsiglietti and Melbourne [33, 34] for log-concave densities (which corresponds to the $s = 0$ case).

Let $s \in [-\infty, \infty]$. A function $f : \mathbb{R}^d \to [0, \infty)$ is called s-concave if the inequality

$$f((1 - \lambda)x + \lambda y) \geq ((1 - \lambda)f(x)^s + \lambda f(y)^s)^{1/s} \qquad (4.5)$$

holds for all $x, y \in \mathbb{R}^d$ such that $f(x)f(y) > 0$ and $\lambda \in (0, 1)$. For $s \in \{-\infty, 0, \infty\}$, the RHS of (4.5) is understood in the limiting sense; that is $\min\{f(x), f(y)\}$ for $s = -\infty$, $f(x)^{1-\lambda}f(y)^\lambda$ for $s = 0$, and $\max\{f(x), f(y)\}$ for $s = \infty$. The case $s = 0$ corresponds to log-concave functions. The study of measures with s-concave densities was initiated by Borell in the seminal work [13, 14]. One can think of s-concave densities, in particular log-concave densities, as functional versions of convex sets. There has been a recent stream of research on a formal parallel relation between functional inequalities of s-concave densities and geometric inequalities of convex sets.

Theorem 4.2 *For any $s \in (-1/d, 0)$ and $r \in (-sd, 1)$, there exists $c = c(s, r, d, n)$ such that for all independent random vectors $X_1, \cdots, X_n$ with s-*

concave densities in $\mathbb{R}^d$, we have

$$N_r(X_1 + \cdots + X_n) \geq c \sum_{i=1}^{n} N_r(X_i).$$

In particular, one can take

$$c = r^{\frac{1}{1-r}} \left(1 + \frac{1}{n|r'|}\right)^{1+n|r'|} \left(\prod_{k=1}^{d} \frac{(1+ks)^{|r'|(n-1)}(1+\frac{ks}{r})^{1+|r'|}}{(1+ks(1+\frac{1}{n|r'|}))^{1+n|r'|}}\right)^{\frac{2}{d}},$$

where $r' = r/(r-1)$ is the Hölder conjugate of r.

Theorem 4.3 *Given $s \in (-1/d, 0)$, there exist $0 < r_0 < 1$ and $\alpha = \alpha(s, r, d)$ such that for $r \in (r_0, 1)$ and independent random vectors X and Y in $\mathbb{R}^d$ with s-concave densities,*

$$N_r(X+Y)^\alpha \geq N_r(X)^\alpha + N_r(Y)^\alpha.$$

In particular, one can take

$$r_0 = \left(1 - \frac{2}{1+\sqrt{3}}\left(1 + \frac{1}{sd}\right)\right)^{-1}$$

$$\alpha = \left(1 + \frac{\log r + (r+1)\log\frac{r+1}{2r} + C(s)}{(1-r)\log 2}\right)^{-1},$$

where

$$C(s) = \frac{2}{d} \sum_{k=1}^{d} \left(\log\left(1 + \frac{ks}{r}\right) + r\log(1+ks) - (r+1)\log\left(1 + \frac{ks(r+1)}{2r}\right)\right).$$

Owing to the convexity, random vectors with s-concave densities also satisfy a reverse EPI, which was first proved by Bobkov and Madiman [9]. This can be seen as the functional lifting of Milman's well known reverse Brunn–Minkowski inequality [35]. Motivated by Busemann's theorem [17] in convex geometry, Ball et al. [2] conjectured that the following reverse EPI

$$N(X+Y)^{1/2} \leq N(X)^{1/2} + N(Y)^{1/2} \tag{4.6}$$

holds for any symmetric log-concave random vector $(X, Y) \in \mathbb{R}^2$. The r-Rényi entropy analogue was asked in [30], and the $r = 2$ case was soon verified in [27]. It was also observed in [27] that the r-Rényi entropy analogue is equivalent to the convexity of p-cross-section body in convex geometry introduced by Gardner and

Giannopoulos [23]. The equivalent linearization of (4.6) reads as follows. Let (X, Y) be a symmetric log-concave random vector in $\mathbb{R}^2$ such that $h(X) = h(Y)$. Then for any $\lambda \in [0, 1]$ we have

$$h((1 - \lambda)X + \lambda Y) \le h(X).$$

Cover and Zhang [20] proved the above inequality under the stronger assumption that X and Y have the same log-concave distribution. They also showed that this provides a characterization of log-concave distributions on the real line. The following theorem extends Cover and Zhang's result from log-concave densities to a more general class of s-concave densities. This gives an entropic characterization of s-concave densities and implies a reverse Rényi EPI for random vectors with the same s-concave density.

Theorem 4.4 *Let $r > 1 - 1/d$. Let f be a probability density function on $\mathbb{R}^d$. For any fixed integer $n \ge 2$, the identity*

$$\sup_{X_i \sim f} h_r \left(\sum_{i=1}^{n} \lambda_i X_i \right) = h_r(X_1)$$

holds for all $\lambda_i \ge 0$ such that $\sum_{i=1}^{n} \lambda_i = 1$ if and only if the density f is $(r-1)$-concave.

The paper is organized as follows. In Sect. 4.2, we explore the convergence along the CLT for r-Rényi entropies. For $r > 1$, the convergence is fully characterized for densities on $\mathbb{R}^d$, while for $r \in (0, 1)$ sufficient conditions are obtained for a large class of densities. More precisely, we prove the convergence for log-concave densities and for compactly supported, spherically symmetric and unimodal densities. As an application, we prove in Sect. 4.3 that a general r-Rényi EPI fails when $r \in (0, 1)$, thus establishing Theorem 4.1. We also complement this result by proving Theorems 4.2 and 4.3. In the last section, we provide an entropic characterization of the class of s-concave densities, and include a reverse Rényi EPI as an immediate consequence.

4.2 Convergence Along the CLT for Rényi Entropies

Let $\{X_n\}_{n \in \mathbb{N}}$ be a sequence of independent identically distributed (henceforth, i.i.d.) centered random vectors in $\mathbb{R}^d$ with finite covariance matrix. We denote by Z_n the normalized sum

$$Z_n = \frac{X_1 + \cdots + X_n}{\sqrt{n}}. \tag{4.7}$$

An important tool used to prove various forms of CLT is the characteristic function. Recall that the characteristic function of a random vector X is defined by

$$\varphi_X(t) = \mathbb{E}\left[e^{i\langle t, X\rangle}\right], \quad t \in \mathbb{R}^d.$$

Before providing sufficient conditions for the convergence along the CLT for Rényi entropy of order $r \in (0, 1)$, we first extend [11, Theorem 1.1] to higher dimensions.

Theorem 4.5 *Let* $r > 1$. *Let* $X_1, \cdots, X_n$ *be i.i.d. centered random vectors in* $\mathbb{R}^d$. *We denote by* ρ_n *the density of* Z_n *defined in* (4.7). *The following statements are equivalent.*

1. $h_r(Z_n) \to h_r(Z)$ *as* $n \to +\infty$, *where* Z *is a Gaussian random vector with mean* 0 *and the same covariance matrix as* X_1.
2. $h_r(Z_{n_0})$ *is finite for some integer* n_0.
3. $\int_{\mathbb{R}^d} |\varphi_{X_1}(t)|^\nu \, dt < +\infty$ *for some* $\nu \geq 1$.
4. Z_{n_0} *has a bounded density* ρ_{n_0} *for some integer* n_0.

Proof $1 \Longrightarrow 2$: Assume that $h_r(Z_n) \to h_r(Z)$ as $n \to +\infty$. Then there exists an integer n_0 such that

$$h_r(Z) - 1 < h_r(Z_{n_0}) < h_r(Z) + 1.$$

Since $h_r(Z)$ is finite, we conclude that $h_r(Z_{n_0})$ is finite as well.

$2 \Longrightarrow 3$: Assume that $h_r(Z_{n_0})$ is finite for some integer n_0. Then Z_{n_0} has a density $\rho_{n_0} \in L^r(\mathbb{R}^d)$.

Case 1 If $r \geq 2$, we have $\rho_{n_0} \in L^2(\mathbb{R}^d)$. Using Plancherel's identity, we have $\varphi_{Z_{n_0}} \in L^2(\mathbb{R}^d)$. It follows that

$$\int_{\mathbb{R}^d} |\varphi_{Z_{n_0}}(t)|^2 \, dt = \int_{\mathbb{R}^d} |\varphi_{X_1}\left(t/\sqrt{n_0}\right)|^{2n_0} \, dt < +\infty.$$

For $\nu = 2n_0$, we have

$$\int_{\mathbb{R}^d} |\varphi_{X_1}(t)|^\nu \, dt < +\infty.$$

Case 2 If $r \in (1, 2)$, we apply the Hausdorff–Young inequality to obtain

$$\|\varphi_{Z_{n_0}}\|_{L^{r'}} \leq \frac{1}{(2\pi)^{d/r'}} \|\rho_{n_0}\|_{L^r},$$

where r' is the conjugate of r such that $1/r + 1/r' = 1$. Hence, for $\nu = r'n_0$, we have

$$\int_{\mathbb{R}^d} |\varphi_{X_1}(t)|^\nu \, dt < +\infty.$$

$3 \implies 4$: Since $\int_{\mathbb{R}^d} |\varphi_{X_1}(t)|^\nu \, dt < +\infty$ for some $\nu \geq 1$, one may apply Gnedenko's local limit theorems (see [24]), which is valid in arbitrary dimensions (see [5]). In particular, we have

$$\lim_{n \to +\infty} \sup_{x \in \mathbb{R}^d} |\rho_n(x) - \phi_\Sigma(x)| = 0, \tag{4.8}$$

where ϕ_Σ denotes the density of a Gaussian random vector with mean 0 and the same covariance matrix as X_1. We deduce that there exists an integer n_0 and a constant $M > 0$ such that $\rho_n \leq M$ for all $n \geq n_0$.

$4 \implies 1$: Since ρ_{n_0} is bounded, then $\rho_{n_0} \in L^2$, and we deduce by Plancherel's identity that $\int_{\mathbb{R}^d} |\varphi_{X_1}(t)|^\nu \, dt < +\infty$ for $\nu = 2n_0$. Hence, (4.8) holds and there exists $M > 0$ such that $\rho_n \leq M$ for all $n \geq n_0$. Let us show that $\int_{\mathbb{R}^d} \rho_n(x)^r dx \to \int_{\mathbb{R}^d} \phi_\Sigma(x)^r dx$ as $n \to +\infty$, where ϕ_Σ denotes the density of a Gaussian random vector with mean 0 and the same covariance matrix as X_1. By the CLT, for any $\varepsilon > 0$, there exists $T > 0$ such that for all n large enough,

$$\int_{|x|>T} \rho_n(x) dx < \varepsilon,$$

which implies that

$$\int_{|x|>T} \rho_n(x)^r dx \leq M^{r-1} \int_{|x|>T} \rho_n(x) dx < M^{r-1}\varepsilon.$$

The function ϕ_Σ satisfies similar inequalities. Hence, for any $\delta > 0$, there exists $T > 0$ such that for all n large enough,

$$\left| \int_{|x|>T} \rho_n(x)^r dx - \int_{|x|>T} \phi_\Sigma(x)^r dx \right| < \delta.$$

On the other hand, by (4.8), for all $T > 0$, the function $\rho_n^r(x)\mathbf{1}_{\{|x| \leq T\}}$ converges everywhere to $\phi_\Sigma^r(x)\mathbf{1}_{\{|x| \leq T\}}$ as $n \to +\infty$. Since $\rho_n^r(x)\mathbf{1}_{\{|x| \leq T\}}$ is dominated by the integrable function $M^r \mathbf{1}_{\{|x| \leq T\}}$, one may use the Lebesgue dominated theorem to conclude that

$$\lim_{n \to +\infty} \left| \int_{|x| \leq T} \rho_n(x)^r dx - \int_{|x| \leq T} \phi_\Sigma(x)^r dx \right| = 0.$$

$\square$

Remark 4.6 Theorem 4.5 fails for $r \in (0, 1)$. For example, one can consider i.i.d. random vectors with a bounded density $\rho(x)$ such that $\int_{\mathbb{R}^d} \rho(x)^r dx = +\infty$ (e.g., Cauchy-type distributions). The implication $4 \implies 2$ (and thus $4 \implies 1$) will not hold since by Jensen inequality $h_r(Z_n) \geq h_r(X_1/\sqrt{n}) = \infty$ for all $n \geq 1$. As observed by Barron [3], the implication $1 \implies 4$ does not necessarily hold in the Shannon entropy case $r = 1$.

The following result yields a sufficient condition for convergence along the CLT to hold for Rényi entropies of order $r \in (0, 1)$ for a large class of random vectors in $\mathbb{R}^d$.

Theorem 4.7 *Let* $r \in (0, 1)$. *Let* $X_1, \cdots, X_n$ *be i.i.d. centered log-concave random vectors in* $\mathbb{R}^d$. *Then we have* $h_r(Z_n) < +\infty$ *for all* $n \geq 1$, *and*

$$\lim_{n \to \infty} h_r(Z_n) = h_r(Z),$$

where Z_n *is the normalized sum in* (4.7) *and* Z *is a Gaussian random vector with mean* 0 *and the same covariance matrix as* X_1.

Proof Since log-concavity is preserved under independent sum, Z_n is log-concave for all $n \geq 1$. Hence, for all $n \geq 1$, Z_n has a bounded log-concave density ρ_n, which satisfies

$$\rho_n(x) \leq e^{-a_n|x|+b_n},$$

for all $x \in \mathbb{R}^d$, and for some constants $a_n > 0$, $b_n \in \mathbb{R}$ possibly depending on the dimension (see, e.g., [16]). Hence, for all $n \geq 1$, we have

$$\int_{\mathbb{R}^d} \rho_n(x)^r \, dx \leq \int_{\mathbb{R}^d} e^{-r(a_n|x|+b_n)} \, dx < +\infty.$$

We deduce that $h_r(Z_n) < +\infty$ for all $n \geq 1$.

The boundedness of ρ_n implies that (4.8) holds, and thus there exists an integer n_0 such that for all $n \geq n_0$,

$$\rho_n(0) > \frac{1}{2}\phi_\Sigma(0),$$

where Σ is the covariance matrix of X_1 (and thus does not depend on n). Moreover, since ρ_n is log-concave, one has for all $x \in \mathbb{R}^d$ that

$$\rho_n(rx) = \rho_n((1-r)0 + rx) \geq \rho_n(0)^{1-r}\rho_n(x)^r \geq \frac{1}{2^{1-r}}\phi_\Sigma(0)^{1-r}\rho_n(x)^r.$$

Hence, for all $T > 0$, we have

$$\int_{|x|>T} \rho_n(x)^r \, dx \leq \frac{2^{1-r}}{\phi_\Sigma(0)^{1-r}} \int_{|x|>T} \rho_n(rx) \, dx$$

$$= \frac{2^{1-r}}{r^d \phi_\Sigma(0)^{1-r}} \mathbb{P}(|Z_n| > rT)$$

$$\leq \frac{1}{T^2} \frac{2^{1-r}\mathbb{E}[|X_1|^2]}{r^{d+2}\phi_\Sigma(0)^{1-r}},$$

where the last inequality follows from Markov's inequality and the fact that

$$\mathbb{E}[|Z_n|^2] = \frac{\mathbb{E}[|X_1|^2] + \cdots + \mathbb{E}[|X_n|^2]}{n} = \mathbb{E}[|X_1|^2].$$

Hence, for every $\varepsilon > 0$, one may choose a positive number T such that for all n large enough,

$$\int_{|x|>T} \rho_n(x)^r dx < \varepsilon, \qquad \int_{|x|>T} \phi_\Sigma(x)^r dx < \varepsilon,$$

and hence

$$\left| \int_{|x|>T} \rho_n(x)^r dx - \int_{|x|>T} \phi_\Sigma(x)^r dx \right| < \varepsilon.$$

On the other hand, from (4.8), we conclude as in the proof of Theorem 4.5 that for all $T > 0$,

$$\lim_{n \to +\infty} \left| \int_{|x| \leq T} \rho_n(x)^r dx - \int_{|x| \leq T} \phi_\Sigma(x)^r dx \right| = 0.$$

$\square$

A function $f \colon \mathbb{R}^d \to \mathbb{R}$ is called unimodal if the super-level sets $\{x \in \mathbb{R}^d : f(x) > t\}$ are convex for all $t \in \mathbb{R}$. Next, we provide a convergence result for random vectors in $\mathbb{R}^d$ with unimodal densities under additional symmetry assumptions. First, we need the following stability result.

Proposition 4.8 *The class of spherically symmetric and unimodal random variables is stable under convolution.*

Proof Let f_1 and f_2 be two spherically symmetric and unimodal densities. By assumption, f_i satisfy that $f_i(Tx) = f_i(x)$ for an orthogonal map T and $|x| \leq |y|$ implies $f_i(x) \geq f_i(y)$. By the layer cake decomposition, we write

$$f_i(x) = \int_0^\infty \mathbf{1}_{\{(u,v):f_i(u)>v\}}(x, \lambda) d\lambda.$$

Apply Fubini's theorem to obtain

$$f_1 \star f_2(x) = \int_{\mathbb{R}^d} f_1(x - y) f_2(y) dy$$

$$= \int_0^\infty \int_0^\infty \left(\int_{\mathbb{R}^d} \mathbf{1}_{\{(u,v):f_1(u)>v\}}(x - y, \lambda_1) \mathbf{1}_{\{(u,v):f_2(u)>v\}}(y, \lambda_2) dy \right)$$

$$\times d\lambda_1 d\lambda_2. \tag{4.9}$$

Notice that by the spherical symmetry and decreasingness of f_i, the super-level set

$$L_{\lambda_i} = \{u : f_i(u) > \lambda_i\}$$

is an origin symmetric ball. Thus we can write the integrand in (4.9) as

$$\int_{\mathbb{R}^d} \mathbf{1}_{L_{\lambda_1}}(x - y)\mathbf{1}_{L_{\lambda_2}}(y)dy = \mathbf{1}_{L_{\lambda_1}} \star \mathbf{1}_{L_{\lambda_2}}(x).$$

This quantity is clearly dependent only on $|x|$, giving spherical symmetry. In addition, as the convolution of two log-concave functions, $\mathbf{1}_{L_{\lambda_1}} \star \mathbf{1}_{L_{\lambda_2}}$ is log-concave as well. It follows that for every λ_1, λ_2, and $|x| \leq |y|$ we have

$$\mathbf{1}_{L_{\lambda_1}} \star \mathbf{1}_{L_{\lambda_2}}(x) \geq \mathbf{1}_{L_{\lambda_1}} \star \mathbf{1}_{L_{\lambda_2}}(y).$$

Integrating this inequality completes the proof. $\qquad\square$

Let us establish large deviation and pointwise inequalities for compactly supported, spherically symmetric and unimodal densities.

Theorem 4.9 (Hoeffding [25]) *Let $X_1, \cdots , X_n$ be independent random variables with mean 0 and bounded in (a_i, b_i), respectively. One has for all $T > 0$,*

$$\mathbb{P}\left(\sum_{i=1}^n X_i > T\right) \leq \exp\left(-\frac{2T^2}{\sum_{i=1}^n (b_i - a_i)^2}\right).$$

The following result is Hoeffding's inequality in higher dimensions.

Lemma 4.10 *Let $X_1, \cdots , X_n$ be centered independent random vectors in $\mathbb{R}^d$ satisfying $\mathbb{P}(|X_i| > R) = 0$ for some $R > 0$. One has for all $T > 0$ that*

$$\mathbb{P}\left(\left|\frac{X_1 + \cdots + X_n}{\sqrt{n}}\right| > T\right) \leq 2d \exp\left(-\frac{T^2}{2d^2 R^2}\right).$$

Proof Let $X_{i,j}$ be the j-th coordinate of the random vector X_i. Then we have

$$\mathbb{P}\left(\left|\frac{X_1 + \cdots + X_n}{\sqrt{n}}\right| > T\right) \leq \mathbb{P}\left(\bigcup_{j=1}^d \left\{|X_{1,j} + \cdots + X_{n,j}| > \frac{T\sqrt{n}}{d}\right\}\right) \quad (4.10)$$

$$\leq \sum_{j=1}^d \mathbb{P}\left(|X_{1,j} + \cdots + X_{n,j}| > \frac{T\sqrt{n}}{d}\right) \quad (4.11)$$

$$\leq 2d \exp\left(-\frac{T^2}{2d^2 R^2}\right), \quad (4.12)$$

where inequality (4.10) follows from the pigeon-hole principle, (4.11) from a union bound, and (4.12) follows from applying Theorem 4.9 to $X_{1,j} + \cdots + X_{n,j}$ and $(-X_{1,j}) + \cdots + (-X_{n,j})$. □

We deduce the following pointwise estimate for unimodal spherically symmetric and bounded random variables.

Corollary 4.11 *Let $X_1, \cdots, X_n$ be i.i.d. random vectors with spherically symmetric, unimodal density supported on the Euclidean ball $B_R = \{x : |x| \leq R\}$ for some $R > 0$. Let ρ_n denote the density of the normalized sum Z_n. Then there exists $c_d > 0$ such that for all $n \geq 1$ and $|x| > 2$,*

$$\rho_n(x) \leq c_d \exp\left(-\frac{(|x|-1)^2}{2d^2 R^2}\right).$$

Proof Stating Lemma 4.10 in terms of ρ_n, we have

$$\int_{|w|>T} \rho_n(w)dw \leq 2d \exp\left(-\frac{T^2}{2d^2 R^2}\right). \tag{4.13}$$

Since the class of spherically symmetric unimodal random variables is stable under independent summation by Proposition 4.8, ρ_n is spherically symmetric and unimodal, so that

$$\rho_n(x) \leq \frac{\int_{B_{|x|}\setminus B_{|x|-1}} \rho_n(w)dw}{\mathrm{Vol}(B_{|x|}\setminus B_{|x|-1})}$$

$$\leq \frac{\int_{|w|\geq|x|-1} \rho_n(w)dw}{(2^d - 1)\omega_d} \tag{4.14}$$

where $B_{|x|}$ represents the Euclidean ball of radius $|x|$ centered at the origin and ω_d is the volume of the unit ball. Note that

$$\mathrm{Vol}(B_{|x|}\setminus B_{|x|-1}) = (|x|^d - (|x|-1)^d)\omega_d \geq (2^d - 1)\omega_d,$$

since $t \mapsto t^d - (t-1)^d$ is increasing, so that (4.14) follows. Now applying (4.13) we have

$$\rho_n(x) \leq \frac{\int_{|w|\geq|x|-1} \rho_n(w)dw}{(2^d - 1)\omega_d}$$

$$\leq \frac{2d}{(2^d - 1)\omega_d} \exp\left(-\frac{(|x|-1)^2}{2d^2 R^2}\right)$$

and our result holds with

$$c_d = \frac{2d}{(2^d - 1)\omega_d}.$$

□

We are now ready to establish a convergence result for bounded spherically symmetric unimodal random vectors.

Theorem 4.12 *Let* $r \in (0, 1)$. *Let* $X_1, \cdots, X_n$ *be i.i.d. random vectors in* $\mathbb{R}^d$ *with a spherically symmetric unimodal density with compact support. Then we have*

$$\lim_{n \to \infty} h_r(Z_n) = h_r(Z),$$

where Z_n *is the normalized sum in* (4.7) *and* Z *is a Gaussian random vector with mean* 0 *and the same covariance matrix as* X_1.

Proof Let us denote by ρ_n the density of Z_n. Since ρ_1 is bounded, one may apply (4.8) together with Lebesgue dominated convergence to conclude that for all $T > 0$,

$$\lim_{n \to +\infty} \left| \int_{|x| \leq T} \rho_n(x)^r dx - \int_{|x| \leq T} \phi_\Sigma(x)^r dx \right| = 0.$$

On the other hand, by Corollary 4.11, one may choose $T > 0$ such that for all $n \geq 1$,

$$\int_{|x| > T} \rho_n(x)^r dx < \varepsilon, \qquad \int_{|x| > T} \phi_\Sigma(x)^r dx < \varepsilon,$$

and hence

$$\left| \int_{|x| > T} \rho_n(x)^r dx - \int_{|x| > T} \phi_\Sigma(x)^r dx \right| < \varepsilon.$$

□

4.3 Rényi EPIs of Order $r \in (0, 1)$

A striking difference between Rényi EPIs of orders $r \in (0, 1)$ and $r \geq 1$ is the lack of an absolute constant. Indeed, it was shown in [8] that for $r \geq 1$ Rényi EPI of the form (4.2) holds for generic independent random vectors with an absolute constant $c \geq \frac{1}{e} r^{\frac{1}{r-1}}$. In the following subsection, we show that such a Rényi EPI does not hold for $r \in (0, 1)$.

4.3.1   Failure of a Generic Rényi EPI

Definition 4.13 For $r \in [0, \infty]$, we define c_r as the largest number such that for all $n, d \geq 1$ and any independent random vectors $X_1, \cdots, X_n$ in $\mathbb{R}^d$, we have

$$N_r(X_1 + \cdots + X_n) \geq c_r \sum_{i=1}^{n} N_r(X_i). \tag{4.15}$$

Then we can rephrase Theorem 4.1 as follows.

Theorem 4.14 *For* $r \in (0, 1)$, *the constant* c_r *defined in* (4.15) *satisfies* $c_r = 0$.

The motivating observation for this line of argument is the fact that for $r \in (0, 1)$, there exist distributions with finite covariance matrices and infinite r-Rényi entropies. One might anticipate that this could contradict the existence of an r-Rényi EPI, as the CLT forces the normalized sum of i.i.d. random vectors $X_1, \cdots, X_n$ drawn from such a distribution to become "more Gaussian". Heuristically, one anticipates that $N_r(X_1 + \cdots + X_n)/n = N_r(Z_n)$ should approach $N_r(Z)$ for large n, where Z_n is the normalized sum in (4.7) and Z is a Gaussian vector with the same covariance matrix as X_1, while $\sum_{i=1}^{n} N_r(X_i)/n = N_r(X_1)$ is infinite.

Proof of Theorem 4.14 Let us consider the following density

$$f_{R,p,d}(x) = C_R(1 + |x|)^{-p} \mathbf{1}_{B_R}(x) \quad x \in \mathbb{R}^d,$$

with $p, R > 0$ and C_R implicitly determined to make $f_{R,p,d}$ a density. Since the density is spherically symmetric, its covariance matrix can be rewritten as $\sigma_R^2 I$ for some $\sigma_R > 0$, where I is the identity matrix. Computing in spherical coordinates one can check that $\lim_{R \to \infty} C_R$ is finite for $p > d$, and we can thus define a density $f_{\infty,p,d}$. What is more, when $p > d + 2$, the limiting density $f_{\infty,p,d}$ has a finite covariance matrix, and has finite Rényi entropy if and only if $p > d/r$.

For fixed $r \in (0, 1)$, we take $p \in (d^* + 2, d^*/r]$, where $d^* = \min\{d \in \mathbb{N} : d > 2r/(1 - r)\}$ guarantees the existence of such p. In this case, the limiting density f_{∞,p,d^*} is well defined and it has finite covariance matrix $\sigma_\infty^2 I$, but the corresponding r-Rényi entropy is infinite. Now we select independent random vectors $X_1, \cdots, X_n$ from the distribution f_{R,p,d^*}. Since f_{R,p,d^*} is a spherically symmetric and unimodal density with compact support, we can apply Theorem 4.12 to conclude that

$$\lim_{n \to \infty} N_r(Z_n) = \sigma_R^2 N_r(Z_{Id}),$$

where Z_n is the normalized sum in (4.7) and Z_{Id} is the standard d-dimensional Gaussian. Since $\lim_{R \to \infty} \sigma_R = \sigma_\infty < \infty$, we can take R large enough such that

$|\sigma_R^2 - \sigma_\infty^2| \le 1$. Then we can take n large enough such that

$$N_r(Z_n) \le (\sigma_\infty^2 + 2)N_r(Z_{Id}).\qquad(4.16)$$

Since the limiting density $f_{\infty, p, d*}$ has infinite r-Rényi entropy, given $M > 0$, we can take R large enough such that

$$N_r(X_1) \ge M.\qquad(4.17)$$

Combining (4.16) and (4.17), we conclude that for inequality (4.15) to hold we must have

$$c_r \le \frac{(\sigma_\infty^2 + 2)N_r(Z_{Id})}{M}$$

for all $M > 0$. Then the statement follows from taking the limit $M \to \infty$. □

Remark 4.15 Random vectors in our proof has identical s-concave density with $s \le -r/d$. In the following section, we provide a complementary result by showing that Rényi EPI of order $r \in (0, 1)$ does hold for s-concave densities when $-r/d < s < 0$.

4.3.2 Rényi EPIs for s-Concave Densities

As showed above, a generic Rényi EPI of the form (4.2) fails for $r \in (0, 1)$. In this part, we establish Rényi EPIs of the forms (4.2) and (4.3) for an important class of random vectors with s-concave densities (see (4.5)).

Following Lieb [29], we prove Theorems 4.2 and 4.3 by showing their equivalent linearizations. The following linearization of (4.2) and (4.3) is due to Rioul [37]. The $c = 1$ case was used in [27].

Theorem 4.16 ([37]) *Let* $X_1, \cdots, X_n$ *be independent random vectors in* $\mathbb{R}^d$. *The following statements are equivalent.*

1. There exist a constant $c > 0$ *and an exponent* $\alpha > 0$ *such that*

$$N_r^\alpha\left(\sum_{i=1}^n X_i\right) \ge c\sum_{i=1}^n N_r^\alpha(X_i).\qquad(4.18)$$

2. For any $\lambda_1, \cdots, \lambda_n \ge 0$ *such that* $\sum_{i=1}^n \lambda_i = 1$, *one has*

$$h_r\left(\sum_{i=1}^n \sqrt{\lambda_i}X_i\right) - \sum_{i=1}^n \lambda_i h_r(X_i) \ge \frac{d}{2}\left(\frac{\log c}{\alpha} + \left(\frac{1}{\alpha} - 1\right)H(\lambda)\right),\qquad(4.19)$$

where $H(\lambda) \triangleq H(\lambda_1, \cdots, \lambda_n)$ is the discrete entropy defined as

$$H(\lambda) = -\sum_{i=1}^{n} \lambda_i \log \lambda_i.$$

Inequality (4.19) is the linearized form of inequality (4.18). One of the ingredients used to establish (4.19) is Young's sharp convolution inequality [4, 15]. Its information-theoretic formulation was given in [21], which we recall below. We denote by r' the Hölder conjugate of r such that $1/r + 1/r' = 1$.

Theorem 4.17 ([15, 21]) *Let $r > 0$. Let $\lambda_1, \cdots, \lambda_n \geq 0$ such that $\sum_{i=1}^{n} \lambda_i = 1$, and let $r_1, \cdots, r_n$ be positive reals such that $\lambda_i = r'/r_i'$. For any independent random vectors $X_1, \cdots, X_n$ in $\mathbb{R}^d$, one has*

$$h_r\left(\sum_{i=1}^{n} \sqrt{\lambda_i} X_i\right) - \sum_{i=1}^{n} \lambda_i h_{r_i}(X_i) \geq \frac{d}{2} r' \left(\frac{\log r}{r} - \sum_{i=1}^{n} \frac{\log r_i}{r_i}\right). \tag{4.20}$$

The second ingredient is a comparison between Rényi entropies h_r and h_{r_i}. When $r > 1$, we have $1 < r_i < r$, and Jensen's inequality implies that $h_r \leq h_{r_i}$. In this case, one can deduce (4.19) from (4.20) with h_{r_i} replaced by h_r. However, when $r \in (0, 1)$, the order of r and r_i are reversed, i.e., $0 < r < r_i < 1$, and we need a reverse entropy comparison inequality. The so-called s-concave densities do satisfy such a reverse entropy comparison inequality. The following result of Fradelizi et al. [22] serves this purpose.

Theorem 4.18 ([22]) *Let $s \in \mathbb{R}$. Let $f : \mathbb{R}^d \to [0, +\infty)$ be an integrable s-concave function. The function*

$$G(r) = C(r) \int_{\mathbb{R}^d} f(x)^r \, dx$$

is log-concave for $r > \max\{0, -sd\}$, where

$$C(r) = (r + s) \cdots (r + sd). \tag{4.21}$$

We deduce the following Rényi entropic comparison for random vectors with s-concave densities.

Corollary 4.19 *Let X be a random vector in $\mathbb{R}^d$ with a s-concave density. For $-sd < r < q < 1$, we have*

$$h_q(X) \geq h_r(X) + \log \frac{C(r)^{\frac{1}{1-r}} C(1)^{\frac{q-r}{(1-q)(1-r)}}}{C(q)^{\frac{1}{1-q}}}.$$

Proof Write $q = (1 - \lambda) \cdot r + \lambda \cdot 1$. Using the log-concavity of the function G in Theorem 4.18, we have

$$G(q) \geq G(r)^{1-\lambda}G(1)^{\lambda} = G(r)^{\frac{1-q}{1-r}}G(1)^{\frac{q-r}{1-r}}.$$

The above inequality can be rewritten in terms of entropy power as follows

$$C(q)^{\frac{2}{d}\cdot\frac{1}{1-q}}N_q(X) \geq C(r)^{\frac{2}{d}\cdot\frac{1-q}{1-r}\cdot\frac{1}{1-q}}N_r(X)C(1)^{\frac{2}{d}\cdot\frac{q-r}{1-r}\cdot\frac{1}{1-q}}.$$

The desired statement follows from taking the logarithm of both sides of the above inequality.   □

Theorem 4.17 together with Corollary 4.19 yields the following Rényi EPI with a single Rényi parameter $r \in (0, 1)$ for s-concave densities.

Theorem 4.20 *Let* $s \in (-1/d, 0)$ *and* $r \in (-sd, 1)$. *Let* $X_1, \cdots , X_n$ *be independent random vectors in* $\mathbb{R}^d$ *with* s-*concave densities. For all* $\lambda = (\lambda_1, \cdots , \lambda_n) \in [0, 1]^n$ *such that* $\sum_{i=1}^n \lambda_i = 1$, *we have*

$$h_r\left(\sum_{i=1}^n \sqrt{\lambda_i}X_i\right) - \sum_{i=1}^n \lambda_i h_r(X_i) \geq \frac{d}{2}A(\lambda) + \sum_{k=1}^d g_k(\lambda),$$

where

$$A(\lambda) = r'\left(\left(1 - \frac{1}{r'}\right)\log\left(1 - \frac{1}{r'}\right) - \sum_{i=1}^n \left(1 - \frac{\lambda_i}{r'}\right)\log\left(1 - \frac{\lambda_i}{r'}\right)\right),$$

$$g_k(\lambda) = (1 - n)r'\log(1 + ks) + (1 - r')\log\left(1 + \frac{ks}{r}\right) + r'\sum_{i=1}^n\left(1 - \frac{\lambda_i}{r'}\right)$$

$$\times \log\left(1 + ks\left(1 - \frac{\lambda_i}{r'}\right)\right).$$

Proof Let r_i be defined by $\lambda_i = r'/r_i'$, where r' and r_i' are Hölder conjugates of r and r_i, respectively. Combining Theorem 4.17 with Corollary 4.19, we have

$$h_r\left(\sum_{i=1}^n \sqrt{\lambda_i}X_i\right) - \sum_{i=1}^n \lambda_i h_r(X_i) \geq \frac{d}{2}r'\left(\frac{\log r}{r} - \sum_{i=1}^n \frac{\log r_i}{r_i}\right)$$

$$+ \sum_{i=1}^n \lambda_i \log \frac{C(r)^{\frac{1}{1-r}}C(1)^{\frac{r_i-r}{(1-r_i)(1-r)}}}{C(r_i)^{\frac{1}{1-r_i}}}. \tag{4.22}$$

Notice that $C(r) = r^d D(r)$, where $C(r)$ is given in (4.21) and $D(r) = (1 + s/r) \cdots (1 + sd/r)$. Thus,

$$\sum_{i=1}^{n} \lambda_i \log \frac{C(r)^{\frac{1}{1-r}} C(1)^{\frac{r_i-r}{(1-r_i)(1-r)}}}{C(r_i)^{\frac{1}{1-r_i}}}$$

$$= \sum_{i=1}^{n} \lambda_i \left(\frac{\log D(r)}{1-r} + \left(\frac{1}{1-r_i} - \frac{1}{1-r} \right) \log D(1) - \frac{\log D(r_i)}{1-r_i} \right)$$

$$+ d \left(\frac{\log r}{1-r} - \sum_{i=1}^{n} \lambda_i \frac{\log r_i}{1-r_i} \right). \tag{4.23}$$

Using the identities $1/(1-r) = 1 - r'$ and $\lambda_i/(1-r_i) = \lambda_i - r'$, we have

$$\sum_{i=1}^{n} \lambda_i \left(\frac{\log D(r)}{1-r} + \left(\frac{1}{1-r_i} - \frac{1}{1-r} \right) \log D(1) - \frac{\log D(r_i)}{1-r_i} \right)$$

$$= (1 - r') \log D(r) + (1 - n)r' \log D(1) + \sum_{k=1}^{d} \sum_{i=1}^{n} (r' - \lambda_i) \log \left(1 + \frac{ks}{r_i} \right)$$

$$= \sum_{k=1}^{d} \left((1 - r') \log \left(1 + \frac{ks}{r} \right) + (1 - n)r' \log(1 + ks) \right.$$

$$\left. + \sum_{i=1}^{n} (r' - \lambda_i) \log \left(1 + \frac{ks}{r_i} \right) \right) = \sum_{k=1}^{d} g_k(\lambda). \tag{4.24}$$

The last identity follows from $1/r_i = 1 - \lambda_i/r'$. Using (4.24) and (4.23), the RHS of (4.22) can be written as

$$\frac{d}{2} r' \left(\frac{\log r}{r} - \sum_{i=1}^{n} \frac{\log r_i}{r_i} \right) + d \left(\frac{\log r}{1-r} - \sum_{i=1}^{n} \lambda_i \frac{\log r_i}{1-r_i} \right)$$

$$+ \sum_{k=1}^{d} g_k(\lambda) = \frac{d}{2} A(\lambda) + \sum_{k=1}^{d} g_k(\lambda).$$

This concludes the proof. □

Having Theorems 4.16 and 4.20 at hand, we are ready to prove Theorems 4.2 and 4.3.

4.3.2.1 Proof of Theorem 4.2

Put Theorems 4.16 and 4.20 together. Then it suffices to find c such that the following inequality

$$\frac{d}{2} A(\lambda) + \sum_{k=1}^{d} g_k(\lambda) \geq \frac{d}{2} \log c$$

holds for all $\lambda = (\lambda_1, \cdots, \lambda_n) \in [0, 1]^n$ such that $\sum_{i=1}^{n} \lambda_i = 1$. Hence, we can set

$$c = \inf_{\lambda} \exp \left(A(\lambda) + \frac{2}{d} \sum_{k=1}^{d} g_k(\lambda) \right),$$

where the infimum runs over all $\lambda = (\lambda_1, \cdots, \lambda_n) \in [0, 1]^n$ such that $\sum_{i=1}^{n} \lambda_i = 1$. For fixed r, both $A(\lambda)$ and $g_k(\lambda)$ are sum of one-dimensional convex functions of the form $(1 + x) \log(1 + x)$. Furthermore, both $A(\lambda)$ and $g_k(\lambda)$ are permutation invariant. Hence, the minimum is achieved at $\lambda = (1/n, \cdots, 1/n)$. This yields the numerical value of c in Theorem 4.2.

4.3.2.2 Proof of Theorem 4.3

The following lemma in [33] serves us in the proof of Theorem 4.3.

Lemma 4.21 ([33]) *Let $c > 0$. Let $L, F : [0, c] \to [0, \infty)$ be twice differentiable on $(0, c]$, continuous on $[0, c]$, such that $L(0) = F(0) = 0$ and $L'(c) = F'(c) = 0$. Let us also assume that $F(x) > 0$ for $x > 0$, that F is strictly increasing, and that F' is strictly decreasing. Then $\frac{L''}{F''}$ increasing on $(0, c)$ implies that $\frac{L}{F}$ is increasing on $(0, c)$ as well. In particular,*

$$\max_{x \in [0,c]} \frac{L(x)}{F(x)} = \frac{L(c)}{F(c)}.$$

Proof of Theorem 4.3 Apply Theorems 4.16 and 4.20 with $n = 2$. Then it suffices to find α such that for all $\lambda \in [0, 1]$ we have

$$\frac{d}{2} A(\lambda) + \sum_{k=1}^{d} g_k(\lambda) \geq \frac{d}{2} \left(\frac{1}{\alpha} - 1 \right) H(\lambda),$$

where

$$A(\lambda) = r' \left(\left(1 - \frac{1}{r'}\right) \log \left(1 - \frac{1}{r'}\right) - \left(1 - \frac{\lambda}{r'}\right) \log \left(1 - \frac{\lambda}{r'}\right) - \left(1 - \frac{1 - \lambda}{r'}\right) \right.$$
$$\left. \times \log \left(1 - \frac{1 - \lambda}{r'}\right) \right),$$

$$g_k(\lambda) = (1 - r') \log \left(1 + \frac{ks}{r}\right) - r' \log(1 + ks)$$
$$+ r' \left(\left(1 - \frac{\lambda}{r'}\right) \log \left(1 + ks \left(1 - \frac{\lambda}{r'}\right)\right) + \left(1 - \frac{1 - \lambda}{r'}\right) \right.$$
$$\left. \times \log \left(1 + ks \left(1 - \frac{1 - \lambda}{r'}\right)\right) \right).$$

We can set

$$\alpha = \left(1 - \sup_{0 \le \lambda \le 1} \left(-\frac{A(\lambda)}{H(\lambda)} - \frac{2}{d} \sum_{k=1}^{d} \frac{g_k(\lambda)}{H(\lambda)}\right)\right)^{-1}. \tag{4.25}$$

We will show that the optimal value is achieved at $\lambda = 1/2$. Since the function is symmetric about $\lambda = 1/2$, it suffices to show that

$$-\frac{A(\lambda)}{H(\lambda)} - \frac{2}{d} \sum_{k=1}^{n} \frac{g_k(\lambda)}{H(\lambda)} \tag{4.26}$$

is increasing on $[0, 1/2]$. It has been shown in [27] that $-A(\lambda)/H(\lambda)$ is increasing on $[0, 1/2]$. We will show that for each $k = 1, \cdots, n$ the function $-g_k(\lambda)/H(\lambda)$ is also increasing on $[0, 1/2]$. One can check that $-g_k(\lambda)$ and $H(\lambda)$ satisfy the conditions in Lemma 4.21. Hence, it suffices to show that $-g_k''(\lambda)/H''(\lambda)$ is increasing on $[0, 1/2]$. Elementary calculation yields that

$$H''(\lambda) = -\frac{1}{\lambda(1 - \lambda)}.$$

Define $x = \frac{\lambda}{|r'|}$ and $y = \frac{1 - \lambda}{|r'|} = \frac{1}{|r'|} - x$. Then one can check that

$$-g_k''(\lambda) = \frac{ks}{|r'|} \left(\frac{1}{1 + ks(1 + x)} + \frac{1}{1 + ks(1 + y)} + \frac{1}{(1 + ks(1 + x))^2} \right.$$
$$\left. + \frac{1}{(1 + ks(1 + y))^2} \right).$$

Hence, we have

$$-\frac{g_k''(\lambda)}{H''(\lambda)} = ksr'W(x),$$

where

$$W(x) = xy\left(\frac{1}{1+ks(1+x)} + \frac{1}{1+ks(1+y)} + \frac{1}{(1+ks(1+x))^2}\right.$$
$$\left. + \frac{1}{(1+ks(1+y))^2}\right).$$

Since $s, r' < 0$, it suffices to show that $W(x)$ is increasing on $[0, \frac{1}{2|r'|}]$. We rewrite W as follows

$$W(x) = W_1(x) + W_2(x),$$

where

$$W_1(x) = xy\left(\frac{1}{1+ks(1+x)} + \frac{1}{1+ks(1+y)}\right),$$
$$W_2(x) = xy\left(\frac{1}{(1+ks(1+x))^2} + \frac{1}{(1+ks(1+y))^2}\right). \qquad (4.27)$$

We will show that both $W_1(x)$ and $W_2(x)$ are increasing on $[0, \frac{1}{2|r'|}]$.

Now let us focus on W_1. Since $y = \frac{1}{|r'|} - x$, one can check that

$$W_1'(x) = \left(\frac{1}{|r'|} - 2x\right)\left(\frac{1}{1+ks(1+x)} + \frac{1}{1+ks(1+y)}\right)$$
$$- ksxy\left(\frac{1}{(1+ks(1+x))^2} - \frac{1}{(1+ks(1+y))^2}\right).$$

Let us denote

$$a \triangleq a(x) = 1 + ks(1+x), \qquad (4.28)$$
$$b \triangleq b(x) = 1 + ks(1+y) = 1 + ks\left(\frac{1}{|r'|} - x + 1\right). \qquad (4.29)$$

The condition $r > -sd$ implies that $a, b \geq 0$. With these notations, we have

$$
W_1'(x) = \left(\frac{1}{a} + \frac{1}{b}\right)\left(\frac{1}{|r'|} - 2x - ksxy\left(\frac{1}{a} - \frac{1}{b}\right)\right)
$$

$$
= \left(\frac{1}{a} + \frac{1}{b}\right)\left(\frac{1}{|r'|} - 2x\right)\left(1 - (ks)^2\frac{xy}{ab}\right).
$$

The last identity follows from

$$
\frac{1}{a} - \frac{1}{b} = \frac{ks}{ab}\left(\frac{1}{|r'|} - 2x\right).
$$

Since $a, b \geq 0$ and $x \in [0, \frac{1}{2|r'|}]$, it suffices to show that

$$
ab - (ks)^2 xy \geq 0.
$$

Using (4.28) and (4.29), we have

$$
ab - (ks)^2 xy = (1 + ks)\left(1 + \frac{ks}{r}\right).
$$

Then the desired statement follows from that $s > -1/d$ and $r > -sd$. We conclude that W_1 is increasing on $[0, \frac{1}{2|r'|}]$.

It remains to show that $W_2(x)$ is increasing on $[0, \frac{1}{2|r'|}]$. Recall the definition of $W_2(x)$ in (4.27), one can check that

$$
W_2'(x) = \left(\frac{1}{|r'|} - 2x\right)\left(\frac{1}{a^2} + \frac{1}{b^2}\right) - 2ksxy\left(\frac{1}{a^3} - \frac{1}{b^3}\right)
$$

$$
= \frac{b - a}{ks}\left(\frac{1}{a^2} + \frac{1}{b^2}\right) - 2ksxy\left(\frac{1}{a^3} - \frac{1}{b^3}\right)
$$

$$
= \frac{b - a}{ksa^3b^3}T(x),
$$

where a and b are defined in (4.28) and (4.29), and

$$
T(x) = ab(a^2 + b^2) - 2k^2s^2xy(a^2 + ab + b^2).
$$

Since

$$
\frac{b - a}{ks} = \frac{1}{|r'|} - 2x \geq 0, \quad x \in [0, \frac{1}{2|r'|}],
$$

it suffices to show that $T(x) \geq 0$ for $[0, \frac{1}{2|r'|}]$. Using the identity

$$a'(x)b(x) + a(x)b'(x) = ks(b - a) = -a(x)a'(x) - b(x)b'(x),$$

one can check that

$$T'(x) = ks(a - b)U(x),$$

where

$$U(x) = a^2 + b^2 + 4ab - 2k^2s^2xy.$$

Notice that $U'(x) \equiv 0$, which implies that $U(x)$ is a constant. Since $a, b \geq 0$, we have

$$U(0) = a^2 + b^2 + 4ab > 0.$$

Hence, $T'(x) \leq 0$, i.e., $T(x)$ is decreasing. Therefore, since $a = b$ when $x = \frac{1}{2|r'|}$, we have

$$T(x) \geq T\left(\frac{1}{2|r'|}\right) = 2a^2(a^2 - 3k^2s^2x^2) \quad \text{at } x = \frac{1}{2|r'|}.$$

It suffices to have

$$a^2 \geq 3k^2s^2x^2, \quad x = \frac{1}{2|r'|},$$

which is equivalent to

$$\frac{1}{|r'|} \leq \frac{2}{1 + \sqrt{3}}\left(\frac{1}{k|s|} - 1\right).$$

This finishes the proof that every $-g_k(\lambda)/H(\lambda)$ is also increasing on $[0, 1/2]$. Then the numerical value of α in Theorem 4.3 follows from setting $\lambda = 1/2$ in (4.25). □

Remark 4.22 Our optimization argument heavily relies on the fact that $-A(\lambda)/H(\lambda)$ and $-g_k(\lambda)/H(\lambda)$ are monotonically increasing for $\lambda \in [0, 1/2]$. As observed in [27], the monotonicity of $-A(\lambda)/H(\lambda)$ does not depend on the value of r. Numerical examples show that $-g_k(\lambda)/H(\lambda)$, even the whole quantity in (4.26), is not monotone when r is small. This is one of the reasons for the restriction $r > r_0$.

Remark 4.23 Note that the condition $r > -sd$ of Theorem 4.18 can be rewritten as

$$\frac{1}{|r'|} < \left(\frac{1}{d|s|} - 1\right).$$

We do not know whether Theorem 4.3 holds when

$$\frac{2}{1 + \sqrt{3}}\left(\frac{1}{d|s|} - 1\right) < \frac{1}{|r'|} < \left(\frac{1}{d|s|} - 1\right).$$

4.4   An Entropic Characterization of s-Concave Densities

Let X and Y be real-valued random variables (possibly dependent) with the identical density f. Cover and Zhang [20] proved that

$$h(X + Y) \le h(2X)$$

holds for every coupling of X and Y if and only if f is log-concave. This yields an entropic characterization of one-dimensional log-concave densities. We will extend Cover and Zhang's result to Rényi entropies of random vectors with s-concave densities (defined in (4.5)), which particularly include log-concave densities as a special case. This was previously proved in [28] when f is continuous.

Firstly, we introduce some classical variations of convexity and concavity which will be needed in our proof.

Definition 4.24 Let $\lambda \in (0, 1)$ be fixed. A function $f : \mathbb{R}^d \to \mathbb{R}$ with convex support is called almost λ-convex if the following inequality

$$f((1 - \lambda)x + \lambda y) \le (1 - \lambda)f(x) + \lambda f(y) \tag{4.30}$$

holds for almost every pair x, y in the domain of f. We say that f is λ-convex if the above inequality holds for every pair x, y in the domain of f. Particularly, for $\lambda = 1/2$, it is usually called mid-convex or Jensen convex. We say that f is convex if f is λ-convex for any $\lambda \in (0, 1)$.

One can define almost λ-concavity, λ-concavity and concavity by reversing inequality (4.30). Adamek [1, Theorem 1] showed that an almost λ-convex function is identical to a λ-convex function except on a set of Lebesgue measure 0. (To apply the theorem there, one can take the ideals $\mathcal{I}_1$ and $\mathcal{I}_2$ as the family of sets with Lebesgue measure 0 in $\mathbb{R}^d$ and $\mathbb{R}^{2d}$, respectively). In general, λ-convexity is not equivalent to convexity, as it is not a strong enough notion to imply continuity, at least not in a logical framework that accepts the axiom of choice. Indeed, counterexamples can be constructed using a Hamel basis for $\mathbb{R}$ as a vector space over $\mathbb{Q}$. However, in the case that f is Lebesgue measurable, a classical result of

Blumberg [6] and Sierpinski [40] (see also [18] in more general setting) shows that
λ-convexity implies continuity, and thus convexity.

Theorem 4.25 *Let $s > -1/d$ and we define $r = 1 + s$. Let f be a probability density on $\mathbb{R}^d$. The following statements are equivalent.*

1. *The density f is s-concave.*
2. *For any $\lambda \in (0, 1)$, we have $h_r(\lambda X + (1 - \lambda)Y) \leq h_r(X)$ for any random vectors X and Y with the identical density f.*
3. *We have $h_r\left(\frac{X+Y}{2}\right) \leq h_r(X)$ for any random vectors X and Y with the identical density f.*

Proof We only prove the statement for $s > 0$, or equivalently $r > 1$. The proof for $-1/d < s < 0$, or equivalently $1 - 1/d < r < 1$, is similar and sketched below.

$1 \implies 2$: The proof is taken from [28]. We include it for completeness. Let g be the density of $\lambda X + (1 - \lambda)Y$. Then we have

$$h_r(X) = \frac{1}{1-r} \log \mathbb{E} f^{r-1}(X)$$

$$= \frac{1}{1-r} \log(\lambda \mathbb{E} f^{r-1}(X) + (1 - \lambda)\mathbb{E} f^{r-1}(Y)) \tag{4.31}$$

$$\geq \frac{1}{1-r} \log \mathbb{E} f^{r-1}(\lambda X + (1 - \lambda)Y) \tag{4.32}$$

$$= \frac{1}{1-r} \log \int_{\mathbb{R}^d} f(x)^{r-1} g(x) dx$$

$$\geq \frac{1}{1-r} \log \left(\int_{\mathbb{R}^d} f(x)^r dx\right)^{1-\frac{1}{r}} \left(\int_{\mathbb{R}^d} g(x)^r dx\right)^{\frac{1}{r}} \tag{4.33}$$

$$= \frac{r-1}{r} h_r(X) + \frac{1}{r} h_r(\lambda X + (1 - \lambda)Y).$$

This is equivalent to the desired statement. Identity (4.31) follows from the assumption that X and Y have the same distribution. In inequality (4.32), we use the concavity of f^{r-1} and the fact that $\frac{1}{1-r} \log x$ is decreasing when $r > 1$. Inequality (4.33) follows from Hölder's inequality and the fact that $\frac{1}{1-r} \log x$ is decreasing when $r > 1$. For $1 - 1/d < r < 1$, the statement follows from the same argument in conjunction with the convexity of f^{r-1}, the converse of Hölder's inequality and the fact that $\frac{1}{1-r} \log x$ is increasing when $0 < r < 1$.

$2 \implies 3$: Obvious by taking $\lambda = \frac{1}{2}$.

$3 \implies 1$: We will prove the statement by contradiction. We first show an example borrowed from Cover and Zhang [20] to illustrate the "mass transferring" argument used in our proof. Consider the density $f(x) = 3/2$ in the intervals $(0, 1/3)$ and $(2/3, 1)$. It is clear that f is not $(r-1)$-concave. The joint distribution of (X, Y) with $Y \equiv X$ is supported on the diagonal line $y = x$. The Radon-Nikodym derivative g with respect to the one-dimensional Lebesgue measure on the line $y = x$ exists and is shown in Fig. 4.1. We remove some "mass" from the diagonal line $y = x$ to

Fig. 4.1 g

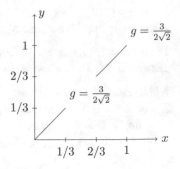

Fig. 4.2 $\hat{g}$

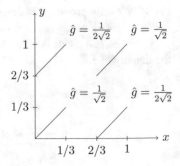

the lines $y = x - 2/3$ and $y = x + 2/3$. The new Radon–Nikodym derivative $\hat{g}$ is shown in Fig. 4.2. Let $(\hat{X}, \hat{Y})$ be a pair of random variables whose joint distribution possesses this new Radon–Nikodym derivative. It is easy to see that $\hat{X}$ and $\hat{Y}$ still have the same density f. But $\hat{X} + \hat{Y}$ is uniformly distributed on $(0, 2)$, and thus $h_r(\hat{X} + \hat{Y}) = \log 2$. One can check that $h_r(2X) = \log(4/3)$.

Now we turn to the general case. Suppose that f is not $(r - 1)$-concave, i.e., f^{r-1} is not concave (for $r > 1$). We claim that there exists a set $A \subseteq \mathbb{R}^{2d}$ of positive Lebesgue measure on $\mathbb{R}^{2d}$ such that the inequality

$$2 f^{r-1} \left(\frac{x + y}{2} \right) < f^{r-1}(x) + f^{r-1}(y) \tag{4.34}$$

holds for all $(x, y) \in A$. Otherwise, the converse of (4.34) holds for almost every pair (x, y), and thus f^{r-1} is an almost mid-concave function (i.e., 1/2-concave). By Theorem 1 in [1], f^{r-1} is identical to a mid-concave function except on a set of Lebesgue measure 0. Without changing the distribution, we can modify f such that f^{r-1} is mid-concave. Using the equivalence of mid-concavity and concavity (under the Lebesgue measurability), after modification, f^{r-1} is concave, i.e., f is $(r - 1)$-concave. This contradicts our assumption. Hence, there exists such a set A with positive Lebesgue measure on $\mathbb{R}^{2d}$. Then there exists y such that (4.34) holds for a set of x with positive Lebesgue measure on $\mathbb{R}^d$. We rephrase this statement in a form suitable for our purpose. There is $x_0 \neq 0$ such that the set

$$\Lambda = \left\{ x \in \mathbb{R}^d : 2f(x)^{r-1} < f(x + x_0)^{r-1} + f(x - x_0)^{r-1} \right\} \tag{4.35}$$

has positive Lebesgue measure on $\mathbb{R}^d$. For $\epsilon > 0$, we denote by $\Lambda(\epsilon)$ a ball of radius ϵ whose intersection with Λ has positive Lebesgue measure on $\mathbb{R}^d$. Consider (X, Y) such that $X \equiv Y$, where X and Y have the identical density f. Let $g(x, y)$ be the Radon-Nikodym derivative of (X, Y) with respect to the d-dimensional Lebesgue measure on the "diagonal line" $y = x$. Now we build a new density $\hat{g}$ by translating a small amount of "mass" from "diagonal points" $(x-x_0, x-x_0)$ and $(x+x_0, x+x_0)$ to "off-diagonal points" $(x - x_0, x + x_0)$ and $(x + x_0, x - x_0)$. To be more precise, we define the new joint density $\hat{g}$ as

$$\hat{g}(x, y) = g(x, y)\mathbf{1}_{\{x=y\}} - \sqrt{d/2}\delta(\mathbf{1}_{\{(x-x_0,x-x_0):x\in\Lambda(\epsilon)\}} + \mathbf{1}_{\{(x+x_0,x+x_0):x\in\Lambda(\epsilon)\}})$$
$$+\sqrt{d/2}\delta(\mathbf{1}_{\{(x-x_0,x+x_0):x\in\Lambda(\epsilon)\}} + \mathbf{1}_{\{(x+x_0,x-x_0):x\in\Lambda(\epsilon)\}}),$$

where $\delta > 0$ and $\mathbf{1}_S$ is the indicator function of the set S. The function $\hat{g}$ is supported on the "diagonal line" $y = x$ and "off-diagonal segments" $\{(x - x_0, x + x_0) : x \in \Lambda(\epsilon)\}$ and $\{(x + x_0, x - x_0) : x \in \Lambda(\epsilon)\}$, which are disjoint for sufficiently small $\epsilon > 0$. (This is similar to Fig. 4.2.) When $\delta > 0$ is small enough, $\hat{g}(x, y)$ is non-negative everywhere. Furthermore, our construction preserves the "total mass". Hence, the function $\hat{g}(x, y)$ is indeed a probability density with respect to the d-dimensional Lebesgue measure on the "diagonal line" and two "off-diagonal segments". Let $(\hat{X}, \hat{Y})$ be a pair with the joint density $\hat{g}(x, y)$. The marginals $\hat{X}$ and $\hat{Y}$ have the same distribution as that of X, since the "positive mass" on "off-diagonal points" complements the "mass deficit" on "diagonal points" when we project in the x and y directions. We claim that $\frac{\hat{X}+\hat{Y}}{2}$ has larger entropy than $\hat{X}$. One can check that the density of $\frac{\hat{X}+\hat{Y}}{2}$ is

$$\hat{f}(x) = f(x) + \delta(2\mathbf{1}_{\Lambda(\epsilon)} - \mathbf{1}_{\Lambda(\epsilon)+x_0} - \mathbf{1}_{\Lambda(\epsilon)-x_0}).$$

Let Ω denote the union of $\Lambda(\epsilon)$, $\Lambda(\epsilon) + x_0$ and $\Lambda(\epsilon) - x_0$. Then we have

$$h_r\left(\frac{\hat{X} + \hat{Y}}{2}\right) = \frac{1}{1-r}\log\left(\int_\Omega \hat{f}(x)^r dx + \int_{\Omega^c} f^r(x)dx\right). \qquad (4.36)$$

Since $x_0 \neq 0$, for $\epsilon > 0$ small enough, Ω is the union of disjoint translates of $\Lambda(\epsilon)$. When $\delta > 0$ is sufficiently small, we have

$$\int_\Omega \hat{f}(x)^r dx = \int_{\Lambda(\epsilon)} \left[(f(x) + 2\delta)^r + (f(x + x_0) - \delta)^r + (f(x - x_0) - \delta)^r\right] dx$$

$$< \int_{\Lambda(\epsilon)} \left[f(x)^r + f(x + x_0)^r + f(x - x_0)^r\right] dx \qquad (4.37)$$

$$= \int_\Omega f(x)^r dx, \qquad (4.38)$$

where inequality (4.37) follows from the observation that for $x \in \Lambda(\epsilon) \subset \Lambda$ (see (4.35)) the derivative of the integrand at $\delta = 0$ is

$$r[2f(x)^{r-1} - f(x - x_0)^{r-1} - f(x + x_0)^{r-1}] < 0. \tag{4.39}$$

Since $r > 1$, (4.36) together with (4.38) implies that

$$h_r\left(\frac{\hat{X} + \hat{Y}}{2}\right) > \frac{1}{1 - r} \log\left(\int_\Omega f(x)^r dx + \int_{\Omega^c} f(x)^r dx\right) = h(X) = h(\hat{X}).$$

This is contradictory to our assumption. Hence, f has to be $(r - 1)$-concave. For $1 - 1/d < r < 1$, we redefine the set Λ by reversing inequality (4.35), and inequality (4.37) will be also reversed. We will arrive at the same conclusion. □

Remark 4.26 The proof of $1 \implies 2$ is an immediate consequence of Theorem 3.36 in [30]. The theorem there draws heavily on the ideas of [42], where a related study, deriving the Schur convexity of Rényi entropies under the assumption of exchangeability and s-concavity of the random variables, generalizing Yu's results in [43] on the entropies of sums of i.i.d. log-concave random variables. Although we state Theorem 4.25 for two random vectors, the argument also works for more than two random vectors. Hence, it implies the seemingly stronger Theorem 4.4.

As an immediate consequence of Theorem 4.25, we have the following reverse Rényi EPI for random vectors with the same distribution.

Corollary 4.27 *Let $s > -1/d$ and let $r = 1 + s$. Let X and Y be (possibly dependent) random vectors in $\mathbb{R}^d$ with the same density f being s-concave. Then we have*

$$N_r(X + Y) \leq 4N_r(X).$$

References

1. M. Adamek, Almost λ-convex and almost Wright-convex functions. Math. Slovaca **53**(1), 67–73 (2003)
2. K. Ball, P. Nayar, T. Tkocz, A reverse entropy power inequality for log-concave random vectors. Stud. Math. **235**(1), 17–30 (2016)
3. A.R. Barron, Entropy and the central limit theorem. Ann. Probab. **14**, 336–342 (1986)
4. W. Beckner, Inequalities in Fourier analysis. Ann. Math. **102**(1), 159–182 (1975)
5. R.N. Bhattacharya, R. Ranga Rao, *Normal Approximation and Asymptotic Expansions*. (Wiley/Society for Industrial and Applied Mathematics, Hoboken/Philadelphia, 2010/1976)
6. H. Blumberg, On convex functions. Trans. Am. Math. Soc. **20**, 40–44 (1919)
7. S.G. Bobkov, G.P. Chistyakov, Bounds for the maximum of the density of the sum of independent random variables. Zap. Nauchn. Sem. S.-Peterburg. Otdel. Mat. Inst. Steklov. **408**(18), 62–73, 324 (2012)

8. S.G. Bobkov, G.P. Chistyakov, Entropy power inequality for the Rényi entropy. IEEE Trans. Inform. Theory **61**(2), 708–714 (2015)
9. S. Bobkov, M. Madiman, Reverse Brunn-Minkowski and reverse entropy power inequalities for convex measures. J. Funct. Anal. **262**, 3309–3339 (2012)
10. S.G. Bobkov, A. Marsiglietti, Variants of the entropy power inequality. IEEE Trans. Inform. Theory **63**(12), 7747–7752 (2017)
11. S.G. Bobkov, A. Marsiglietti, Asymptotic behavior of Rényi entropy in the central limit theorem (2018). Preprint. arXiv:1802.10212
12. S.G. Bobkov, G.P. Chistyakov, F. Götze, Rényi divergence and the central limit theorem. Ann. Probab. **47**(1), 270–323 (2019)
13. C. Borell, Convex measures on locally convex spaces. Ark. Mat. **12**, 239–252 (1974)
14. C. Borell, Convex set functions in d-space. Period. Math. Hungar. **6**(2), 111–136 (1975)
15. H.J. Brascamp, E.H. Lieb, Best constants in Young's inequality, its converse, and its generalization to more than three functions. Adv. Math. **20**(2), 151–173 (1976)
16. S. Brazitikos, A. Giannopoulos, P. Valettas, B.-H. Vritsiou, *Geometry of isotropic convex bodies*. AMS-Mathematical Surveys and Monographs, vol. 196 (American Mathematical Society, Providence, 2014)
17. H. Busemann, A theorem on convex bodies of the Brunn-Minkowski type. Proc. Natl. Acad. Sci. U. S. A. **35**, 27–31 (1949)
18. A. Chademan, F. Mirzapour, Midconvex functions in locally compact groups. Proc. Am. Math. Soc. **127**, 2961–2968 (1999)
19. M.H.M. Costa, A new entropy power inequality. IEEE Trans. Inform. Theory **31**(6), 751–760 (1985)
20. T.M. Cover, Z. Zhang, On the maximum entropy of the sum of two dependent random variables. IEEE Trans. Inform. Theory **40**(4), 1244–1246 (1994)
21. A. Dembo, T.M. Cover, J.A. Thomas, Information-theoretic inequalities. IEEE Trans. Inform. Theory **37**(6), 1501–1518 (1991)
22. M. Fradelizi, J. Li, M. Madiman, Concentration of information content for convex measures (2015). Preprint. arXiv:1512.01490
23. R.J. Gardner, A. Giannopoulos, p-cross-section bodies. Indiana Univ. Math. J. **48**(2), 593–614 (1999)
24. B.V. Gnedenko, A.N. Kolmogorov, *Limit distributions for sums of independent random variables*. Translated from the Russian, annotated, and revised by K. L. Chung. With appendices by J. L. Doob and P. L. Hsu. Revised edition. (Addison-Wesley Publishing, Reading, 1968)
25. W. Hoeffding, Probability inequalities for sums of bounded random variables. J. Am. Stat. Assoc. **58**, 13–30 (1963)
26. O. Johnson, *Information theory and the central limit theorem* (Imperial College Press, London, 2004)
27. J. Li, Rényi entropy power inequality and a reverse. Stud. Math. **242**, 303–319 (2018)
28. J. Li, J. Melbourne, Further investigations of the maximum entropy of the sum of two dependent random variables, in *Proceedings of IEEE International Symposium on Information Theory*, Vail (2018), pp. 1969–1972
29. E.H. Lieb, Proof of an entropy conjecture of Wehrl. Commun. Math. Phys. **62**(1), 35–41 (1978)
30. M. Madiman, J. Melbourne, P. Xu, Forward and reverse entropy power inequalities in convex geometry. Convexity Concentration **161**, 427–485 (2017)
31. M. Madiman, J. Melbourne, P. Xu, Rogozin's convolution inequality for locally compact groups (2017). Preprint. arXiv:1705.00642
32. M. Madiman, J. Melbourne, P. Xu, Infinity-Rényi entropy power inequalities, in *2017 IEEE International Symposium on Information Theory (ISIT)*, Aachen (2017), pp. 2985–2989
33. A. Marsiglietti, J. Melbourne, On the entropy power inequality for the Rényi entropy of order [0, 1]. IEEE Trans. Inform. Theory **65**, 1387–1396 (2018). https://doi.org/10.1109/TIT.2018.2877741

34. A. Marsiglietti, J. Melbourne, A Rényi entropy power inequality for log-concave vectors and parameters in [0, 1], in *Proceedings of IEEE International Symposium on Information Theory*, Vail (2018), pp. 1964–1968
35. V.D. Milman, Inégalité de Brunn-Minkowski inverse et applications à la théorie locale des espaces normés. C. R. Acad. Sci. Paris Sér. I Math. **302**(1), 25–28 (1986)
36. E. Ram, I. Sason, On Rényi entropy power inequalities. IEEE Trans. Inform. Theory **62**(12), 6800–6815 (2016)
37. O. Rioul, Rényi entropy power inequalities via normal transport and rotation. Entropy **20**(9), 641 (2018)
38. G. Savaré, G. Toscani, The concavity of Rényi entropy power. IEEE Trans. Inform. Theory **60**(5), 2687–2693 (2014)
39. C.E. Shannon, A mathematical theory of communication. Bell Syst. Tech. J. **27**, 379–423 623–656 (1948)
40. W. Sierpinski, Sur les fonctions convexes mesurables. Fund. Math. **1**, 125–129 (1920)
41. A.J. Stam, Some inequalities satisfied by the quantities of information of Fisher and Shannon. Inf. Control. **2**, 101–112 (1959)
42. P. Xu, J. Melbourne, M. Madiman, Reverse entropy power inequalities for s-concave densities, in *Proceedings of International Symposium on Information Theory*, Barcelona (2016), pp. 2284–2288
43. Y. Yu, Letter to the editor: on an inequality of Karlin and Rinott concerning weighted sums of i.i.d. random variables. Adv. Appl. Probab. **40**(4), 1223–1226 (2008)

Chapter 5
Small Ball Probability for the Condition Number of Random Matrices

Alexander E. Litvak, Konstantin Tikhomirov,
and Nicole Tomczak-Jaegermann

Abstract Let A be an $n \times n$ random matrix with i.i.d. entries of zero mean, unit variance and a bounded sub-Gaussian moment. We show that the condition number $s_{\max}(A)/s_{\min}(A)$ satisfies the small ball probability estimate

$$\mathbb{P}\big\{ s_{\max}(A)/s_{\min}(A) \leq n/t \big\} \leq 2\exp(-ct^2), \quad t \geq 1,$$

where $c > 0$ may only depend on the sub-Gaussian moment. Although the estimate can be obtained as a combination of known results and techniques, it was not noticed in the literature before. As a key step of the proof, we apply estimates for the singular values of A, $\mathbb{P}\big\{ s_{n-k+1}(A) \leq ck/\sqrt{n} \big\} \leq 2\exp(-ck^2)$, $1 \leq k \leq n$, obtained (under some additional assumptions) by Nguyen.

5.1 Introduction

We say that a random variable ξ has sub-Gaussian moment bounded above by $K > 0$ if

$$\mathbb{P}\{|\xi| \geq t\} \leq \exp\big(1 - t^2/(2K^2)\big), \quad t \geq 0.$$

AMS 2010 Classification 60B20, 15B52, 46B06, 15A18

A. E. Litvak (✉) · N. Tomczak-Jaegermann
Department of Mathematical and Statistical Sciences, University of Alberta, Edmonton, AB, Canada
e-mail: nicole.tomczak@ualberta.ca

K. Tikhomirov
School of Mathematics, Georgia Institute of Technology, Atlanta, GA, USA
e-mail: ktikhomirov6@gatech.edu

© Springer Nature Switzerland AG 2020
B. Klartag, E. Milman (eds.), *Geometric Aspects of Functional Analysis*,
Lecture Notes in Mathematics 2266,
https://doi.org/10.1007/978-3-030-46762-3_5

Let A be an $n \times n$ random matrix with i.i.d. entries of zero mean, unit variance and sub-Gaussian moment bounded above by K, and denote by $s_i(A)$, $1 \leq i \leq n$, its singular values arranged in non-increasing order. We will write $s_{\max}(A)$ and $s_{\min}(A)$ for $s_1(A)$ and $s_n(A)$, respectively. Estimating the magnitude of the condition number,

$$\kappa(A) = s_{\max}(A)/s_{\min}(A),$$

is a well studied problem, with connections to numerical analysis and computation of the limiting distribution of the matrix spectrum; we refer, in particular, to [20] for discussion. Since the largest singular value $s_{\max}(A)$ is strongly concentrated (see the proof of Corollary 5.1.2 below), estimating $\kappa(A)$ is essentially reduced to estimating $s_{\min}(A)$ from above and below.

The main result of [12] provides small ball probability estimates for $s_{\min}(A)$ of the form

$$\mathbb{P}\{s_{\min}(A) \leq t/\sqrt{n}\} \leq Ct + e^{-cn}, \quad t \leq 1,$$

for some $C, c > 0$ depending only on the sub-Gaussian moment. It seems natural to investigate the complementary regime—the large deviation estimates for $s_{\min}(A)$. It was shown in [13] that

$$\mathbb{P}\{s_{\min}(A) \geq t/\sqrt{n}\} \leq \frac{C \ln t}{t} + e^{-cn}, \quad t \geq 2$$

(see also [21] for an extension of this result to distributions with no assumptions on moments higher than 2). The probability estimate was improved in [10] to

$$\mathbb{P}\{s_{\min}(A) \geq t/\sqrt{n}\} \leq e^{-ct}, \quad t \geq 2,$$

for $c > 0$ depending only on the sub-Gaussian moment. The existing results on the distribution of the singular values of random Gaussian matrices [4, 18] suggest that the optimal dependence on t in the exponent on the right hand side is quadratic, i.e. the variable $\sqrt{n}\, s_{\min}(A)$ is sub-Gaussian. Specifically, it is shown in [18] that $s_{\min}(G)$ for the standard $n \times n$ Gaussian matrix G satisfies two-sided estimates

$$\exp(-Ct^2) \leq \mathbb{P}\{s_{\min}(G) \geq t/\sqrt{n}\} \leq \exp(-ct^2), \quad t \geq C_1,$$

where $C, C_1, c > 0$ are some universal constants. The main result of our note provides matching upper estimate for matrices with sub-Gaussian entries:

Theorem 5.1.1 *Let A be an $n \times n$ random matrix with i.i.d. entries of zero mean, unit variance, and sub-Gaussian moment bounded above by $K > 0$. Then the smallest singular value $s_{\min}(A)$ satisfies*

$$\mathbb{P}\{s_{\min}(A) \geq t/\sqrt{n}\} \leq 2\exp(-ct^2), \quad t \geq 1,$$

where $c > 0$ is a constant depending only on K.

As a simple corollary of the theorem, we obtain small ball probability estimates for the condition number:

Corollary 5.1.2 *Let A be an $n \times n$ random matrix with i.i.d. entries of zero mean, unit variance, and sub-Gaussian moment bounded above by $K > 0$. Then the condition number $\kappa(A)$ satisfies*

$$\mathbb{P}\{\kappa(A) \leq n/t\} \leq 2\exp(-ct^2), \quad t \geq 1,$$

where $c > 0$ is a constant depending only on K.

Theorem 5.1.1 is a consequence of the following theorem, which is of independent interest.

Theorem 5.1.3 *Under conditions of Theorem 5.1.1 one has*

$$\mathbb{P}\{\|A^{-1}\|_{HS} \leq \min(n/t, \sqrt{n/t})\} \leq 2\exp(-ct^2), \quad t \geq 0,$$

where $c > 0$ is a constant depending only on K.

The proof of Theorem 5.1.3 uses, as a main step, the estimates

$$\mathbb{P}\{s_{n-k+1}(A) \leq ck/\sqrt{n}\} \leq 2\exp(-ck^2), \quad 1 \leq k \leq n,$$

for the singular values of the matrix A. These estimates, based on the restricted invertibility of matrices and certain averaging arguments, were recently obtained by Nguyen [9] under some additional assumptions (which will be discussed in the next section).

5.2 Preliminaries

Given a matrix A, it singular values $s_i = s_i(A)$, $i \geq 1$, are square roots of eigenvalues of AA^*. We always assume that $s_1 \geq s_2 \geq \ldots$ By $\|A\|$ and $\|A\|_{HS}$ we denote the operator $\ell_2 \to \ell_2$ norm of A (also called the spectral norm) and the Hilbert–Schmidt norm respectively. Note that

$$\|A\| = s_1 \quad \text{and} \quad \|A\|_{HS}^2 = \sum_{i \geq 1} s_i^2.$$

The columns and rows of A are denoted by $\mathbf{C}_i(A)$ and $\mathbf{R}_i(A)$, $i \geq 1$, respectively. Given $J \subset [m]$, the coordinate projection in $\mathbb{R}^m$ onto $\mathbb{R}^J$ is denoted by P_J. For convenience, we often write A_J instead of AP_J. Given $m \geq 1$, the identity operator $\mathbb{R}^\ell \to \mathbb{R}^\ell$ we denote by I_m. Given $x, y \in \mathbb{R}^n$ by $\langle x, \cdot \rangle y$ we denote the operator

$z \mapsto \langle x, z \rangle y$ (in the literature it is often denoted by $x \otimes y$ or $yx^\top$). The canonical Euclidean norm in $\mathbb{R}^m$ is denoted by $\| \cdot \|_2$ and the unit Euclidean sphere by S^{m-1}.

As the most important part of our argument, we will use the following result.

Theorem 5.2.1 *Let A be an $n \times n$ random matrix with i.i.d. entries of zero mean, unit variance, and sub-Gaussian moment bounded above by $K > 0$. Then for any $1 \le k \le n$ one has*

$$\mathbb{P}\{s_{n-k+1}(A) \le ck/\sqrt{n}\} \le 2\exp(-ck^2),$$

where $c > 0$ is a constant depending only on K.

The above theorem, up to some minor modifications, was proved by Nguyen in [9]. Specifically, in the case $k \ge C \log n$, the theorem follows from [9, Theorem 1.7] (or [9, Corollary 1.8]) if one additionally assumes either that the entries of A are uniformly bounded by a constant, or that the distribution density of the entries is bounded. Removing these conditions requires a minor change of the proof in [9]. Further, in the case $k \le C \log n$, the above result (in fact, in a stronger form) is stated as formula (4) in [9, Theorem 1.4]. However, [9, Theorem 3.6], which is used to derive [9, formula (4)], provides a non-trivial probability estimate only for the event $\{s_{n-k+1}(A) \le c_\gamma k^{1-\gamma}/\sqrt{n}\}$ (for any given $\gamma \in (0, 1)$ and c_γ depending on γ), see [9, formula (31)]. Again, a minor update of the argument of [9] provides the result needed for our purposes. In view of the above and for the reader's convenience, we provide a proof of Theorem 5.2.1 in the last section.

The following result was proved in [17] as an extension of the classical Bourgain–Tzafriri restricted invertibility theorem [2]. With worse dependence on ε, the theorem was earlier proved in [22]. See also recent papers [1, 8] for further improvements and discussions.

Theorem 5.2.2 ([17]) *Let T be $n \times n$ matrix. Then for any $\varepsilon \in (0, 1)$ there is a set $J \subset [n]$ such that*

$$\ell := |J| \ge \left\lfloor \frac{\varepsilon^2 \|T\|_{HS}^2}{\|T\|^2} \right\rfloor \qquad and \qquad s_\ell(T_J) \ge \frac{(1-\varepsilon)\|T\|_{HS}}{\sqrt{n}}.$$

We will use two following results by Rudelson–Verhsynin. The first one was one of the key ingredients in estimating the smallest singular value of rectangular matrices. The second one is an immediate consequence of the Hanson–Wright inequality [5, 23] generalized in [15].

Theorem 5.2.3 ([14, Theorem 4.1]) *Let X be a vector in $\mathbb{R}^n$, whose coordinates are i.i.d. mean-zero, sub-Gaussian random variables with unit variance. Let F be a random subspace in $\mathbb{R}^n$ spanned by $n - \ell$ vectors, $1 \le \ell \le c'n$, whose coordinates are i.i.d. mean-zero, sub-Gaussian random variables with unit variance, jointly independent with X. Then, for every $\varepsilon > 0$, one has*

$$\mathbb{P}\{\text{dist}(X, F) \le \varepsilon\sqrt{\ell}\} \le (C\varepsilon)^\ell + \exp(-cn).$$

where $C > 0$, c, $c' \in (0, 1)$ are constants depending only on the sub-Gaussian moments.

Theorem 5.2.4 ([15, Corollary 3.1]) *Let X be a vector in $\mathbb{R}^n$, whose coordinates are i.i.d. mean-zero random variables with unit variance and with sub-Gaussian moment bounded by K. Let F be a fixed subspace in $\mathbb{R}^n$ of dimension $n - \ell$. Then, for every $t > 0$, one has*

$$\mathbb{P}\{|\mathrm{dist}(X, F) - \sqrt{\ell}| \geq t\} \leq 2\exp(-ct^2/K^4).$$

where $c > 0$ is an absolute constant.

We will also need the following standard claim, which can be proved by integrating the indicator functions (see e.g., [9, Claim 3.4], cf. [7, Claim 4.9]).

Claim 5.2.5 Let $\alpha, p \in (0, 1)$. Let $\mathcal{E}$ be an event. Let Z be a finite index set, and $\{\mathcal{E}_z\}_{z \in Z}$ be a collection of $|Z|$ events satisfying $\mathbb{P}(\mathcal{E}_z) \leq p$ for every $z \in Z$. Assume that at least $\alpha|Z|$ of events $\mathcal{E}_z$ hold whenever the event $\mathcal{E}$ occurs. Then $\mathbb{P}(\mathcal{E}) \leq p/\alpha$.

5.3 Proofs of Main Results

Proof of Theorem 5.1.1 In the case $t > n$ we have

$$\mathbb{P}\{s_{\min}(A) \geq t/\sqrt{n}\} = \mathbb{P}\{s_1(A^{-1}) \leq \sqrt{n}/t\} \leq \mathbb{P}\left\{\sum_{i=1}^{n} s_i(A^{-1})^2 \leq n^2/t^2\right\}$$

and the result follows from Theorem 5.1.3.

Now we consider the case $1 \leq t \leq n$. Let $L \geq 1$ be a parameter which we will choose later. Then

$$\mathbb{P}\{s_{\min}(A) \geq t/\sqrt{n}\} = \mathbb{P}\{s_1(A^{-1}) \leq \sqrt{n}/t\}$$

$$\leq \mathbb{P}\left\{s_1(A^{-1})^2 \leq n/t^2 \text{ and } \sum_{i \geq \lceil t \rceil} s_i(A^{-1})^2 \geq Ln/t\right\}$$

$$+ \mathbb{P}\left\{s_1(A^{-1})^2 \leq n/t^2 \text{ and } \sum_{i \geq \lceil t \rceil} s_i(A^{-1})^2 < Ln/t\right\}$$

$$\leq \mathbb{P}\left\{\sum_{i \geq \lceil t \rceil} s_i(A^{-1})^2 \geq Ln/t\right\}$$

$$+ \mathbb{P}\left\{\sum_{i=1}^{n} s_i(A^{-1})^2 \leq n/t + Ln/t\right\}.$$

For the first summand in the last expression, we apply Theorem 5.2.1. Since $\sum_{i=\lfloor t \rfloor}^{\infty} \frac{1}{i^2} \leq \frac{2}{t}$, we obtain

$$\mathbb{P}\left\{ \sum_{i=\lceil t \rceil}^{n} s_i(A^{-1})^2 \geq Ln/t \right\} \leq \sum_{i=\lceil t \rceil}^{n} \mathbb{P}\left\{ s_i(A^{-1})^2 \geq Ln/(2i^2) \right\}$$

$$= \sum_{i=\lceil t \rceil}^{n} \mathbb{P}\left\{ s_{n-i+1}(A) \leq \sqrt{2}i/\sqrt{Ln} \right\}.$$

Choosing L so that $\sqrt{2/L}$ is equal to the constant from Theorem 5.2.1, we get

$$\sum_{i=\lfloor t \rfloor}^{n} \mathbb{P}\left\{ s_{n-i+1}(A) \leq \sqrt{2}i/\sqrt{Ln} \right\} \leq 2 \sum_{i=\lfloor t \rfloor}^{n} \exp(-ci^2) \leq 3\exp(-c't^2)$$

for some $c' > 0$ depending only on K. The bound on the second summand follows from Theorem 5.1.3 applied with $t/(L+1)$ instead of t. This completes the proof. $\square$

Proof of Corollary 5.1.2 Theorem 5.2.4 implies that there exists an absolute constant $c_1 > 0$ depending only on K such that for every $i \leq n$

$$\mathbb{P}(\|C_i(A)\|_2 \leq \sqrt{n}/2) \leq \exp(-c_1 n)$$

(this can be shown by direct calculations as well, see e.g. Fact 2.5 in [6]). Since the entries of A are independent, we obtain

$$\mathbb{P}(\|A\| \leq \sqrt{n}/2) \leq \prod_{i=1}^{n} \mathbb{P}(\|C_i(A)\|_2 \leq \sqrt{n}/2) \leq \exp(-c_1 n^2).$$

Note that if $\|A\| \geq \sqrt{n}/2$ and $\kappa(A) \leq n/2t$ then $s_n(A) = \|A\|/\kappa(A) \geq t/\sqrt{n}$. Therefore, by Theorem 5.1.1,

$$\mathbb{P}\{\kappa(A) \leq n/2t\} \leq 2\exp(-ct^2) + \exp(-c_1 n^2).$$

By adjusting constants, this implies the conclusion for $t \leq n$. Since $\kappa(A) \geq 1$, the case $t > n$ is trivial. $\square$

Proof of Theorem 5.1.3 Adjusting the constant in the exponent if needed, without loss of generality, we assume that $t \geq C_0$, where $C_0 > 0$ is a large enough constant depending only on K. Denote

$$\mathcal{E}_0 := \left\{ \sum_{i=1}^{n} s_i(A^{-1})^2 \leq n/t \right\}.$$

We first consider the case $t \leq n$. Applying the negative second moment identity (see e.g. Exercise 2.7.3 in [19]),

$$\sum_{i=1}^{n} s_i(A^{-1})^2 = \sum_{i=1}^{n} \text{dist}(\mathbf{C}_i(A), \text{span}\{\mathbf{C}_j(A), \ j \neq i\})^{-2},$$

we observe that on the event $\mathcal{E}_0$,

$$\left|\{i \leq n : \ \text{dist}(\mathbf{C}_i(A), \text{span}\{\mathbf{C}_j(A), \ j \neq i\}) \geq \sqrt{t/2}\}\right| \geq n/2.$$

For each subset $I \subset [n]$ of cardinality $k \leq n/2$ (the actual value of k will be defined later), let $\mathbf{1}_I$ be the indicator of the event

$$\{\text{dist}(\mathbf{C}_i(A), \text{span}\{\mathbf{C}_j(A), \ j \in [n] \setminus I\}) \geq \sqrt{t/2} \text{ for all } i \in I\}.$$

Then, in view of the above, everywhere on the event $\mathcal{E}_0$ we have

$$\sum_{I \subset [n], \ |I|=k} \mathbf{1}_I \geq \binom{\lceil n/2 \rceil}{k} \geq \left(\frac{n}{2k}\right)^k \geq (2e)^{-k} \binom{n}{k}.$$

Hence, by Markov's inequality and permutation invariance of the matrix distribution,

$$\mathbb{P}(\mathcal{E}_0) \leq (2e)^k \, \mathbb{E} \, \mathbf{1}_{[k]}.$$

As the last step of the proof, we estimate the expectation of $\mathbf{1}_{[k]}$ (with a suitable choice of k). In view of independence and equidistribution of the matrix columns, we have

$$\mathbb{E} \, \mathbf{1}_{[k]} = \left(\mathbb{P}\{\text{dist}(\mathbf{C}_1(A), \text{span}\{\mathbf{C}_j(A), \ j \in [n] \setminus [k]\}) \geq \sqrt{t/2}\}\right)^k.$$

Choose $k := \lfloor t/4 \rfloor \leq n/2$ and denote

$$D := \text{dist}(\mathbf{C}_1(A), \text{span}\{\mathbf{C}_j(A), \ j \in [n] \setminus [k]\}).$$

Using independence of columns of the matrix A and applying Theorem 5.2.4 with $\ell = k$ and $F = \text{span}\{\mathbf{C}_j(A), \ j \in [n] \setminus [k]\}$, we obtain

$$\mathbb{P}\left\{D \geq \sqrt{t/2}\right\} \leq \mathbb{P}\left\{D - \sqrt{k} \geq (\sqrt{2} - 1) \sqrt{t/4}\right\} \leq 2 \exp(-\bar{c} t)$$

for some $\bar{c} > 0$ depending only on K. Hence,

$$\mathbb{P}(\mathcal{E}_0) \leq (2e)^k \, 2^k \, \exp(-\bar{c} t \, k) \leq \exp(-\bar{c} t^2/16),$$

provided that t is larger than a certain constant depending only on K. This implies the desired result for $t \leq n$.

In the case $t > n$ we essentially repeat the argument along the same lines. Define

$$\mathcal{E}_0' := \left\{ \sum_{i=1}^{n} s_i(A^{-1})^2 \leq n^2/t^2 \right\}.$$

Observe that on the event $\mathcal{E}_0'$,

$$\left| \left\{ i \leq n : \text{dist}\big(\mathbf{C}_i(A), \text{span}\{\mathbf{C}_j(A), \ j \neq i\}\big) \geq t/\sqrt{2n} \right\} \right| \geq n/2.$$

Repeating the above computations with the same notation and with $k = \lfloor n/4 \rfloor$ we obtain

$$\mathbb{P}\left\{ D \geq t/\sqrt{2n} \right\} \leq \mathbb{P}\left\{ D - \sqrt{k} \geq t/(5\sqrt{n}) \right\} \leq 2\exp(-\bar{c}\, t^2/n),$$

which leads to

$$\mathbb{P}(\mathcal{E}_0') \leq (2e)^k \, 2^k \, \exp(-\bar{c}\, kt^2/n) \leq \exp(-\bar{c} t^2/16),$$

provided that $t > Cn$ for large enough C depending only on K. For $n < t \leq Cn$ the result follows by adjusting the absolute constants. $\square$

5.4 Small Ball Estimates for Singular Values

The goal of this section is to prove Theorem 5.2.1. As we have noted, the argument essentially reproduces that of [9]. An important part of the proof is the use of restricted invertibility (see also [3] and [11] for some recent applications of restricted invertibility in the context of random matrices).

We will use a construction from [9]. Given an integer k and an $n \times n$ matrix A define a $k \times n$ matrix $Z = Z(A, k)$ in the following way. Consider singular value decomposition $A = \sum_{i=1}^{n} s_i \langle v_i, \cdot \rangle w_i$, where $s_i = s_i(A)$ are singular values of A (arranged in non-increasing order) and $\{v_i\}_i$, $\{w_i\}_i$ are two orthonormal systems in $\mathbb{R}^n$. For $i \leq k$ denote $z_i = v_{n-i+1}$. Let Z be the matrix whose rows are $\mathbf{R}_i(Z) = z_i$. Clearly, the rows of Z are orthonormal and for every $i \leq k$,

$$\|A z_i\|_2 = s_{n-i+1} \leq s_{n-k+1}. \tag{5.1}$$

Moreover,

$$\|Z\| = 1 \qquad \text{and} \qquad \|Z\|_{HS} = \sqrt{k}.$$

The matrix Z is not uniquely defined when some of the k smallest singular values of A have non-trivial multiplicity; we will however assume that for each realization of A, a single admissible Z is chosen in such a way that Z is a (measurable) random matrix.

5.4.1 Proof of Theorem 5.2.1, the Case $k \geq \ln n$

Let C, c, c' be constants from Theorem 5.2.3. Let $\gamma = \sqrt{c'}$. Note that C, c, c', γ depend only on K. Let $Z = Z(A, k)$ be the $k \times n$ matrix constructed above. Applying Theorem 5.2.2 to Z (one can add zero rows to make it an $n \times n$ matrix), there exists $J \subset [n]$ such that

$$|J| = \ell := \lfloor \gamma^2 k \rfloor \leq c'k \qquad \text{and} \qquad s_\ell(Z_J) \geq (1 - \gamma)\sqrt{k/n}.$$

Fix a (small enough, depending on K) constant $c_0 > 0$. Define the event

$$\mathcal{E}_k := \{s_{n-k+1}(A) \leq c_0 k/\sqrt{n}\}.$$

Consider the $n \times k$ matrix $B = AZ^\top$. Using property (5.1), on the event $\mathcal{E}_k$, we have for every $i \leq k$,

$$\|\mathbf{C}_i(B)\|_2 = \|Az_i\|_2 \leq c_0 k/\sqrt{n},$$

hence $\|B\|_{HS} \leq c_0 k^{3/2}/\sqrt{n}$. Now, since $s_\ell(Z_J) > 0$, there exists a $k \times \ell$ matrix M such that $Z_J^\top M = I_\ell$. Then

$$\|M\| = 1/s_\ell(Z) \leq (1 - \gamma)^{-1}\sqrt{n/k}.$$

Therefore,

$$\|BM\|_{HS} \leq \|B\|_{HS} \|M\| \leq c_0(1 - \gamma)^{-1}k.$$

Writing $B = A_J(Z_J)^\top + A_{J^c}(Z_{J^c})^\top$, we also have $BM = A_J + A_{J^c}(Z_{J^c})^\top M$. Next denote

$$F = F(A, J) := \text{span}\{\mathbf{C}_i(A_{J^c})\}_{i \in J^c},$$

and let P be the orthogonal projection on $F^\perp$. Then, on the event $\mathcal{E}_k$,

$$c_0^2(1 - \gamma)^{-2}k^2 \geq \|PBM\|_{HS}^2 \geq \|PA_J\|_{HS}^2 = \sum_{i \in J} \|P\,\mathbf{C}_i(A_J)\|_2^2$$

$$= \sum_{i \in J} \text{dist}^2(\mathbf{C}_i(A), F).$$

Therefore, for at least $\ell/2$ indices $i \in J$, one has

$$\text{dist}(\mathbf{C}_i(A), F) \le \sqrt{2}c_0(1-\gamma)^{-1}k/\sqrt{\ell} \le 2c_0\sqrt{\ell}/((1-\gamma)\gamma^2).$$

Note that the subspace F is spanned by $n - \ell$ random vectors, it is independent of columns $\mathbf{C}_i(A)$, $i \in J$, and that columns of A are independent. Therefore, by Theorem 5.2.3 and the union bound we obtain

$$\mathbb{P}(\mathcal{E}_k) \le \sum_{\substack{J \subset [n] \\ |J|=\ell}} \sum_{\substack{J_1 \subset J \\ |J_1|=\lceil \ell/2 \rceil}} \mathbb{P}\left\{ \forall i \in J_1 \ \ \text{dist}(\mathbf{C}_i(A), F) \le 2c_0\sqrt{\ell}/((1-\gamma)\gamma^2) \right\}$$

$$\le \binom{n}{\ell} 2^\ell \left((2Cc_0/((1-\gamma)\gamma^2))^\ell + \exp(-cn) \right)^{\ell/2}$$

$$\le \left(\frac{4en}{\ell} \max\left\{ \left(\frac{\sqrt{2Cc_0}}{\gamma\sqrt{1-\gamma}} \right)^\ell, \exp(-cn/2) \right\} \right)^\ell$$

Choosing small enough c_0 and using $k \ge \ln n$, we obtain $\mathbb{P}(\mathcal{E}_k) \le \exp(-c_3\ell^2)$, where $c_3 > 0$ depends only on K. By adjusting constants this proves the desired result for $k \ge \ln n$. $\square$

5.4.2   Proof of Theorem 5.2.1, the Case $k \le \ln n$

Let A be as in Theorem 5.2.1. It is well known (see e.g. Fact 2.4 in [6]) that there is an absolute constant $C_1 > 0$ such that

$$\mathbb{P}\{\|A\| \le C_1 K\sqrt{n}\} \ge 1 - e^{-n}. \tag{5.2}$$

Let $\mathcal{E}_{bd}$ denote the event from this equation. Further, from [16, Theorem 1.5] one infers that for any $\gamma > 0$ there are $\gamma_1, \gamma_2, \gamma_3 > 0$ depending only on γ and K such that, denoting

$$\mathcal{E}_{inc}(\gamma) := \left\{ \forall x \in S^{n-1} \text{ with } \|Ax\|_2 \le \gamma_1\sqrt{n}, \ \forall I \subset [n] \right.$$

$$\left. \text{with } |I| \ge \gamma n \text{ one has } \|P_I x\|_2 \ge \gamma_2 \right\},$$

the event satisfies

$$\mathbb{P}(\mathcal{E}_{inc}(\gamma)) \ge 1 - 2e^{-\gamma_3 n}. \tag{5.3}$$

The following statement was proved by Nguyen ([9, Corollary 3.8]).

Proposition 5.4.1 *For any $K > 0$ there are $C, c_1, c_2, \gamma > 0$ depending only on K with the following property. Let A be an $n \times n$ random matrix with i.i.d. entries of zero mean, unit variance, and sub-Gaussian moment bounded above by K. Let $2 \le k \le n/(C \ln n)$, and let the random $k \times n$ matrix $Z = Z(A, k)$ be defined as above. Then everywhere on the event $\{s_{n-k+1}(A) \le c_1 k/\sqrt{n}\} \cap \mathcal{E}_{inc}(\gamma) \cap \mathcal{E}_{bd}$ one has*

$$\left|\left\{J \subset [n]: |J| = \lfloor k/2\rfloor, \ s_{\lfloor k/2\rfloor}(Z_J) \ge c_1\sqrt{k/n}\right\}\right| \ge c_2^{k \ln k} n^{\lfloor k/2\rfloor}.$$

Now assume that $k \le \ln n$. Without loss of generality we may also assume that k is bounded below by a large constant. Let C, c, c' be constants from Theorem 5.2.3 and c_1, c_2, γ from Proposition 5.4.1. Fix for a moment any realization of A from the event $\{s_{n-k+1}(A) \le c_0 k/\sqrt{n}\} \cap \mathcal{E}_{inc}(\gamma) \cap \mathcal{E}_{bd}$, where $c_0 \in (0, c_1]$ will be chosen later. Let $\ell := \lfloor k/2\rfloor$ and

$$\mathcal{J} := \left\{J \subset [n]: |J| = \lfloor k/2\rfloor, \ s_{\lfloor k/2\rfloor}(Z_J) \ge c_1\sqrt{k/n}\right\}.$$

Fix $J \in \mathcal{J}$ and repeat the procedure used in Sect. 5.4.1 with J and ℓ. We obtain that for at least $\ell/2$ indices $i \in J$, one has

$$\mathrm{dist}(\mathbf{C}_i(A), F) \le \sqrt{2}c_0 k/(c_1\sqrt{\ell}) \le 4c_0\sqrt{\ell}/c_1, \tag{5.4}$$

where $F = \mathrm{span}\{\mathbf{C}_i(A_{J^c})\}_{i \in J^c}$. For any fixed subset $J \subset [n]$ of cardinality ℓ consider the event

$$\mathcal{E}_J := \{\text{for at least } \ell/2 \text{ indices } i \in J \text{ inequality (5.4) holds}\}.$$

Applying Theorem 5.2.3 and the union bound we observe

$$\mathbb{P}(\mathcal{E}_J) \le 2^\ell \left((4c_0C/c_1)^\ell + \exp(-cn)\right)^{\ell/2}$$

$$\le \left(4 \max\left\{(4c_0C/c_1)^\ell, \ \exp(-cn)\right\}\right)^{\ell/2}.$$

Choosing c_0 to be small enough we obtain that $\mathbb{P}(\mathcal{E}_J) \le \exp(-c_4 k^2)$, where $c_4 > 0$ depends only on K. Combining this with Claim 5.2.5 and Proposition 5.4.1 we obtain

$$\mathbb{P}\left(\{s_{n-k+1}(A) \le c_0 k/\sqrt{n}\} \cap \mathcal{E}_{inc}(\gamma) \cap \mathcal{E}_{bd}\right) \le c_2^{-k \ln k} \exp(-c_4 k^2) \le \exp(-c_5 k^2)$$

provided that $k \ge C_2$, where $C_2 \ge 1 \ge c_5 > 0$ are constants depending on on K only. By Eqs. (5.2) and (5.3) this completes the proof in the case $k \le \ln n$. $\qquad\square$

Acknowledgements The authors are grateful to the anonymous referee for careful reading and valuable suggestions that have helped to improve the presentation. The second named author would like to thank the Department of Mathematical and Statistical Sciences, University of Alberta for ideal working conditions.

References

1. J.D. Batson, D.A. Spielman, N. Srivastava, Twice-Ramanujan sparsifiers, in *Proceedings of the 2009 ACM International Symposium on Theory of Computing (STOC'09)* (ACM, New York, 2009), pp. 255–262. MR2780071
2. J. Bourgain, L. Tzafriri, Invertibility of "large" submatrices with applications to the geometry of Banach spaces and harmonic analysis. Israel J. Math. **57**(2), 137–224 (1987). MR0890420
3. N. Cook, Lower bounds for the smallest singular value of structured random matrices. Ann. Prob. (to appear). arXiv:1608.07347
4. A. Edelman, Eigenvalues and condition numbers of random matrices. SIAM J. Matrix Anal. Appl. **9**(4), 543–560 (1988). MR0964668
5. D.L. Hanson, F.T. Wright, *A bound on tail probabilities for quadratic forms in independent random variables*. Ann. Math. Stat. **42**, 1079–1083 (1971). MR0279864
6. A.E. Litvak, A. Pajor, M. Rudelson, N. Tomczak-Jaegermann, Smallest singular value of random matrices and geometry of random polytopes. Adv. Math. **195**(2), 491–523 (2005). MR2146352
7. A.E. Litvak, A. Lytova, K. Tikhomirov, N. Tomczak-Jaegermann, P. Youssef, Adjacency matrices of random digraphs: singularity and anti-concentration. J. Math. Anal. Appl. **445**(2), 1447–1491 (2017). MR3545253
8. A. Naor, P. Youssef, Restricted invertibility revisited, in *A journey through discrete mathematics* (Springer, Cham, 2017), pp. 657–691. MR3726618
9. H.H. Nguyen, Random matrices: overcrowding estimates for the spectrum. J. Funct. Anal. **275**(8), 2197–2224 (2018). MR3841540
10. H.H. Nguyen, V.H. Vu, Normal vector of a random hyperplane. Int. Math. Res. Not. **2018**(6), 1754–1778 (2018). MR3800634
11. M. Rudelson, K. Tikhomirov, The sparse circular law under minimal assumptions. Geom. Funct. Anal. **29**(2), 561–637 (2019)
12. M. Rudelson, R. Vershynin, The Littlewood-Offord problem and invertibility of random matrices. Adv. Math. **218**(2), 600–633 (2008). MR2407948
13. M. Rudelson, R. Vershynin, The least singular value of a random square matrix is $O(n^{-1/2})$. C. R. Math. Acad. Sci. Paris **346**(15–16), 893–896 (2008). MR2441928
14. M. Rudelson, R. Vershynin, Smallest singular value of a random rectangular matrix. Commun. Pure Appl. Math. **62**(12), 1707–1739 (2009). MR2569075
15. M. Rudelson, R. Vershynin, Hanson-Wright inequality and sub-Gaussian concentration. Electron. Commun. Probab. **18**(82), 9 pp. (2013). MR3125258
16. M. Rudelson, R. Vershynin, No-gaps delocalization for general random matrices. Geom. Funct. Anal. **26**(6), 1716–1776 (2016). MR3579707
17. D.A. Spielman, N. Srivastava, An elementary proof of the restricted invertibility theorem. Israel J. Math. **190**, 83–91 (2012). MR2956233
18. S.J. Szarek, Condition numbers of random matrices. J. Complexity **7**(2), 131–149 (1991). MR1108773
19. T. Tao, *Topics in random matrix theory*. Graduate Studies in Mathematics, vol. 132 (American Mathematical Society, Providence, 2012)
20. T. Tao, V.H. Vu, Inverse Littlewood-Offord theorems and the condition number of random discrete matrices. Ann. Math. **169**(2), 595–632 (2009). MR2480613

21. K. Tatarko, An upper bound on the smallest singular value of a square random matrix. J. Complexity **48**, 119–128 (2018). MR3828841
22. R. Vershynin, John's decompositions: selecting a large part. Israel J. Math. **122**, 253–277 (2001). MR1826503
23. F.T. Wright, A bound on tail probabilities for quadratic forms in independent random variables whose distributions are not necessarily symmetric. Ann. Probab. **1**(6), 1068–1070 (1973). MR0353419

Chapter 6
Concentration of the Intrinsic Volumes of a Convex Body

Martin Lotz, Michael B. McCoy, Ivan Nourdin, Giovanni Peccati, and Joel A. Tropp

Abstract The intrinsic volumes are measures of the content of a convex body. This paper applies probabilistic and information-theoretic methods to study the sequence of intrinsic volumes. The main result states that the intrinsic volume sequence concentrates sharply around a specific index, called the central intrinsic volume. Furthermore, among all convex bodies whose central intrinsic volume is fixed, an appropriately scaled cube has the intrinsic volume sequence with maximum entropy.

M. Lotz
Mathematics Institute, University of Warwick, Coventry, UK
e-mail: martin.lotz@warwick.ac.uk

M. B. McCoy
Department of Computing and Mathematical Sciences, California Institute of Technology, Pasadena, CA, USA

Cruise LLC, San Francisco, CA, USA
e-mail: mike.mccoy@getcruise.com

I. Nourdin · G. Peccati
Faculty of Science, Technology and Medicine, University of Luxembourg, Esch-sur-Alzette, Luxembourg

Department of Mathematics, University of Luxembourg, Esch-sur-Alzette, Luxembourg
e-mail: ivan.nourdin@uni.lu; giovanni.peccati@uni.lu

J. A. Tropp (✉)
Department of Computing and Mathematical Sciences, California Institute of Technology, Pasadena, CA, USA
e-mail: jtropp@cms.caltech.edu

© Springer Nature Switzerland AG 2020
B. Klartag, E. Milman (eds.), *Geometric Aspects of Functional Analysis*,
Lecture Notes in Mathematics 2266,
https://doi.org/10.1007/978-3-030-46762-3_6

6.1   Introduction and Main Results

Intrinsic volumes are the fundamental measures of content for a convex body. Some of the most celebrated results in convex geometry describe the properties of the intrinsic volumes and their interrelationships. In this paper, we identify several new properties of the sequence of intrinsic volumes by exploiting recent results from information theory and geometric functional analysis. In particular, we establish that the mass of the intrinsic volume sequence concentrates sharply around a specific index, which we call the *central intrinsic volume*. We also demonstrate that a scaled cube has the maximum-entropy distribution of intrinsic volumes among all convex bodies with a fixed central intrinsic volume.

6.1.1   Convex Bodies and Volume

For each natural number m, the Euclidean space $\mathbb{R}^m$ is equipped with the ℓ_2 norm $\|\cdot\|$, the associated inner product, and the canonical orthonormal basis. The origin of $\mathbb{R}^m$ is written as $\mathbf{0}_m$.

Throughout the paper, n denotes a fixed natural number. A *convex body* in $\mathbb{R}^n$ is a compact and convex subset, possibly empty. Throughout this paper, K will denote a *nonempty* convex body in $\mathbb{R}^n$. The dimension of the convex body, $\dim \mathsf{K}$, is the dimension of the affine hull of K; the dimension takes values in the range $\{0, 1, 2, \ldots, n\}$. When K has dimension j, we define the j-dimensional volume $\mathrm{Vol}_j(\mathsf{K})$ to be the Lebesgue measure of K, computed relative to its affine hull. If K is zero-dimensional (i.e., a single point), then $\mathrm{Vol}_0(\mathsf{K}) = 1$.

For sets $\mathsf{C} \subset \mathbb{R}^n$ and $\mathsf{D} \subset \mathbb{R}^m$, we define the *orthogonal* direct product

$$\mathsf{C} \times \mathsf{D} := \{(x, y) \in \mathbb{R}^{n+m} : x \in \mathsf{C} \text{ and } y \in \mathsf{D}\}.$$

To be precise, the concatenation $(x, y) \in \mathbb{R}^{n+m}$ places $x \in \mathbb{R}^n$ in the first n coordinates and $y \in \mathbb{R}^m$ in the remaining $(n-m)$ coordinates. In particular, $\mathsf{K} \times \{\mathbf{0}_m\}$ is the natural embedding of K into $\mathbb{R}^{n+m}$.

Several convex bodies merit special notation. The unit-volume cube is the set $\mathsf{Q}_n := [0, 1]^n \subset \mathbb{R}^n$. We write $\mathsf{B}_n := \{x \in \mathbb{R}^n : \|x\| \leq 1\}$ for the Euclidean unit ball. The volume κ_n and the surface area ω_n of the Euclidean ball are given by the formulas

$$\kappa_n := \mathrm{Vol}_n(\mathsf{B}_n) = \frac{\pi^{n/2}}{\Gamma(1 + n/2)} \quad \text{and} \quad \omega_n := n\kappa_n = \frac{2\pi^{n/2}}{\Gamma(n/2)}. \tag{6.1.1}$$

As usual, Γ denotes the gamma function.

6.1.2 The Intrinsic Volumes

In this section, we introduce the intrinsic volumes, their properties, and connections to other geometric functionals. A good reference for this material is [32]. Intrinsic volumes are basic tools in stochastic and integral geometry [33], and they appear in the study of random fields [2].

We begin with a geometrically intuitive definition.

Definition 6.1.1 (Intrinsic Volumes) For each index $j = 0, 1, 2, \ldots, n$, let $P_j \in \mathbb{R}^{n \times n}$ be the orthogonal projector onto a fixed j-dimensional subspace of $\mathbb{R}^n$. Draw a rotation matrix $Q \in \mathbb{R}^{n \times n}$ uniformly at random (from the Haar measure on the compact, homogeneous group of $n \times n$ orthogonal matrices with determinant one). The *intrinsic volumes* of the nonempty convex body $K \subset \mathbb{R}^n$ are the quantities

$$V_j(K) := \binom{n}{j} \frac{\kappa_n}{\kappa_j \kappa_{n-j}} \mathbb{E}_Q \left[\mathrm{Vol}_j(P_j Q K) \right]. \tag{6.1.2}$$

We write $\mathbb{E}$ for expectation and $\mathbb{E}_X$ for expectation with respect to a specific random variable X. The intrinsic volumes of the empty set are identically zero: $V_j(\emptyset) = 0$ for each index j.

Up to scaling, the jth intrinsic volume is the average volume of a projection of the convex body onto a j-dimensional subspace, chosen uniformly at random. Following Federer [11], we have chosen the normalization in (6.1.2) to remove the dependence on the dimension in which the convex body is embedded. McMullen [25] introduced the term "intrinsic volumes". In her work, Chevet [10] called V_j the j-*ième épaisseur* or the "jth thickness".

Example 6.1.2 (The Euclidean Ball) We can easily calculate the intrinsic volumes of the Euclidean unit ball because each projection is simply a Euclidean unit ball of lower dimension. Thus,

$$V_j(\mathsf{B}_n) = \binom{n}{j} \frac{\kappa_n}{\kappa_{n-j}} \quad \text{for } j = 0, 1, 2, \ldots, n.$$

Example 6.1.3 (The Cube) We can also determine the intrinsic volumes of a cube:

$$V_j(\mathsf{Q}_n) = \binom{n}{j} \quad \text{for } j = 0, 1, 2, \ldots, n.$$

See Sect. 6.5 for the details of the calculation. A classic reference is [30, pp. 224–227].

6.1.2.1 Geometric Functionals

The intrinsic volumes are closely related to familiar geometric functionals. The intrinsic volume V_0 is called the *Euler characteristic*; it takes the value zero for the empty set and the value one for each nonempty convex body. The intrinsic volume V_1 is proportional to the *mean width*, scaled so that $V_1([0,1] \times \{0_{n-1}\}) = 1$. Meanwhile, V_{n-1} is half the surface area, and V_n coincides with the ordinary volume measure, Vol_n.

6.1.2.2 Properties

The intrinsic volumes satisfy many important properties. Let $\mathsf{C}, \mathsf{K} \subset \mathbb{R}^n$ be nonempty convex bodies. For each index $j = 0, 1, 2, \ldots, n$, the intrinsic volume V_j is...

1. **Nonnegative:** $V_j(\mathsf{K}) \geq 0$.
2. **Monotone:** $\mathsf{C} \subset \mathsf{K}$ implies $V_j(\mathsf{C}) \leq V_j(\mathsf{K})$.
3. **Homogeneous:** $V_j(\lambda \mathsf{K}) = \lambda^j V_j(\mathsf{K})$ for each $\lambda \geq 0$.
4. **Invariant:** $V_j(T\mathsf{K}) = V_j(\mathsf{K})$ for each *proper rigid motion T*. That is, T acts by rotation and translation.
5. **Intrinsic:** $V_j(\mathsf{K}) = V_j(\mathsf{K} \times \{0_m\})$ for each natural number m.
6. **A Valuation:** $V_j(\emptyset) = 0$. If $\mathsf{C} \cup \mathsf{K}$ is also a convex body, then

$$V_j(\mathsf{C} \cap \mathsf{K}) + V_j(\mathsf{C} \cup \mathsf{K}) = V_j(\mathsf{C}) + V_j(\mathsf{K}).$$

7. **Continuous:** If $\mathsf{K}_m \to \mathsf{K}$ in the Hausdorff metric, then $V_j(\mathsf{K}_m) \to V_j(\mathsf{K})$.

With sufficient energy, one may derive all of these facts directly from Definition 6.1.1. See the books [14, 20, 30, 32, 33] for further information about intrinsic volumes and related matters.

6.1.2.3 Hadwiger's Characterization Theorems

Hadwiger [15–17] proved several wonderful theorems that characterize the intrinsic volumes. To state these results, we need a short definition. A valuation F on $\mathbb{R}^n$ is *simple* if $F(\mathsf{K}) = 0$ whenever $\dim \mathsf{K} < n$.

Fact 6.1.4 (Uniqueness of Volume) *Suppose that F is a simple, invariant, continuous valuation on convex bodies in $\mathbb{R}^n$. Then F is a scalar multiple of the intrinsic volume V_n.*

Fact 6.1.5 (The Basis of Intrinsic Volumes) *Suppose that F is an invariant, continuous valuation on convex bodies in $\mathbb{R}^n$. Then F is a linear combination of the intrinsic volumes $V_0, V_1, V_2, \ldots, V_n$.*

Together, these theorems demonstrate the fundamental importance of intrinsic volumes in convex geometry. They also construct a bridge to the field of integral geometry, which provides explicit formulas for geometric functionals defined by integrating over geometric groups (e.g., the family of proper rigid motions).

6.1.2.4 Quermassintegrals

With a different normalization, the mean projection volume appearing in (6.1.2) is also known as a *quermassintegral*. The relationship between the quermassintegrals and the intrinsic volumes is

$$\binom{n}{j} W_j^{(n)}(\mathsf{K}) := \kappa_j V_{n-j}(\mathsf{K}) \quad \text{for } j = 0, 1, 2, \dots, n.$$

The notation reflects the fact that the quermassintegral $W_j^{(n)}$ depends on the ambient dimension n, while the intrinsic volume does not.

6.1.3 The Intrinsic Volume Random Variable

In view of Example 6.1.3, we see that the intrinsic volume sequence of the cube Q_n is sharply peaked (around index $n/2$). Example 6.1.2 shows that intrinsic volumes of the Euclidean ball B_n drop off quickly (starting around index $\sqrt{2\pi n}$). This observation motivates us to ask whether the intrinsic volumes of a general convex body also exhibit some type of concentration.

It is natural to apply probabilistic methods to address this question. To that end, we first need to normalize the intrinsic volumes to construct a probability distribution.

Definition 6.1.6 (Normalized Intrinsic Volumes) The *total intrinsic volume* of the convex body K, also known as the *Wills functional* [18, 25, 37], is the quantity

$$W(\mathsf{K}) := \sum_{j=0}^{n} V_j(\mathsf{K}). \tag{6.1.3}$$

The *normalized intrinsic volumes* compose the sequence

$$\tilde{V}_j(\mathsf{K}) := \frac{V_j(\mathsf{K})}{W(\mathsf{K})} \quad \text{for } j = 0, 1, 2, \dots, n.$$

In particular, the sequence $\{\tilde{V}_j(\mathsf{K}) : j = 0, 1, 2, \dots, n\}$ forms a probability distribution.

In spite of the similarity of notation, the total intrinsic volume W should *not* be confused with a quermassintegral.

We may now construct a random variable that reflects the distribution of the intrinsic volumes of a convex body.

Definition 6.1.7 (Intrinsic Volume Random Variable) The *intrinsic volume random variable* Z_K associated with a convex body K takes nonnegative integer values according to the distribution

$$\mathbb{P}\{Z_K = j\} = \tilde{V}_j(K) \quad \text{for } j = 0, 1, 2, \ldots, n. \tag{6.1.4}$$

The mean of the intrinsic volume random variable plays a special role in the analysis, so we exalt it with its own name and notation.

Definition 6.1.8 (Central Intrinsic Volume) The *central intrinsic volume* of the convex body K is the quantity

$$\Delta(K) := \mathbb{E}\, Z_K = \sum_{j=0}^{n} j \cdot \tilde{V}_j(K). \tag{6.1.5}$$

Equivalently, the central intrinsic volume is the centroid of the sequence of intrinsic volumes.

Since the intrinsic volume sequence of a convex body $K \subset \mathbb{R}^n$ is supported on $\{0, 1, 2, \ldots, n\}$, it is immediate that the central intrinsic volume satisfies $\Delta(K) \in [0, n]$. The extreme n is unattainable (because a nonempty convex body has Euler characteristic $V_0(K) = 1$). But it is easy to construct examples that achieve values across the rest of the range.

Example 6.1.9 (The Scaled Cube) Fix $s \in [0, \infty)$. Using Example 6.1.3 and the homogeneity of intrinsic volumes, we see that total intrinsic volume of the scaled cube is

$$W(sQ_n) = \sum_{j=0}^{n} \binom{n}{j} \cdot s^j = (1+s)^n.$$

The central intrinsic volume of the scaled cube is

$$\Delta(sQ_n) = \frac{1}{(1+s)^n} \sum_{j=0}^{n} j \cdot \binom{n}{j} \cdot s^j = \sum_{j=0}^{n} j \cdot \binom{n}{j} \cdot \left(\frac{s}{1+s}\right)^j \left(1 - \frac{s}{1+s}\right)^{n-j}$$

$$= \frac{ns}{1+s}.$$

We recognize the mean of the random variable $\text{BIN}(s/(1+s), n)$ to reach the last identity. Note that the quantity $\Delta(sQ_n) = ns/(1+s)$ sweeps through the interval $[0, n)$ as we vary $s \in [0, \infty)$.

Example 6.1.10 (Large Sets) More generally, we can compute the limits of the normalized intrinsic volumes of a growing set:

$$\lim_{s\to\infty} \tilde{V}_j(sK) \to 0 \quad \text{for } j < \dim K;$$

$$\lim_{s\to\infty} \tilde{V}_j(sK) \to 1 \quad \text{for } j = \dim K.$$

This point follows from the homogeneity of intrinsic volumes, noted in Sect. 6.1.2.2.

6.1.4 Concentration of Intrinsic Volumes

Our main result states that the intrinsic volume random variable concentrates sharply around the central intrinsic volume.

Theorem 6.1.11 (Concentration of Intrinsic Volumes) *Let* $K \subset \mathbb{R}^n$ *be a nonempty convex body with intrinsic volume random variable* Z_K. *The variance satisfies*

$$\mathrm{Var}[Z_K] \leq 4n.$$

Furthermore, in the range $0 \leq t \leq \sqrt{n}$, *we have the tail inequality*

$$\mathbb{P}\{|Z_K - \mathbb{E}\, Z_K| \geq t\sqrt{n}\} \leq 2e^{-3t^2/28}.$$

To prove this theorem, we first convert questions about the intrinsic volume random variable into questions about metric geometry (Sect. 6.2). We reinterpret the metric geometry formulations in terms of the information content of a log-concave probability density. Then we can control the variance (Sect. 6.3) and concentration properties (Sect. 6.4) of the intrinsic volume random variable using the analogous results for the information content random variable.

A general probability distribution on $\{0, 1, 2, \ldots, n\}$ can have variance higher than $n^2/3$. In contrast, the intrinsic volume random variable has variance no greater than $4n$. Moreover, the intrinsic volume random variable behaves, at worst, like a normal random variable with mean $\mathbb{E}\, Z_K$ and variance less than $5n$. Thus, most of the mass of the intrinsic volume sequence is concentrated on an interval of about $O(\sqrt{n})$ indices.

Looking back to Example 6.1.3, concerning the unit-volume cube Q_n, we see that Theorem 6.1.11 gives a qualitatively accurate description of the intrinsic volume sequence. On the other hand, the bounds for scaled cubes sQ_n can be quite poor; see Sect. 6.5.3.

6.1.5 Concentration of Conic Intrinsic Volumes

Theorem 6.1.11 and its proof parallel recent developments in the theory of *conic* intrinsic volumes, which appear in the papers [3, 13, 22]. Using the concentration of conic intrinsic volumes, we were able to establish that random configurations of convex cones exhibit striking phase transitions; these facts have applications in signal processing [3, 21, 23, 24]. We are confident that extending the ideas in the current paper will help us discover new phase transition phenomena in Euclidean integral geometry.

6.1.6 Maximum-Entropy Convex Bodies

The probabilistic approach to the intrinsic volume sequence suggests other questions to investigate. For instance, we can study the entropy of the intrinsic volume random variable, which reflects the dispersion of the intrinsic volume sequence.

Definition 6.1.12 (Intrinsic Entropy) Let $K \subset \mathbb{R}^n$ be a nonempty convex body. The intrinsic entropy of K is the entropy of the intrinsic volume random variable Z_K:

$$\mathrm{IntEnt}(K) := \mathrm{Ent}[Z_K] = - \sum_{j=0}^{n} \tilde{V}_j(K) \cdot \log \tilde{V}_j(K).$$

We have the following extremal result.

Theorem 6.1.13 (Cubes Have Maximum Entropy) *Fix the ambient space $\mathbb{R}^n$, and let $d \in [0, n)$. There is a scaled cube whose central intrinsic volume equals d:*

$$\Delta(s_{d,n} Q_n) = d \quad when \quad s_{d,n} = \frac{d}{n-d}.$$

Among convex bodies with central intrinsic volume d, the scaled cube $s_{d,n} Q_n$ has the maximum intrinsic entropy. Among all convex bodies, the unit-volume cube has the maximum intrinsic entropy. In symbols,

$$\max\{\mathrm{IntEnt}(K) : \Delta(K) = d\} = \mathrm{IntEnt}(s_{d,n} Q_n) \leq \mathrm{IntEnt}(Q_n).$$

The maximum takes place over all nonempty convex bodies $K \subset \mathbb{R}^n$.

The proof of Theorem 6.1.13 also depends on recent results from information theory, as well as some deep properties of the intrinsic volume sequence. This analysis appears in Sect. 6.6.

Theorem 6.1.13 joins a long procession of results on the extremal properties of the cube. In particular, the cube solves the (affine) reverse isoperimetric problem for symmetric convex bodies [5]. That is, every symmetric convex body $K \subset \mathbb{R}^n$ has an affine image whose volume is one and whose surface area is not greater than $2n$, the surface area of Q_n. See Sect. 6.1.7.2 for an equivalent statement.

Remark 6.1.14 (Minimum Entropy) The convex body consisting of a single point $x_0 \in \mathbb{R}^n$ has the minimum intrinsic entropy: $\mathrm{IntEnt}(\{x_0\}) = 0$. Very large convex bodies also have negligible entropy:

$$\lim_{s \to \infty} \mathrm{IntEnt}(sK) = 0 \quad \text{for each nonempty convex body } K \subset \mathbb{R}^n.$$

The limit is a consequence of Example 6.1.10.

6.1.7 Other Inequalities for Intrinsic Volumes

The classic literature on convex geometry contains a number of prominent inequalities relating the intrinsic volumes, and this topic continues to arouse interest. This section offers a short overview of the main results of this type. Our presentation is influenced by [26, 28]. See [32, Chap. 7] for a comprehensive treatment.

Remark 6.1.15 (Unrelated Work) Although the title of the paper [1] includes the phrase "concentration of intrinsic volumes," the meaning is quite different. Indeed, the focus of that work is to study hyperplane arrangements via the intrinsic volumes of a random sequence associated with the arrangement.

6.1.7.1 Ultra-Log-Concavity

The Alexandrov–Fenchel inequality (AFI) is a profound result on the behavior of mixed volumes; see [32, Sec. 7.3] or [34]. We can specialize the AFI from mixed volumes to the particular case of quermassintegrals. In this instance, the AFI states that the quermassintegrals of a convex body $K \subset \mathbb{R}^n$ compose a log-concave sequence:

$$W_j^{(n)}(K)^2 \geq W_{j+1}^{(n)}(K) \cdot W_{j-1}^{(n)}(K) \quad \text{for } j = 1, 2, 3, \dots, n-1. \tag{6.1.6}$$

As Chevet [10] and McMullen [26] independently observed, the log-concavity (6.1.6) of the quermassintegral sequence implies that the intrinsic volumes form an ultra-log-concave (ULC) sequence:

$$j \cdot V_j(K)^2 \geq (j+1) \cdot V_{j+1}(K) \cdot V_{j-1}(K) \quad \text{for } j = 1, 2, 3, \dots, n-1. \tag{6.1.7}$$

This fact plays a key role in the proof of Theorem 6.1.13. For more information on log-concavity and ultra-log-concavity, see the survey article [31].

From (6.1.7), Chevet and McMullen both deduce that all of the intrinsic volumes are controlled by the first one, and they derive an estimate for the total intrinsic volume:

$$V_j(\mathsf{K}) \le \frac{1}{j!} V_1(\mathsf{K})^j \quad \text{for } j = 1, 2, 3, \ldots, n, \quad \text{hence} \quad W(\mathsf{K}) \le e^{V_1(\mathsf{K})}.$$

This estimate implies some growth and decay properties of the intrinsic volume sequence. An interesting application appears in Vitale's paper [35], which derives concentration for the supremum of a Gaussian process from the foregoing bound on the total intrinsic volume.

It is possible to establish a concentration result for intrinsic volumes as a direct consequence of (6.1.7). Indeed, it is intuitive that a ULC sequence should concentrate around its centroid. This point follows from Caputo et al. [9, Sec. 3.2], which transcribes the usual semigroup proof of a log-Sobolev inequality to the discrete setting. When applied to intrinsic volumes, this method gives concentration on the scale of the mean width $V_1(\mathsf{K})$ of the convex body K. This result captures a phenomenon different from Theorem 6.1.11, where the scale for the concentration is the dimension n.

6.1.7.2 Isoperimetric Ratios

Another classical consequence of the AFI is a sequence of comparisons for the *isoperimetric ratios* of the volume of a convex body $\mathsf{K} \subset \mathbb{R}^n$, relative to the Euclidean ball B_n:

$$\left(\frac{V_n(\mathsf{K})}{V_n(\mathsf{B}_n)} \right)^{1/n} \le \left(\frac{V_{n-1}(\mathsf{K})}{V_{n-1}(\mathsf{B}_n)} \right)^{1/(n-1)} \le \cdots \le \frac{V_1(\mathsf{K})}{V_1(\mathsf{B}_n)}. \tag{6.1.8}$$

The first inequality is the isoperimetric inequality, and the inequality between V_n and V_1 is called Urysohn's inequality [32, Sec. 7.2]. Isoperimetric ratios play a prominent role in asymptotic convex geometry; for example, see [4, 6, 29].

Some of the inequalities in (6.1.8) can be inverted by applying affine transformations. For example, Ball's reverse isoperimetric inequality [5] states that K admits an affine image $\hat{\mathsf{K}}$ for which

$$\left(\frac{V_{n-1}(\hat{\mathsf{K}})}{V_{n-1}(\mathsf{B}_n)} \right)^{1/(n-1)} \le \text{const}_n \cdot \left(\frac{V_n(\hat{\mathsf{K}})}{V_n(\mathsf{B}_n)} \right)^{1/n}.$$

The sharp value for the constant is known; equality holds when K is a simplex. If we restrict our attention to symmetric convex bodies, then the cube is extremal.

The recent paper [28] of Paouris et al. contains a more complete, but less precise, set of reversals. Suppose that K is a symmetric convex body. Then there is a parameter $\beta_\star := \beta_\star(\mathsf{K})$ for which

$$\frac{V_1(\mathsf{K})}{V_1(\mathsf{B}_n)} \leq \left[1 + \text{const} \cdot \left(\beta_\star j \log \left(\frac{\mathrm{e}}{j\beta_\star} \right) \right)^{1/2} \right] \cdot \left(\frac{V_j(\mathsf{K})}{V_j(\mathsf{B}_n)} \right)^{1/j}$$

$$\text{for } j = 1, 2, 3, \ldots, \text{const}/\beta_\star. \tag{6.1.9}$$

The constants here are universal but unspecified. This result implies that the prefix of the sequence of isoperimetric ratios is roughly constant. The result (6.1.9) leaves open the question about the behavior of the sequence beyond the distinguished point.

It would be interesting to reconcile the work of Paouris et al. [28] with Theorem 6.1.11. In particular, it is unclear whether the isoperimetric ratios remain constant, or whether they exhibit some type of phase transition. We believe that our techniques have implications for this question.

6.2 Steiner's Formula and Distance Integrals

The first step in our program is to convert questions about the intrinsic volume random variable into questions in metric geometry. We can accomplish this goal using Steiner's formula, which links the intrinsic volumes of a convex body to its expansion properties. We reinterpret Steiner's formula as a distance integral, and we use this result to compute moments of the intrinsic volume random variable. This technique, which appears to be novel, drives our approach.

6.2.1 Steiner's Formula

The Minkowski sum of a nonempty convex body and a Euclidean ball is called a *parallel body*. Steiner's formula gives an explicit expansion for the volume of the parallel body in terms of the intrinsic volumes of the convex body.

Fact 6.2.1 (Steiner's Formula) *Let* $\mathsf{K} \subset \mathbb{R}^n$ *be a nonempty convex body. For each* $\lambda \geq 0$,

$$\mathrm{Vol}_n(\mathsf{K} + \lambda \mathsf{B}_n) = \sum_{j=0}^{n} \lambda^{n-j} \kappa_{n-j} V_j(\mathsf{K}).$$

In other words, the volume of the parallel body is a *polynomial* function of the expansion radius. Moreover, the coefficients depend only on the intrinsic volumes of the convex body. The proof of Fact 6.2.1 is fairly easy; see [14, 32].

Remark 6.2.2 (Steiner and Kubota) Steiner's formula can be used to *define* the intrinsic volumes. The definition we have given in (6.1.2) is usually called *Kubota's formula*; it can be derived as a consequence of Fact 6.2.1 and Cauchy's formula for surface area. For example, see [4, Sec. B.5].

6.2.2   Distance Integrals

The parallel body can also be expressed as the set of points within a fixed distance of the convex body. This observation motivates us to introduce the distance to a convex set.

Definition 6.2.3 (Distance to a Convex Body)   The distance to a nonempty convex body K is the function

$$\text{dist}(x, K) := \min \left\{ \|y - x\| : y \in K \right\} \quad \text{where } x \in \mathbb{R}^n.$$

It is not hard to show that the distance, $\text{dist}(\cdot, K)$, and its square, $\text{dist}^2(\cdot, K)$, are both convex functions.

Here is an alternative statement of Steiner's formula in terms of distance integrals [18].

Proposition 6.2.4 (Distance Integrals)   *Let $K \subset \mathbb{R}^n$ be a nonempty convex body. Let $f : \mathbb{R}_+ \to \mathbb{R}$ be an absolutely integrable function. Provided that the integrals on the right-hand side converge,*

$$\int_{\mathbb{R}^n} f(\text{dist}(x, K)) \, dx = f(0) \cdot V_n(K) + \sum_{j=0}^{n-1} \left(\omega_{n-j} \int_0^\infty f(r) \cdot r^{n-j-1} \, dr \right) \cdot V_j(K).$$

This result is equivalent to Fact 6.2.1.

Proof   For $r > 0$, Steiner's formula gives an expression for the volume of the locus of points within distance r of the convex body:

$$\text{Vol}_n \{ x \in \mathbb{R}^n : \text{dist}(x, K) \leq r \} = \sum_{j=0}^{n} r^{n-j} \kappa_{n-j} V_j(K).$$

The rate of change in this volume satisfies

$$\frac{d}{dr} \text{Vol}_n \{ x \in \mathbb{R}^n : \text{dist}(x, K) \leq r \} = \sum_{j=0}^{n-1} r^{n-j-1} \omega_{n-j} V_j(K). \tag{6.2.1}$$

We have used the relation (6.1.1) that $\omega_{n-j} = (n - j) \kappa_{n-j}$.

Let $\mu_\sharp$ be the push-forward of the Lebesgue measure on $\mathbb{R}^n$ to $\mathbb{R}_+$ by the function $\operatorname{dist}(\cdot; \mathsf{K})$. That is,

$$\mu_\sharp(\mathsf{A}) := \operatorname{Vol}_n\{x \in \mathbb{R}^n : \operatorname{dist}(x; \mathsf{K}) \in \mathsf{A}\} \quad \text{for each Borel set } \mathsf{A} \subset \mathbb{R}_+.$$

This measure clearly satisfies $\mu_\sharp(\{0\}) = V_n(\mathsf{K})$. Beyond that, when $0 < a < b$,

$$\mu_\sharp((a, b]) = \operatorname{Vol}_n\{x \in \mathbb{R}^n : a < \operatorname{dist}(x; \mathsf{K}) \le b\}$$
$$= \operatorname{Vol}_n\{x \in \mathbb{R}^n : \operatorname{dist}(x; \mathsf{K}) \le b\} - \operatorname{Vol}_n\{x \in \mathbb{R}^n : \operatorname{dist}(x; \mathsf{K}) \le a\}$$
$$= \int_a^b \frac{\mathrm{d}}{\mathrm{d}r} \operatorname{Vol}_n\{x \in \mathbb{R}^n : \operatorname{dist}(x; \mathsf{K}) \le r\} \, \mathrm{d}r.$$

Therefore, by definition of the push-forward,

$$\int_{\mathbb{R}^n} f(\operatorname{dist}(x; \mathsf{K})) \, \mathrm{d}x = \int_{\mathbb{R}_+} f(r) \, \mathrm{d}\mu_\sharp(r)$$
$$= f(0) \cdot V_n(\mathsf{K})$$
$$+ \int_0^\infty f(r) \cdot \frac{\mathrm{d}}{\mathrm{d}r} \operatorname{Vol}_n\{x \in \mathbb{R}^n : \operatorname{dist}(x; \mathsf{K}) \le r\} \, \mathrm{d}r.$$

Introduce (6.2.1) into the last display to arrive at the result. □

6.2.3 Moments of the Intrinsic Volume Sequence

We can compute moments (i.e., linear functionals) of the sequence of intrinsic volumes by varying the function f in Proposition 6.2.4. To that end, it is helpful to make another change of variables.

Corollary 6.2.5 (Distance Integrals II) *Let $\mathsf{K} \subset \mathbb{R}^n$ be a nonempty convex body. Let $g : \mathbb{R}_+ \to \mathbb{R}$ be an absolutely integrable function. Provided the integrals on the right-hand side converge,*

$$\int_{\mathbb{R}^n} g(\pi \operatorname{dist}^2(x, \mathsf{K})) \cdot \mathrm{e}^{-\pi \operatorname{dist}^2(x, \mathsf{K})} \, \mathrm{d}x$$

$$= g(0) \cdot V_n(\mathsf{K}) + \sum_{j=0}^{n-1} \left(\frac{1}{\Gamma((n-j)/2)} \int_0^\infty g(r) \cdot r^{-1+(n-j)/2} \mathrm{e}^{-r} \, \mathrm{d}r \right) \cdot V_j(\mathsf{K}).$$

Proof Set $f(r) = g(\pi r^2) \cdot \mathrm{e}^{-\pi r^2}$ in Proposition 6.2.4 and invoke (6.1.1). □

We are now prepared to compute some specific moments of the intrinsic volume sequence by making special choices of g in Corollary 6.2.5.

Example 6.2.6 (Total Intrinsic Volume) Consider the case where $g(r) = 1$. We obtain the appealing formula

$$\int_{\mathbb{R}^n} e^{-\pi \operatorname{dist}^2(x, K)} \, dx = \sum_{j=0}^{n} V_j(K) = W(K).$$

The total intrinsic volume $W(K)$ was defined in (6.1.3). This identity appears in [18, 25].

Example 6.2.7 (Central Intrinsic Volume) The choice $g(r) = 2r/W(K)$ yields

$$\frac{1}{W(K)} \int_{\mathbb{R}^n} 2\pi \operatorname{dist}^2(x, K) \cdot e^{-\pi \operatorname{dist}^2(x, K)} \, dx = \frac{1}{W(K)} \sum_{j=0}^{n} (n-j) \cdot V_j(K) = n - \mathbb{E} \, Z_K.$$

We have recognized the total intrinsic volume (6.1.3) and the central intrinsic volume (6.1.5).

Example 6.2.8 (Generating Functions) We can also develop an expression for the generating function of the intrinsic volume sequence by selecting $g(r) = e^{(1-\lambda^2)r}$. Thus,

$$\int_{\mathbb{R}^n} e^{-\lambda^2 \pi \operatorname{dist}^2(x, K)} \, dx = \lambda^{-n} \sum_{j=0}^{n} \lambda^j V_j(K). \tag{6.2.2}$$

This expression is valid for all $\lambda > 0$. See [18] or [33, Lem. 14.2.1].

We can reframe the relation (6.2.2) in terms of the moment generating function of the intrinsic volume random variable Z_K. To do so, we make the change of variables $\lambda = e^\theta$ and divide by the total intrinsic volume $W(K)$:

$$\mathbb{E} \, e^{\theta(Z_K - n)} = \frac{1}{W(K)} \int_{\mathbb{R}^n} e^{-e^{2\theta} \pi \operatorname{dist}^2(x, K)} \, dx. \tag{6.2.3}$$

This expression remains valid for all $\theta \in \mathbb{R}$.

Remark 6.2.9 (Other Moments) In fact, we can compute *any* moment of the intrinsic volume sequence by selecting an appropriate function f in Proposition 6.2.4. Corollary 6.2.5 is designed to produce gamma integrals. Beta integrals also arise naturally and lead to other striking relations. For instance,

$$\int_{\mathbb{R}^n} \frac{dx}{(1 + \lambda \operatorname{dist}(x, K))^{n+1}} = \kappa_n \lambda^{-n} \sum_{j=0}^{n} \lambda^j \frac{V_j(K)}{V_j(B_n)} \quad \text{for } \lambda > 0.$$

The intrinsic volumes of the Euclidean ball are computed in Example 6.1.2. Isoperimetric ratios appear naturally in convex geometry (see Sect. 6.1.7.2), so this type of result may have independent interest.

6.3 Variance of the Intrinsic Volume Random Variable

Let us embark on our study of the intrinsic volume random variable. The main result of this section states that the variance of the intrinsic volume random variable is significantly smaller than its range. This is a more precise version of the variance bound in Theorem 6.1.11.

Theorem 6.3.1 (Variance of the Intrinsic Volume Random Variable) *Let* $\mathsf{K} \subset \mathbb{R}^n$ *be a nonempty convex body with intrinsic volume random variable* Z_K. *We have the inequalities*

$$\mathrm{Var}[Z_\mathsf{K}] \leq 2(n + \mathbb{E}\, Z_\mathsf{K}) \leq 4n.$$

The proof of Theorem 6.3.1 occupies the rest of this section. We make a connection between the distance integrals from Sect. 6.2 and the information content of a log-concave probability measure. By using recent results on the variance of information, we can develop bounds for the distance integrals. These results, in turn, yield bounds on the variance of the intrinsic volume random variable. A closely related argument, appearing in Sect. 6.4, produces exponential concentration.

Remark 6.3.2 (An Alternative Argument) Theorem 6.3.1 can be sharpened using variance inequalities for log-concave densities. Indeed, it holds that

$$\mathrm{Var}[Z_\mathsf{K}] \leq 2(n - \mathbb{E}\, Z_\mathsf{K}).$$

To prove this claim, we apply the Brascamp–Lieb inequality [8, Thm. 4.1] to a perturbation of the log-concave density (6.3.4) described below. It is not clear whether similar ideas lead to normal concentration (because the density is not *strongly* log-concave), so we have chosen to omit this development.

6.3.1 The Varentropy of a Log-Concave Distribution

First, we outline some facts from information theory about the information content in a log-concave random variable. Let $\mu : \mathbb{R}^n \to \mathbb{R}_+$ be a log-concave probability density; that is, a probability density that satisfies the inequalities

$$\mu(\tau x + (1 - \tau)y) \geq \mu(x)^\tau \mu(y)^{1-\tau} \quad \text{for } x, y \in \mathbb{R}^n \text{ and } \tau \in [0, 1].$$

We define the *information content* I_μ of a random point drawn from the density μ to be the random variable

$$I_\mu := -\log \mu(y) \quad \text{where} \quad y \sim \mu. \tag{6.3.1}$$

The symbol $\sim$ means "has the distribution." The terminology is motivated by the operational interpretation of the information content of a *discrete* random variable as the number of bits required to represent a random realization using a code with minimal average length [7].

The expected information content $\mathbb{E}\, I_\mu$ is usually known as the *entropy* of the distribution μ. The *varentropy* of the distribution is the variance of information content:

$$\mathrm{VarEnt}[\mu] := \mathrm{Var}[I_\mu] = \mathbb{E}\,(I_\mu - \mathbb{E}\, I_\mu)^2. \tag{6.3.2}$$

Here and elsewhere, nonlinear functions bind before the expectation.

Bobkov and Madiman [7] showed that the varentropy of a log-concave distribution on $\mathbb{R}^n$ is not greater than a constant multiple of n. Other researchers quickly determined the optimal constant. The following result was obtained independently by Nguyen [27] and by Wang [36] in their doctoral dissertations.

Fact 6.3.3 (Varentropy of a Log-Concave Distribution) *Let* $\mu : \mathbb{R}^n \to \mathbb{R}_+$ *be a log-concave probability density. Then*

$$\mathrm{VarEnt}[\mu] \leq n.$$

See Fradelizi et al. [12] for more background and a discussion of this result.

For future reference, note that the varentropy and related quantities exhibit a simple scale invariance. Consider the shifted information content

$$I_{c\mu} := -\log(c\mu(\boldsymbol{y})) \quad \text{where } c > 0 \text{ and } \boldsymbol{y} \sim \mu.$$

It follows from the definition that

$$I_{c\mu} - \mathbb{E}\, I_{c\mu} = I_\mu - \mathbb{E}\, I_\mu \quad \text{for each } c > 0. \tag{6.3.3}$$

In particular, $\mathrm{Var}[I_{c\mu}] = \mathrm{Var}[I_\mu]$.

6.3.2 A Log-Concave Density

Next, we observe that the central intrinsic volume is related to the information content of a log-concave density. For a nonempty convex body $\mathsf{K} \subset \mathbb{R}^n$, define

$$\mu_\mathsf{K}(\boldsymbol{x}) := \frac{1}{W(\mathsf{K})} e^{-\pi\, \mathrm{dist}^2(\boldsymbol{x},\mathsf{K})} \quad \text{for } \boldsymbol{x} \in \mathbb{R}^n. \tag{6.3.4}$$

The density μ_K is log-concave because the squared distance to a convex body is a convex function. The calculation in Example 6.2.6 ensures that μ_K is a probability density.

Introduce the (shifted) information content random variable associated with K:

$$H_K := -\log(W(K) \cdot \mu_K(y)) = \pi \operatorname{dist}^2(y, K) \quad \text{where} \quad y \sim \mu_K. \tag{6.3.5}$$

Up to the presence of the factor $W(K)$, the random variable H_K is the information content of a random draw from the distribution μ_K. In view of (6.3.2) and (6.3.3),

$$\operatorname{Var}[H_K] = \operatorname{Var}[I_{\mu_K}] = \operatorname{VarEnt}[\mu_K]. \tag{6.3.6}$$

More generally, *all* central moments and cumulants of H_K coincide with the corresponding central moments and cumulants of I_{μ_K}:

$$\mathbb{E} f(H_K - \mathbb{E} H_K) = \mathbb{E} f(I_{\mu_K} - \mathbb{E} I_{\mu_K}). \tag{6.3.7}$$

This expression is valid for any function $f : \mathbb{R} \to \mathbb{R}$ such that the expectations exist.

6.3.3 Information Content and Intrinsic Volumes

We are now prepared to connect the moments of the intrinsic volume random variable Z_K with the moments of the information content random variable H_K. These representations allow us to transfer results about information content into data about the intrinsic volumes.

Using the notation from the last section, Example 6.2.7 gives a relation between the expectations:

$$\mathbb{E} Z_K = n - 2 \mathbb{E} H_K. \tag{6.3.8}$$

The next result provides a similar relationship between the variances.

Proposition 6.3.4 (Variance of the Intrinsic Volume Random Variable) *Let* $K \subset \mathbb{R}^n$ *be a nonempty convex body with intrinsic volume random variable* Z_K *and information content random variable* H_K. *We have the variance identity*

$$\operatorname{Var}[Z_K] = 4 \left(\operatorname{Var}[H_K] - \mathbb{E} H_K \right).$$

Proof Apply Corollary 6.2.5 with the function $g(r) = 4r^2/W(\mathsf{K})$ to obtain

$$4 \, \mathbb{E} \, H_\mathsf{K}^2 = \frac{1}{W(\mathsf{K})} \int_{\mathbb{R}^n} 4\pi^2 \, \mathrm{dist}^4(x, \mathsf{K}) \cdot e^{-\pi \, \mathrm{dist}^2(x,\mathsf{K})} \, dx$$

$$= \frac{1}{W(\mathsf{K})} \sum_{j=0}^{n-1} (n-j)((n-j)+2) \cdot V_j(\mathsf{K})$$

$$= \mathbb{E}(n - Z_\mathsf{K})^2 + 2 \, \mathbb{E}[n - Z_\mathsf{K}]$$

$$= \mathrm{Var}[n - Z_\mathsf{K}] + (\mathbb{E}[n - Z_\mathsf{K}])^2 + 2 \, \mathbb{E}[n - Z_\mathsf{K}]$$

$$= \mathrm{Var}[Z_\mathsf{K}] + 4(\mathbb{E} \, H_\mathsf{K})^2 + 4 \, \mathbb{E} \, H_\mathsf{K}.$$

We have used the definition (6.1.4) of the intrinsic volume random variable to express the sum as an expectation. In the last step, we used the relation (6.3.8) twice to pass to the random variable H_K. Finally, rearrange the display to complete the proof. □

6.3.4 Proof of Theorem 6.3.1

We may now establish the main result of this section. Proposition 6.3.4 yields

$$\mathrm{Var}[Z_\mathsf{K}] = 4 \, (\mathrm{Var}[H_\mathsf{K}] - \mathbb{E} \, H_\mathsf{K}) = 4 \, \mathrm{VarEnt}[\mu_\mathsf{K}] - 2(n - \mathbb{E} \, Z_\mathsf{K}) \leq 2n + 2 \, \mathbb{E} \, Z_\mathsf{K} \leq 4n.$$

We have invoked (6.3.6) to replace the variance of H_K with the varentropy and (6.3.8) to replace $\mathbb{E} \, H_\mathsf{K}$ by the central intrinsic volume $\mathbb{E} \, Z_\mathsf{K}$. The inequality is a consequence of Fact 6.3.3, which controls the varentropy of the log-concave density μ_K. We obtain the final bound by noting that $\mathbb{E} \, Z_\mathsf{K} \preceq n$.

Here is an alternative approach to the final bound that highlights the role of the varentropy:

$$\mathrm{Var}[Z_\mathsf{K}] \leq 4 \, \mathrm{Var}[H_\mathsf{K}] = 4 \, \mathrm{VarEnt}[\mu_\mathsf{K}] \leq 4n.$$

The first inequality follows from Proposition 6.3.4, and the second inequality is Fact 6.3.3.

6.4 Concentration of the Intrinsic Volume Random Variable

The square root of the variance of the intrinsic volume random variable Z_K gives the scale for fluctuations about the mean. These fluctuations have size $O(\sqrt{n})$, which is much smaller than the $O(n)$ range of the random variable. This observation motivates us to investigate the concentration properties of Z_K. In this section, we develop a refined version of the tail bound from Theorem 6.1.11.

Theorem 6.4.1 (Tail Bounds for Intrinsic Volumes) *Let* $K \subset \mathbb{R}^n$ *be a nonempty convex body with intrinsic volume random variable* Z_K. *For all* $t \geq 0$, *we have the inequalities*

$$\mathbb{P}\{Z_K - \mathbb{E}\, Z_K \geq t\} \leq \exp\left\{-(n + \mathbb{E}\, Z_K) \cdot \psi^*\left(\frac{t}{n + \mathbb{E}\, Z_K}\right)\right\};$$

$$\mathbb{P}\{Z_K - \mathbb{E}\, Z_K \leq -t\} \leq \exp\left\{-(n + \mathbb{E}\, Z_K) \cdot \psi^*\left(\frac{-t}{n + \mathbb{E}\, Z_K}\right)\right\}.$$

The function $\psi^*(s) := ((1+s)\log(1+s) - s)/2$ *for* $s > -1$.

The proof of this result follows the same pattern as the argument from Theorem 6.3.1. In Sect. 6.4.5, we derive Theorem 6.4.1 as an immediate consequence.

6.4.1 Moment Generating Function of the Information Content

In addition to the variance, one may study other moments of the information content random variable. In particular, bounds for the moment generating function (mgf) of the centered information content lead to exponential tail bounds for the information content. Bobkov and Madiman [7] proved the first result in this direction. More recently, Fradelizi et al. [12] have obtained the optimal bound.

Fact 6.4.2 (Information Content mgf) *Let* $\mu : \mathbb{R}^n \to \mathbb{R}_+$ *be a log-concave probability density. For* $\beta < 1$,

$$\mathbb{E}\, e^{\beta(I_\mu - \mathbb{E}\, I_\mu)} \leq e^{n\varphi(\beta)},$$

where $\varphi(s) := -s - \log(1-s)$ *for* $s < 1$. *The information content random variable* I_μ *is defined in* (6.3.1).

6.4.2 Information Content and Intrinsic Volumes

We extract concentration inequalities for the intrinsic volume random variable Z_K by studying its (centered) exponential moments. Define

$$m_K(\theta) := \mathbb{E}\, e^{\theta(Z_K - \mathbb{E}\, Z_K)} \quad \text{for } \theta \in \mathbb{R}.$$

The first step in the argument is to represent the mgf in terms of the information content random variable H_K defined in (6.3.5).

Proposition 6.4.3 (mgf of Intrinsic Volume Random Variable) *Let $K \subset \mathbb{R}^n$ be a nonempty convex body with intrinsic volume random variable Z_K and information content random variable H_K. For $\theta \in \mathbb{R}$,*

$$m_K(\theta) = e^{-\varphi(\beta)\,\mathbb{E}\,H_K} \cdot \mathbb{E}\,e^{\beta(H_K - \mathbb{E}\,H_K)} \quad where \quad \beta := 1 - e^{2\theta}.$$

The function φ is defined in Fact 6.4.2.

Proof The formula (6.2.3) from Example 6.2.8 yields the identity

$$\mathbb{E}\,e^{\theta(Z_K - n)} = \frac{1}{W(K)} \int_{\mathbb{R}^n} e^{(1-e^{2\theta}) \cdot \pi\,\mathrm{dist}^2(x,K)} \cdot e^{-\pi\,\mathrm{dist}^2(x,K)}\,dx = \mathbb{E}\,e^{(1-e^{2\theta})H_K}.$$

We can transfer this result to obtain another representation for m_K. First, use the identity (6.3.8) to replace $\mathbb{E}\,Z_K$ with $\mathbb{E}\,H_K$. Then invoke the last display to reach

$$\begin{aligned}
m_K(\theta) = \mathbb{E}\,e^{\theta(Z_K - \mathbb{E}\,Z_K)} &= e^{2\theta\,\mathbb{E}\,H_K}\,\mathbb{E}\,e^{\theta(Z_K - n)} \\
&= e^{2\theta\,\mathbb{E}\,H_K}\,\mathbb{E}\,e^{(1-e^{2\theta})H_K} \\
&= e^{(1+2\theta - e^{2\theta})\,\mathbb{E}\,H_K}\,\mathbb{E}\,e^{(1-e^{2\theta})(H_K - \mathbb{E}\,H_K)} \\
&= e^{(\beta + \log(1-\beta))\,\mathbb{E}\,H_K}\,\mathbb{E}\,e^{\beta(H_K - \mathbb{E}\,H_K)}.
\end{aligned}$$

In the last step, we have made the change of variables $\beta = 1 - e^{2\theta}$. Finally, identify the value $-\varphi(\beta)$ in the first exponent. $\qquad\square$

6.4.3 A Bound for the mgf

We are now prepared to bound the mgf m_K. This result will lead directly to concentration of the intrinsic volume random variable.

Proposition 6.4.4 (A Bound for the mgf) *Let $K \subset \mathbb{R}^n$ be a nonempty convex body with intrinsic volume random variable Z_K. For $\theta \in \mathbb{R}$,*

$$m_K(\theta) \le e^{\psi(\theta)(n + \mathbb{E}\,Z_K)},$$

where $\psi(s) := (e^{2s} - 2s - 1)/2$ for $s \in \mathbb{R}$.

Proof For the parameter $\beta = 1 - e^{2\theta}$, Proposition 6.4.3 yields

$$
\begin{aligned}
m_K(\theta) &= e^{-\varphi(\beta)\,\mathbb{E}\,H_K}\,\mathbb{E}\,e^{\beta(H_K - \mathbb{E}\,H_K)} \\
&= e^{-\varphi(\beta)\,\mathbb{E}\,H_K}\,\mathbb{E}\,e^{\beta(I_{\mu_K} - \mathbb{E}\,I_{\mu_K})} \\
&\leq e^{-\varphi(\beta)\,\mathbb{E}\,H_K} \cdot e^{n\varphi(\beta)} \\
&= e^{-\varphi(\beta)(n - \mathbb{E}\,Z_K)/2} \cdot e^{n\varphi(\beta)} = e^{\varphi(\beta)(n + \mathbb{E}\,Z_K)/2}.
\end{aligned}
$$

To reach the second line, we use the equivalence (6.3.7) for the central moments. The inequality is Fact 6.4.2, the mgf bound for the information content I_{μ_K} of the log-concave density μ_K. Afterward, we invoke (6.3.8) to pass from the information content random variable H_K to the intrinsic volume random variable Z_K. The next step is algebraic. The result follows when we return from the variable β to the variable θ, leading to the appearance of the function ψ. $\qquad\square$

6.4.4 Proof of Theorem 6.4.1

The Laplace transform method, combined with the mgf bound from Proposition 6.4.4, produces Bennett-type inequalities for the intrinsic volume random variable. In brief,

$$
\begin{aligned}
\mathbb{P}\{Z_K - \mathbb{E}\,Z_K \geq t\} &\leq \inf_{\theta>0} e^{-\theta t} \cdot m_K(\theta) \\
&\leq \inf_{\theta>0} e^{-\theta t + \psi(\theta)(n + \mathbb{E}\,Z_K)} \\
&= \exp\left\{-(n + \mathbb{E}\,Z_K)\cdot \psi^*\left(\frac{t}{n + \mathbb{E}\,Z_K}\right)\right\}.
\end{aligned}
$$

The Fenchel–Legendre conjugate ψ^* of the function ψ has the explicit form given in the statement of Theorem 6.4.1. The lower tail bound follows from the same argument.

6.4.5 Proof of Theorem 6.1.11

The concentration inequality in the main result, Theorem 6.4.1, follows when we weaken the inequalities obtained in the last section. Comparing derivatives, we can verify that $\psi^*(s) \geq (s^2/4)/(1 + s/3)$ for all $s > -1$. For the interesting range,

$0 \le t \le n$, we have

$$\mathbb{P}\{Z_K - \mathbb{E}\,Z_K \ge t\} \le \exp\left\{\frac{-t^2/4}{n + \mathbb{E}\,Z_K + t/3}\right\};$$

$$\mathbb{P}\{Z_K - \mathbb{E}\,Z_K \le -t\} \le \exp\left\{\frac{-t^2/4}{n + \mathbb{E}\,Z_K - t/3}\right\}.$$

We may combine this pair of inequalities into a single bound:

$$\mathbb{P}\{|Z_K - \mathbb{E}\,Z_K| \ge t\} \le 2\exp\left(\frac{-t^2/4}{n + \mathbb{E}\,Z_K + t/3}\right).$$

Make the estimate $\mathbb{E}\,Z_K \le n$, and bound the denominator using $t \le n$. This completes the argument.

6.5 Example: Rectangular Parallelotopes

In this section, we work out the intrinsic volume sequence of a rectangular parallelotope. This computation involves the generating function of the intrinsic volume sequence. Because of its elegance, we develop this method in more depth than we need to treat the example at hand.

6.5.1 Generating Functions and Intrinsic Volumes

To begin, we collect some useful information about the properties of the generating function of the intrinsic volumes.

Definition 6.5.1 (Intrinsic Volume Generating Function) The *generating function* of the intrinsic volumes of the convex body K is the polynomial

$$G_K(\lambda) := \sum_{j=0}^{n} \lambda^j V_j(K) = W(\lambda K) \quad \text{for } \lambda > 0.$$

We can use the generating function to read off some information about a convex body, including the total intrinsic volume and the central intrinsic volume. This is a standard result [38, Sec. 4.1], so we omit the elementary argument.

Proposition 6.5.2 (Properties of the Generating Function) *For each nonempty convex body* $K \subset \mathbb{R}^n$,

$$W(K) = G_K(1) \quad and \quad \Delta(K) = \frac{G'_K(1)}{G_K(1)} = (\log G_K)'(1).$$

As usual, the prime ' denotes a derivative.

It is usually challenging to compute the intrinsic volumes of a convex body, but the following fact allows us to make short work of some examples.

Fact 6.5.3 (Direct Products) *Let* $C \subset \mathbb{R}^{n_1}$ *and* $K \subset \mathbb{R}^{n_2}$ *be nonempty convex bodies. The generating function of the intrinsic volumes of the convex body* $C \times K \subset \mathbb{R}^{n_1+n_2}$ *takes the form*

$$G_{C\times K}(\lambda) = G_C(\lambda) \cdot G_K(\lambda).$$

For completeness, we include a short proof inspired by Hadwiger [18]; see [33, Lem. 14.2.1].

Proof Abbreviate $n := n_1 + n_2$. For a point $x \in \mathbb{R}^n$, write $x = (x_1, x_2)$ where $x_i \in \mathbb{R}^{n_i}$. Then

$$\mathrm{dist}^2(x, C \times K) = \mathrm{dist}^2(x_1, C) + \mathrm{dist}^2(x_2, K).$$

Invoke the formula (6.2.2) from Example 6.2.8 for the generating function of the intrinsic volumes (three times!). For $\lambda > 0$,

$$\lambda^{-n} \sum_{j=0}^{n} \lambda^j V_j(C \times K) = \int_{\mathbb{R}^n} e^{-\lambda^2 \pi \, \mathrm{dist}^2(x, C\times K)} \, dx$$

$$= \int_{\mathbb{R}^{n_1}} \int_{\mathbb{R}^{n_2}} e^{-\lambda^2 \pi \, \mathrm{dist}^2(x_1, C)} \cdot e^{-\lambda^2 \pi \, \mathrm{dist}^2(x_2, K)} \, dx_1 \, dx_2$$

$$= \left(\lambda^{-n_1} \sum_{j=0}^{n_1} \lambda^j V_j(C) \right) \left(\lambda^{-n_2} \sum_{j=0}^{n_2} \lambda^j V_j(K) \right).$$

Cancel the leading factors of λ to complete the argument.   □

As a corollary, we can derive an expression for the central intrinsic volume of a direct product.

Corollary 6.5.4 (Central Intrinsic Volume of a Product) *Let* $C \subset \mathbb{R}^{n_1}$ *and* $K \subset \mathbb{R}^{n_2}$ *be nonempty convex bodies. Then*

$$\Delta(C \times K) = \Delta(C) + \Delta(K).$$

Proof According to Proposition 6.5.2 and Fact 6.5.3,

$$\Delta(\mathsf{C} \times \mathsf{K}) = (\log G_{\mathsf{C} \times \mathsf{K}})'(1) = (\log(G_{\mathsf{C}} G_{\mathsf{K}}))'(1)$$
$$= (\log G_{\mathsf{C}} + \log G_{\mathsf{K}})'(1) = (\log G_{\mathsf{C}})'(1) + (\log G_{\mathsf{K}})'(1)$$
$$= \Delta(\mathsf{C}) + \Delta(\mathsf{K}).$$

This is what we needed to show.   □

6.5.2 Intrinsic Volumes of a Rectangular Parallelotope

Using Fact 6.5.3, we quickly compute the intrinsic volumes and related statistics for a rectangular parallelotope.

Proposition 6.5.5 (Rectangular Parallelotopes) *For parameters* $s_1, s_2, \ldots, s_n \geq 0$, *construct the rectangular parallelotope*

$$\mathsf{P} := [0, s_1] \times [0, s_2] \times \cdots \times [0, s_n] \subset \mathbb{R}^n.$$

The generating function for the intrinsic volumes of the parallelotope P *satisfies*

$$G_{\mathsf{P}}(\lambda) = \prod_{i=1}^{n} (1 + \lambda s_i).$$

In particular, $V_j(\mathsf{K}) = e_j(s_1, \ldots, s_n)$, *where* e_j *denotes the* j*th elementary symmetric function. The total intrinsic volume and central intrinsic volume satisfy*

$$W(\mathsf{P}) = \prod_{i=1}^{n} (1 + s_i) \quad and \quad \Delta(\mathsf{P}) = \sum_{i=1}^{n} \frac{s_l}{1 + s_i}.$$

Proof Let $s \geq 0$. By direct calculation from Definition 6.1.1, the intrinsic volumes of the interval $[0, s] \subset \mathbb{R}^1$ are $V_0([0, s]) = 1$ and $V_1([0, s]) = s$. Thus,

$$G_{[0,s]}(\lambda) = \sum_{j=0}^{1} \lambda^j V_j([0, s]) = 1 + \lambda s.$$

Fact 6.5.3 implies that the generating function for the intrinsic volumes of the parallelotope P is

$$G_{\mathsf{P}}(\lambda) := \sum_{j=0}^{n} \lambda^j V_j(\mathsf{P}) = \prod_{i=0}^{n} (1 + \lambda s_i).$$

We immediately obtain formulas for the total intrinsic volume and the central intrinsic volume from Proposition 6.5.2. Alternatively, we can compute the central intrinsic volume of an interval $[0, s]$ and use Corollary 6.5.4 to extend this result to the parallelotope P. $\qquad\qquad\square$

6.5.3 Intrinsic Volumes of a Cube

As an immediate consequence of Proposition 6.5.5, we obtain a clean result on the intrinsic volumes of a scaled cube.

Corollary 6.5.6 (Cubes) *Let $\mathbf{Q}_n \subset \mathbb{R}^n$ be the unit cube. For $s \geq 0$, the normalized intrinsic volumes of the scaled cube $s\mathbf{Q}_n$ coincide with a binomial distribution. For each $j = 0, 1, 2, \ldots, n$,*

$$\tilde{V}_j(s\mathbf{Q}_n) = \binom{n}{j} \cdot p^j (1 - p)^{n-j} \quad where \quad p = \frac{s}{1 + s}.$$

In particular, the central intrinsic volume of the scaled cube is

$$\Delta(s\mathbf{Q}_n) = np = \frac{ns}{1 + s}.$$

Corollary 6.5.6 plays a starring role in our analysis of the intrinsic volume sequences that attain the maximum entropy.

We can also use Corollary 6.5.6 to test our results on the variance and concentration properties of the intrinsic volume sequence by comparing them with exact computations for the cube. Fix a number $s \geq 0$, and let $p = s/(1 + s)$. Then

$$\mathrm{Var}[Z_{s\mathbf{Q}_n}] = np(1 - p) = \frac{ns}{(1 + s)^2}.$$

Meanwhile, Theorem 6.3.1 gives the upper bound

$$\mathrm{Var}[Z_{s\mathbf{Q}_n}] \leq 2(n + np) = \frac{2n(1 + 2s)}{1 + s}.$$

For $s = 1$, the ratio of the upper bound to the exact variance is 12. For $s \approx 0$ and $s \to \infty$, the ratio becomes arbitrarily large. Similarly, Theorem 6.4.1 gives a qualitatively good description for $s = 1$, but its predictions are far less accurate for small and large s. There remains more work to do!

6.6 Maximum-Entropy Distributions of Intrinsic Volumes

We have been using probabilistic methods to study the intrinsic volumes of a convex body, and we have seen that the intrinsic volume sequence is concentrated, as reflected in the variance bound (Theorem 6.3.1) and the exponential tail bounds (Theorem 6.4.1). Therefore, it is natural to consider other measures of the dispersion of the sequence. We recall Definition 6.1.12, of the *intrinsic entropy*, which is the entropy of the normalized intrinsic volume sequence. This concept turns out to be interesting.

In this section, we will establish Theorem 6.1.13. This result states that, among all convex bodies with a fixed central intrinsic volume, a scaled cube has the largest entropy. Moreover, the unit-volume cube has the largest intrinsic entropy among all convex bodies in a fixed dimension. We prove this theorem using some recent observations from information theory.

6.6.1 Ultra-Log-Concavity and Convex Bodies

The key step in proving Theorem 6.1.13 is to draw a connection between intrinsic volumes and ultra-log-concave sequences. We begin with an important definition.

Definition 6.6.1 (Ultra-Log-Concave Sequence) A nonnegative sequence $\{a_j : j = 0, 1, 2, \dots\}$ is called *ultra-log-concave*, briefly *ULC*, if it satisfies the relations

$$j \cdot a_j^2 \geq (j+1) \cdot a_{j+1} a_{j-1} \quad \text{for } j = 1, 2, 3, \dots.$$

It is equivalent to say that the sequence $\{j! \, a_j : j = 0, 1, 2, \dots\}$ is log-concave.

Among all finitely supported ULC probability distributions, the binomial distributions have the maximum entropy. This result was obtained by Yaming Yu [39] using methods developed by Oliver Johnson [19] for studying the maximum-entropy properties of Poisson distributions.

Fact 6.6.2 (Binomial Distributions Maximize Entropy) *Let $p \in [0, 1]$, and fix a natural number n. Among all ULC probability distributions with mean pn that are supported on $\{0, 1, 2, \dots, n\}$, the binomial distribution $\mathrm{BIN}(p, n)$ has the maximum entropy.*

These facts are relevant to our discussion because the intrinsic volumes of a convex body form an ultra-log-concave sequence.

Fact 6.6.3 (Intrinsic Volumes are ULC) *The normalized intrinsic volumes of a nonempty convex body in $\mathbb{R}^n$ compose a ULC probability distribution supported on $\{0, 1, 2, \dots, n\}$.*

This statement is a consequence of the Alexandrov–Fenchel inequalities [32, Sec. 7.3]; see the papers of Chevet [10] and McMullen [26].

6.6.2   Proof of Theorem 6.1.13

With this information at hand, we quickly establish the main result of the section. Recall that Q_n denotes the unit-volume cube in $\mathbb{R}^n$. Let $K \subset \mathbb{R}^n$ be a nonempty convex body. Define the number $p \in [0, 1)$ by the relation $pn = \Delta(K)$. According to Corollary 6.5.6, the scaled cube sQ_n satisfies

$$\Delta(sQ_n) = pn = \Delta(K) \quad \text{when} \quad s = \frac{p}{1-p}.$$

Fact 6.6.3 ensures that the normalized intrinsic volume sequence of the convex body K is a ULC probability distribution supported on $\{0, 1, 2, \ldots, n\}$. Since $\mathbb{E} Z_K = \Delta(K) = pn$, Fact 6.6.2 now delivers

$$\text{IntEnt}(K) = \text{Ent}[Z_K] \leq \text{Ent}[\text{BIN}(p, n)] = \text{Ent}[Z_{sQ_n}] = \text{IntEnt}(sQ_n).$$

We have used Corollary 6.5.6 again to see that $Z_{sQ_n} \sim \text{BIN}(p, n)$. The remaining identities are simply the definition of the intrinsic entropy. In other words, the scaled cube has the maximum intrinsic entropy among all convex bodies that share the same central intrinsic volume.

It remains to show that the unit-volume cube has maximum intrinsic entropy among *all* convex bodies. Continuing the analysis in the last display, we find that

$$\text{IntEnt}(K) \leq \text{Ent}[\text{BIN}(p, n)] \leq \text{Ent}[\text{BIN}(1/2, n)] = \text{Ent}[Z_{Q_n}] = \text{IntEnt}(Q_n).$$

Indeed, among the binomial distributions $\text{BIN}(p, n)$ for $p \in [0, 1]$, the maximum entropy distribution is $\text{BIN}(1/2, n)$. But this is the distribution of Z_{Q_n}, the intrinsic volume random variable of the unit cube Q_n. This observation implies the remaining claim in Theorem 6.1.13.

Acknowledgements We are grateful to Emmanuel Milman for directing us to the literature on concentration of information. Dennis Amelunxen, Sergey Bobkov, and Michel Ledoux also gave feedback at an early stage of this project. Ramon Van Handel provided valuable comments and citations, including the fact that ULC sequences concentrate. We thank the anonymous referee for a careful reading and constructive remarks.

Parts of this research were completed at Luxembourg University and at the Institute for Mathematics and its Applications (IMA) at the University of Minnesota. Giovanni Peccati is supported by the internal research project STARS (R-AGR-0502-10) at Luxembourg University. Joel A. Tropp gratefully acknowledges support from ONR award N00014-11-1002 and the Gordon and Betty Moore Foundation.

References

1. K. Adiprasito, R. Sanyal, Whitney numbers of arrangements via measure concentration of intrinsic volumes (2016). http://arXiv.org/abs/1606.09412
2. R.J. Adler, J.E. Taylor, *Random fields and geometry*. Springer Monographs in Mathematics (Springer, New York, 2007)
3. D. Amelunxen, M. Lotz, M.B. McCoy, J.A. Tropp, Living on the edge: phase transitions in convex programs with random data. Inf. Inference **3**(3), 224–294 (2014)
4. S. Artstein-Avidan, A. Giannopoulos, V.D. Milman, *Asymptotic geometric analysis. Part I*. Mathematical Surveys and Monographs, vol. 202 (American Mathematical Society, Providence, 2015)
5. K. Ball, Volume ratios and a reverse isoperimetric inequality. J. Lond. Math. Soc. **44**(2), 351–359 (1991)
6. K. Ball, An elementary introduction to modern convex geometry, in *Flavors of geometry*. Mathematical Sciences Research Institute Publications, vol. 31 (Cambridge Univ. Press, Cambridge, 1997), pp. 1–58
7. S. Bobkov, M. Madiman, Concentration of the information in data with log-concave distributions. Ann. Probab. **39**(4), 1528–1543 (2011)
8. H.J. Brascamp, E.H. Lieb, On extensions of the Brunn-Minkowski and Prékopa-Leindler theorems, including inequalities for log concave functions, and with an application to the diffusion equation. J. Funct. Anal. **22**(4), 366–389 (1976)
9. P. Caputo, P. Dai Pra, G. Posta, Convex entropy decay via the Bochner-Bakry-Emery approach. Ann. Inst. Henri Poincaré Probab. Stat. **45**(3), 734–753 (2009)
10. S. Chevet, Processus Gaussiens et volumes mixtes. Z. Wahrscheinlichkeitstheorie und Verw. Gebiete **36**(1), 47–65 (1976)
11. H. Federer, Curvature measures. Trans. Am. Math. Soc. **93**, 418–491 (1959)
12. M. Fradelizi, M. Madiman, L. Wang, Optimal concentration of information content for log-concave densities, in *High dimensional probability VII*. Progress in Probability, vol. 71 (Springer, Berlin, 2016), pp. 45–60
13. L. Goldstein, I. Nourdin, G. Peccati, Gaussian phase transitions and conic intrinsic volumes: steining the Steiner formula. Ann. Appl. Probab. **27**(1), 1–47 (2017)
14. P.M. Gruber, *Convex and discrete geometry*. Grundlehren der Mathematischen Wissenschaften [Fundamental Principles of Mathematical Sciences], vol. 336 (Springer, Berlin, 2007)
15. H. Hadwiger, Beweis eines Funktionalsatzes für konvexe Körper. Abh. Math. Sem. Univ. Hamburg **17**, 69–76 (1951)
16. H. Hadwiger, Additive Funktionale k-dimensionaler Eikörper. I. Arch. Math. **3**, 470–478 (1952)
17. H. Hadwiger, *Vorlesungen über Inhalt, Oberfläche und Isoperimetrie* (Springer, Berlin, 1957)
18. H. Hadwiger, Das Wills'sche funktional. Monatsh. Math. **79**, 213–221 (1975)
19. O. Johnson, Log-concavity and the maximum entropy property of the Poisson distribution. Stoch. Process. Appl. **117**(6), 791–802 (2007)
20. D.A. Klain, G.-C. Rota, *Introduction to geometric probability*. Lezioni Lincee. [Lincei Lectures] (Cambridge University Press, Cambridge, 1997)
21. M.B. McCoy, *A geometric analysis of convex demixing* (ProQuest LLC, Ann Arbor, 2013). Thesis (Ph.D.)–California Institute of Technology
22. M.B. McCoy, J.A. Tropp, From Steiner formulas for cones to concentration of intrinsic volumes. Discrete Comput. Geom. **51**(4), 926–963 (2014)
23. M.B. McCoy, J.A. Tropp, Sharp recovery bounds for convex demixing, with applications. Found. Comput. Math. **14**(3), 503–567 (2014)
24. M.B. McCoy, J.A. Tropp, The achievable performance of convex demixing. ACM Report 2017-02, California Institute of Technology (2017). Manuscript dated 28 Sep. 2013
25. P. McMullen, Non-linear angle-sum relations for polyhedral cones and polytopes. Math. Proc. Cambridge Philos. Soc. **78**(2), 247–261 (1975)

26. P. McMullen, Inequalities between intrinsic volumes. Monatsh. Math. **111**(1), 47–53 (1991)
27. V.H. Nguyen, Inégalités Fonctionelles et Convexité. Ph.D. Thesis, Université Pierrre et Marie Curie (Paris VI) (2013)
28. G. Paouris, P. Pivovarov, P. Valettas. On a quantitative reversal of Alexandrov's inequality. Trans. Am. Math. Soc. **371**(5), 3309–3324 (2019)
29. G. Pisier, *The volume of convex bodies and Banach space geometry*. Cambridge Tracts in Mathematics, vol. 94 (Cambridge University Press, Cambridge, 1989)
30. L.A. Santaló, *Integral geometry and geometric probability*. Cambridge Mathematical Library, 2nd edn. (Cambridge University Press, Cambridge, 2004). With a foreword by Mark Kac
31. A. Saumard, J.A. Wellner, Log-concavity and strong log-concavity: a review. Stat. Surv. **8**, 45–114 (2014)
32. R. Schneider, *Convex bodies: the Brunn-Minkowski theory*. Encyclopedia of Mathematics and its Applications, vol. 151, expanded edn. (Cambridge University Press, Cambridge, 2014)
33. R. Schneider, W. Weil, *Stochastic and integral geometry*. Probability and its Applications (New York) (Springer, Berlin, 2008)
34. Y. Shenfeld, R. van Handel, Mixed volumes and the Bochner method. Proc. Amer. Math. Soc. **147**, 5385–5402 (2019)
35. R.A. Vitale, The Wills functional and Gaussian processes. Ann. Probab. **24**(4), 2172–2178 (1996)
36. L. Wang, *Heat Capacity Bound, Energy Fluctuations and Convexity* (ProQuest LLC, Ann Arbor, 2014). Thesis (Ph.D.)–Yale University
37. J.M. Wills, Zur Gitterpunktanzahl konvexer Mengen. Elem. Math. **28**, 57–63 (1973)
38. H.S. Wilf, *generatingfunctionology*, 2nd edn. (Academic Press, Boston, 1994)
39. Y. Yu, On the maximum entropy properties of the binomial distribution. IEEE Trans. Inform. Theory **54**(7), 3351–3353 (2008)

Chapter 7
Two Remarks on Generalized Entropy Power Inequalities

Mokshay Madiman, Piotr Nayar, and Tomasz Tkocz

Abstract This note contributes to the understanding of generalized entropy power inequalities. Our main goal is to construct a counter-example regarding monotonicity and entropy comparison of weighted sums of independent identically distributed log-concave random variables. We also present a complex analogue of a recent dependent entropy power inequality of Hao and Jog, and give a very simple proof.

Mokshay Madiman was supported in part by the U.S. National Science Foundation through the grant DMS-1409504. Piotr Nayar was partially supported by the National Science Centre Poland grant 2015/18/A/ST1/00553. The research leading to these results is part of a project that has received funding from the European Research Council (ERC) under the European Union's Horizon 2020 research and innovation programme (grant agreement No 637851). This work was also supported by the NSF under Grant No. 1440140, while the authors were in residence at the Mathematical Sciences Research Institute in Berkeley, California, for the "Geometric and Functional Analysis" program during the fall semester of 2017.

M. Madiman (✉)
University of Delaware, Newark, DE, USA
e-mail: madiman@udel.edu

P. Nayar
University of Warsaw, Warsaw, Poland
e-mail: nayar@mimuw.edu.pl

T. Tkocz
Carnegie Mellon University, Pittsburgh, PA, USA
e-mail: ttkocz@math.cmu.edu

© Springer Nature Switzerland AG 2020
B. Klartag, E. Milman (eds.), *Geometric Aspects of Functional Analysis*,
Lecture Notes in Mathematics 2266,
https://doi.org/10.1007/978-3-030-46762-3_7

169

7.1 Introduction

The differential entropy of a random vector X with density f (with respect to Lebesgue measure on $\mathbb{R}^d$) is defined as

$$h(X) = -\int_{\mathbb{R}^d} f \log f,$$

provided that this integral exists. When the variance of a real-valued random variable X is kept fixed, it is a long known fact [11] that the differential entropy is maximized by taking X to be Gaussian. A related functional is the *entropy power* of X, defined by $N(X) = e^{\frac{2h(X)}{d}}$. As is usual, we abuse notation and write $h(X)$ and $N(X)$, even though these are functionals depending only on the density of X and not on its random realization.

The entropy power inequality is a fundamental inequality in both Information Theory and Probability, stated first by Shannon [34] and proved by Stam [36]. It states that for any two independent random vectors X and Y in $\mathbb{R}^d$ such that the entropies of X, Y and $X + Y$ exist,

$$N(X + Y) \geq N(X) + N(Y).$$

In fact, it holds without even assuming the existence of entropies as long as we set an entropy power to 0 whenever the corresponding entropy does not exist, as noted by Bobkov and Chistyakov [6]. One reason for the importance of this inequality in Probability Theory comes from its close connection to the Central Limit Theorem (see, e.g., [21, 25]). It is also closely related to the Brunn–Minkowski inequality, and thereby to results in Convex Geometry and Geometric Functional Analysis (see, e.g., [7, 31]).

An immediate consequence of the above formulation of the entropy power inequality is its extension to n summands: if $X_1, \ldots, X_n$ are independent random vectors, then $N(X_1 + \cdots + X_n) \geq \sum_{i=1}^n N(X_i)$. Suppose the random vectors X_i are not merely independent but also identically distributed, and that $S_n = \frac{1}{\sqrt{n}} \sum_{i=1}^n X_i$; these are the normalized partial sums that appear in the vanilla version of the Central Limit Theorem. Then one concludes from the entropy power inequality together with the scaling property $N(aX) = a^2 N(X)$ that $N(S_n) \geq N(S_1)$, or equivalently that

$$h(S_n) \geq h(S_1). \tag{7.1}$$

There are several refinements or generalizations of the inequality (7.1) that one may consider. In 2004, Artstein et al. [2] proved (see [13, 26, 35, 38] for simpler proofs and [27, 28] for extensions) that in fact, one has monotonicity of entropy along the Central Limit Theorem, i.e., $h(S_n)$ is a monotonically increasing sequence. If $N(0, 1)$ is the standard normal distribution, Barron [4] had proved much earlier that $h(S_n) \to h(N(0, 1))$ as long as X_1 has mean 0, variance 1, and $h(X_1) > -\infty$. Thus one has the monotone convergence of $h(S_n)$ to the Gaussian entropy, which is the maximum entropy possible under the moment constraints. By standard arguments, the convergence of entropies is equivalent to the relative entropy between the distribution of S_n and the standard Gaussian distribution converging to 0, and this in turn implies not just convergence in distribution but also convergence in total variation. This is the way in which entropy illuminates the Central Limit Theorem.

A different variant of the inequality (7.1) was recently given by Hao and Jog [20], whose paper may be consulted for motivation and proper discussion. A random vector $X = (X_1, \ldots, X_n)$ in $\mathbb{R}^n$ is called *unconditional* if for every choice of signs $\eta_1, \ldots, \eta_n \in \{-1, +1\}$, the vector $(\eta_1 X_1, \ldots, \eta_n X_n)$ has the same distribution as X. Hao and Jog [20] proved that if X is an unconditional random vector in $\mathbb{R}^n$, then $\frac{1}{n} h(X) \leq h\left(\frac{X_1 + \cdots + X_n}{\sqrt{n}}\right)$. If X has independent and identically distributed components instead of being unconditional, this is precisely $h(S_n) \geq h(S_1)$ for real-valued random variables X_i (i.e., in dimension $d = 1$).

The goal of this note is to shed further light on both of these generalized entropy power inequalities. We now explain precisely how we do so.

To motivate our first result, we first recall the notion of Schur-concavity. One vector $a = (a_1, \ldots, a_n)$ in $[0, \infty)^n$ is *majorised* by another one $b = (b_1, \ldots, b_n)$, usually denoted $a \prec b$, if the nonincreasing rearrangements $a_1^* \geq \ldots \geq a_n^*$ and $b_1^* \geq \ldots \geq b_n^*$ of a and b satisfy the inequalities $\sum_{j=1}^{k} a_j^* \leq \sum_{j=1}^{k} b_j^*$ for each $1 \leq k \leq n - 1$ and $\sum_{j=1}^{n} a_j = \sum_{j=1}^{n} b_j$. For instance, any vector a with nonnegative coordinates adding up to 1 is majorised by the vector $(1, 0, \ldots, 0)$ and majorises the vector $(\frac{1}{n}, \frac{1}{n}, \ldots, \frac{1}{n})$. Let $\Phi: \Delta_n \to \mathbb{R}$, where $\Delta_n = \{a \in [0, 1]^n : a_1 + \cdots + a_n = 1\}$ is the standard simplex. We say that Φ is *Schur-concave* if $\Phi(a) \geq \Phi(b)$ when $a \prec b$. Clearly, if Φ is Schur-concave, then one has $\Phi(\frac{1}{n}, \frac{1}{n}, \ldots, \frac{1}{n}) \geq \Phi(a) \geq \Phi(1, 0, \ldots, 0)$ for any $a \in \Delta_n$.

Suppose $X_1, \ldots, X_n$ are i.i.d. copies of a random variable X with finite entropy, and we define

$$\Phi(a) = h\left(\sum \sqrt{a_i} X_i\right) \tag{7.2}$$

for $a \in \Delta_n$. Then the inequality (7.1) simply says that $\Phi(\frac{1}{n}, \frac{1}{n}, \ldots, \frac{1}{n}) \geq \Phi(1, 0, \ldots, 0)$, while the monotonicity of entropy in the Central Limit Theorem says that $\Phi(\frac{1}{n}, \frac{1}{n}, \ldots, \frac{1}{n}) \geq \Phi(\frac{1}{n-1}, \ldots, \frac{1}{n-1}, 0)$. Both these properties would be implied by (but in themselves are strictly weaker than) Schur-concavity. Thus one is led to the natural question: *Is the function Φ defined in (7.2) a Schur-concave function?* For $n = 2$, this would imply in particular that $h(\sqrt{\lambda} X_1 + \sqrt{1 - \lambda} X_2)$ is maximized over $\lambda \in [0, 1]$ when $\lambda = \frac{1}{2}$. The question on the Schur-concavity of Φ had been floating around for at least a decade, until [3] constructed a counterexample showing that Φ cannot be Schur-concave even for $n = 2$. It was conjectured in [3], however, that for $n = 2$, the Schur-concavity should hold if the random variable X has a log-concave distribution, i.e., if X_1 and X_2 are independent, identically distributed, log-concave random variables, the function $\lambda \mapsto h\left(\sqrt{\lambda}X_1 + \sqrt{1-\lambda}X_2\right)$ should be nondecreasing on $[0, \frac{1}{2}]$. More generally, one may ask: *if $X_1, \ldots, X_n$ are n i.i.d. copies of a log-concave random variable X, is it true that $h\left(\sum a_i X_i\right) \geq h\left(\sum b_i X_i\right)$ when $(a_1^2, \ldots, a_n^2) \prec (b_1^2, \ldots, b_n^2)$?* Equivalently, is Φ Schur-concave when X is log-concave?

Our first result implies that the answer to this question is negative. The way we show this is the following: since $(1, \frac{1}{n}, \ldots, \frac{1}{n}, \frac{1}{n}) \prec (1, \frac{1}{n-1}, \ldots, \frac{1}{n-1}, 0)$, if Schur-concavity held, then the sequence $h\left(X_1 + \frac{X_2 + \cdots + X_{n+1}}{\sqrt{n}}\right)$ would be nondecreasing. If we moreover establish convergence of this sequence to $h(X_1 + G)$, where G is an independent Gaussian random variable with the same variance as X_1, we would have in particular that $h\left(X_1 + \frac{X_2 + \cdots + X_{n+1}}{\sqrt{n}}\right) \leq h(X_1 + G)$. We construct examples where the opposite holds.

Theorem 7.1 *There exists a symmetric log-concave random variable X with variance 1 such that if $X_0, X_1, \ldots$ are its independent copies and n is large enough, we have*

$$h\left(X_0 + \frac{X_1 + \cdots + X_n}{\sqrt{n}}\right) > h(X_0 + Z),$$

where Z is a standard Gaussian random variable, independent of the X_i. Moreover, the left hand side of the above inequality converges to $h(X_0 + Z)$ as n tends to infinity. Consequently, even if X is drawn from a symmetric, log-concave distribution, the function Φ defined in (7.2) is not always Schur-concave.

Here by a *symmetric* distribution, we mean one whose density f satisfies $f(-x) = f(x)$ for each $x \in \mathbb{R}$.

In contrast to Theorem 7.1, Φ does turn out to be Schur-concave if the distribution of X is a symmetric Gaussian mixture, as recently shown in [15]. We suspect that Schur-concavity also holds for uniform distributions on intervals (cf. [1]).

Theorem 7.1 can be compared with the afore-mentioned monotonicity of entropy property of the Central Limit Theorem. It also provides an example of two independent symmetric log-concave random variables X and Y with the same variance such that $h(X + Y) > h(X + Z)$, where Z is a Gaussian random variable with the same variance as X and Y, independent of them, which is again in contrast to symmetric Gaussian mixtures (see [15]). The interesting question posed in [15] of whether, for two i.i.d. summands, swapping one for a Gaussian with the same variance increases entropy, remains open.

Our proof of Theorem 7.1 is based on sophisticated and remarkable Edgeworth type expansions recently developed by Bobkov et al. [9] en route to obtaining precise rates of convergence in the entropic central limit theorem, and is detailed in Sect. 7.2.

The **second contribution** of this note is an exploration of a technique to prove inequalities akin to the entropy power inequality by using symmetries and invariance properties of entropy. It is folklore that when X_1 and X_2 are i.i.d. from a symmetric distribution, one can deduce the inequality $h(S_2) \geq h(S_1)$ in an extremely simple fashion (in contrast to any full proof of the entropy power inequality, which tends to require relatively sophisticated machinery– either going through Fisher information or optimal transport or rearrangement theory or functional inequalities). In Sect. 7.3, we will recall this simple proof, and also deduce some variants of the inequality $h(S_2) \geq h(S_1)$ by playing with this basic idea of using invariance, including a complex analogue of a recent entropy power inequality for dependent random variables obtained by Hao and Jog [20].

Theorem 7.2 *Let $X = (X_1, \ldots, X_n)$ be a random vector in $\mathbb{C}^n$ which is complex-unconditional, that is for every complex numbers $z_1, \ldots, z_n$ such that $|z_j| = 1$ for every j, the vector $(z_1 X_1, \ldots, z_n X_n)$ has the same distribution as X. Then*

$$\frac{1}{n} h(X) \leq h\left(\frac{X_1 + \cdots + X_n}{\sqrt{n}}\right).$$

Our proof of Theorem 7.2, which is essentially trivial thanks to the existence of complex Hadamard matrices, is in contrast to the proof given by Hao and Jog [20] for the real case that proves a Fisher information inequality as an intermediary step.

We make some remarks on complementary results in the literature. Firstly, in contrast to the failure of Schur-concavity of Φ implied by Theorem 7.1, the function $\Xi \colon \Delta_n \to \mathbb{R}$ defined by $\Xi(a) = h\left(\sum a_i X_i\right)$ for i.i.d. copies X_i of a random variable X, is actually Schur-convex when X is log-concave [41]. This is an instance of a reverse entropy power inequality, many more of which are discussed in [31]. Note that the weighted sums that appear in the definition of Φ are relevant to the Central Limit Theorem because they have fixed variance, unlike the weighted sums that appear in the definition of Ξ.

Secondly, motivated by the analogies with Convex Geometry mentioned earlier, one may ask if the function $\Psi \colon \Delta_n \to \mathbb{R}$ defined by $\Psi(a) = \mathrm{vol}_d(\sum_{i=1}^n a_i B)$, is Schur-concave for any Borel set $B \subset \mathbb{R}^d$, where vol_d denotes the Lebesgue measure on $\mathbb{R}^d$ and the notation for summation is overloaded as usual to also denote Minkowski summation of sets. (Note that unless B is convex, $(a_1 + a_2)B$ is a subset of, but generally not equal to, $a_1 B + a_2 B$.) The Brunn–Minkowski inequality implies that $\Psi(\frac{1}{n}, \frac{1}{n}, \ldots, \frac{1}{n}) \geq \Psi(1, 0, \ldots, 0)$. The inequality $\Psi(\frac{1}{n}, \frac{1}{n}, \ldots, \frac{1}{n}) \geq \Psi(\frac{1}{n-1}, \ldots, \frac{1}{n-1}, 0)$, which is the geometric analogue of the monotonicity of entropy in the Central Limit Theorem, was conjectured to hold in [8]. However, it was shown in [16] (cf. [17]) that this inequality fails to hold, and therefore Ψ cannot be Schur-concave, for arbitrary Borel sets B. Note that if B is convex, Ψ is trivially Schur-concave, since it is a constant function equal to $\mathrm{vol}_d(B)$.

Finally, it has recently been observed in [32, 33, 40] that majorization ideas are very useful in understanding entropy power inequalities in discrete settings, such as on the integers or on cyclic groups of prime order.

7.2   Failure of Schur-Concavity

Recall that a probability density f on $\mathbb{R}$ is said to be *log-concave* if it is of the form $f = e^{-V}$ for a convex function $V: \mathbb{R} \rightarrow \mathbb{R} \cup \{\infty\}$. Log-concave distributions emerge naturally from the interplay between information theory and convex geometry, and have recently been a very fruitful and active topic of research (see the recent survey [31]).

This section is devoted to a proof of Theorem 7.1, which in particular falsifies the Schur-concavity of Φ defined by (7.2) even when the distribution under consideration is log-concave.

Let us denote

$$Z_n = \frac{X_1 + \cdots + X_n}{\sqrt{n}}$$

and let p_n be the density of Z_n and let φ be the density of Z. Since X_0 is assumed to be log-concave, it satisfies $\mathbb{E}|X_0|^s < \infty$ for all $s > 0$. According to the Edgeworth-type expansion described in [9, (Theorem 3.2 in Chapter 3)], we have (with any $m \leq s < m + 1$)

$$(1 + |x|^m)(p_n(x) - \varphi_m(x)) = o(n^{-\frac{s-2}{2}}) \qquad \text{uniformly in } x,$$

where

$$\varphi_m(x) = \varphi(x) + \sum_{k=1}^{m-2} q_k(x) n^{-k/2}.$$

Here the functions q_k are given by

$$q_k(x) = \varphi(x) \sum H_{k+2j}(x) \frac{1}{r_1! \ldots r_k!} \left(\frac{\gamma_3}{3!}\right)^{r_1} \cdots \left(\frac{\gamma_{k+2}}{(k+2)!}\right)^{r_k},$$

where H_n are Hermite polynomials,

$$H_n(x) = (-1)^n e^{x^2/2} \frac{d^n}{dx^n} e^{-x^2/2},$$

and the summation runs over all nonnegative integer solutions $(r_1, \ldots, r_k)$ to the equation $r_1 + 2r_2 + \cdots + kr_k = k$, and one uses the notation $j = r_1 + \cdots + r_k$. The numbers γ_k are the cumulants of X_0, namely

$$\gamma_k = i^{-k} \frac{d^k}{dt^k} \log \mathbb{E} e^{it X_0}\big|_{t=0}.$$

Let us calculate φ_4. Under our assumption (symmetry of X_0 and $\mathbb{E}X_0^2 = 1$), we have $\gamma_3 = 0$ and $\gamma_4 = \mathbb{E}X_0^4 - 3$. Therefore $q_1 = 0$ and

$$q_2 = \frac{1}{4!}\gamma_4\varphi H_4 = \frac{1}{4!}\gamma_4\varphi^{(4)}, \qquad \varphi_4 = \varphi + \frac{1}{n}\cdot\frac{1}{4!}(\mathbb{E}X_0^4 - 3)\varphi^{(4)}. \qquad (7.3)$$

We get that for any $\varepsilon \in (0, 1)$

$$(1 + x^4)(p_n(x) - \varphi_4(x)) = o(n^{-\frac{3-\varepsilon}{2}}), \qquad \text{uniformly in } x. \qquad (7.4)$$

Let f be the density of X_0. Let us assume that it is of the form $f = \varphi + \delta$, where δ is even, smooth and compactly supported (say, supported in $[-2, -1] \cup [1, 2]$) with bounded derivatives. Moreover, we assume that $\frac{1}{2}\varphi \leq f \leq 2\varphi$, in particular $|\delta| \leq 1/4$. Multiplying δ by a very small constant we can ensure that f is log-concave.

We are going to use Theorem 1.3 from [10]. To check the assumptions of this theorem, we first observe that for any $\alpha > 1$ we have

$$D_\alpha(Z_1||Z) = \frac{1}{\alpha - 1}\log\left(\int\left(\frac{\varphi + \delta}{\varphi}\right)^\alpha \varphi\right) < \infty,$$

since δ has bounded support. We have to show that for sufficiently big $\alpha^\star = \frac{\alpha}{\alpha-1}$ there is

$$\mathbb{E}e^{tX_0} < e^{\alpha^\star t^2/2}, \qquad t \neq 0.$$

Since X_0 is symmetric, we can assume that $t > 0$. Then

$$\mathbb{E}e^{tX_0} = e^{t^2/2} + \sum_{k=1}^\infty \frac{t^{2k}}{(2k)!}\int x^{2k}\delta(x)dx \leq e^{t^2/2} + \sum_{k=1}^\infty \frac{t^{2k}}{(2k)!}2^{2k}\int_{-2}^2 |\delta(x)|dx$$

$$< e^{t^2/2} + \sum_{k=1}^\infty \frac{(2t)^{2k}}{(2k)!} = 1 + \sum_{k=1}^\infty\left(\frac{t^{2k}}{2^k k!} + \frac{(2t)^{2k}}{(2k)!}\right)$$

$$\leq 1 + \sum_{k=1}^\infty\left(\frac{t^{2k}}{k!} + \frac{(2t)^{2k}}{k!}\right) \leq \sum_{k=0}^\infty \frac{t^{2k}4^{2k}}{k!} = e^{16t^2},$$

where we have used the fact that $\int \delta(x)dx = 0$, δ has a bounded support contained in $[-2, 2]$ and $|\delta| \leq 1/4$. We conclude that

$$|p_n(x) - \varphi(x)| \leq \frac{C_0}{n}e^{-x^2/64} \qquad (7.5)$$

for some constant C_0 independent of n. (In this proof, $C_0, C_1, \ldots$ denote sufficiently large constants that may depend on the distribution of X_0.) Thus

$$p_n(x) \leq \varphi(x) + \frac{C_0}{n} e^{-x^2/64} \leq C_1 e^{-x^2/64}. \tag{7.6}$$

Another consequence of (7.5) is the inequality

$$p_n(x) \geq \frac{1}{10} \qquad \text{for } |x| \leq 1 \tag{7.7}$$

and large enough n.

We now prove the convergence part of the theorem. From (7.5) we get that $p_n \to \varphi$ pointwise. Moreover, from (7.6) and from the inequality $f \leq 2\varphi$ we get, by using Lebesgue's dominated convergence theorem, that $f * p_n \to f * \varphi$. In order to show that $\int f * p_n \log f * p_n \to \int f * \varphi \log f * \varphi$ it is enough to bound $f * p_n |\log f * p_n|$ by some integrable function m_0 independent of n and use Lebesgue's dominated convergence theorem. To this end we observe that by (7.6) we have

$$(f * p_n)(x) \leq 2(\varphi * p_n)(x) \leq \frac{2C_1}{\sqrt{2\pi}} \int e^{-t^2/2} e^{-(x-t)^2/64} dt \leq 2C_1 e^{-x^2/66}. \tag{7.8}$$

Moreover, by (7.7)

$$(f * p_n)(x) \geq \frac{1}{2}(\varphi * p_n)(x) \geq \frac{1}{20} \int_{-1}^{1} \varphi(x - t) dt \geq \frac{1}{10} \varphi(|x| + 1). \tag{7.9}$$

Combining (7.8) with (7.9) we get

$$|\log(f * p_n)(x)| \leq \max \left\{ |\log 2C_1|, \frac{1}{10}|\log \varphi(|x| + 1)| \right\} \leq C_2(1 + x^2). \tag{7.10}$$

From (7.10) and (7.8) we see that the function $m_0(x) = 2C_1 C_2 e^{-x^2/66}(1 + x^2)$ is the required majorant.

Let us define $h_n = p_n - \varphi_4$. Note that by (7.3) we have $\varphi_4 = \varphi + \frac{c_1}{n}\varphi^{(4)}$, where $c_1 = \frac{1}{4!}(\mathbb{E}X_0^4 - 3)$. We have

$$\int f * p_n \log f * p_n = \int \left(f * \varphi + \frac{c_1}{n} f * \varphi^{(4)} + f * h_n \right) \log f * p_n$$

$$= \int f * \varphi \log f * p_n + \frac{c_1}{n} \int f * \varphi^{(4)} \log f * p_n$$

$$+ \int f * h_n \log f * p_n$$

$$= I_1 + I_2 + I_3.$$

We first bound I_3. Note that using (7.4) with $\varepsilon = 1/2$ we get

$$|(f * h_n)(x)| \le 2(\varphi * |h_n|)(x) \le C_3 n^{-5/4} \int e^{-y^2/2} \frac{1}{1 + (x - y)^4} dy \qquad (7.11)$$

for sufficiently large n. Assuming without loss of generality that $x > 0$, we have

$$\int e^{-y^2/2} \frac{1}{1 + (x - y)^4} dy \le \int_{y \in [\frac{1}{2}x, 2x]} e^{-y^2/2} \frac{1}{1 + (x - y)^4} dy$$

$$+ \int_{y \notin [\frac{1}{2}x, 2x]} e^{-y^2/2} \frac{1}{1 + (x - y)^4} dy$$

$$\le \int_{y \in [\frac{1}{2}x, 2x]} e^{-x^2/8} dy$$

$$+ \frac{1}{1 + \frac{1}{16}x^4} \int_{y \notin [\frac{1}{2}x, 2x]} e^{-y^2/2} dy$$

$$\le \frac{3}{2} x e^{-x^2/8} + \frac{\sqrt{2\pi}}{1 + \frac{1}{16}x^4} \le \frac{C_4}{1 + x^4}.$$

Combining this with (7.11) one gets for large n

$$|(f * h_n)(x)| \le C_3 C_4 n^{-5/4} \frac{1}{1 + x^4}. \qquad (7.12)$$

Inequalities (7.12) and (7.10) give for large n,

$$|I_3| \le C_3 C_4 C_2 n^{-5/4} \int \frac{1 + x^2}{1 + x^4} dx \le 5 C_3 C_4 C_2 n^{-5/4}. \qquad (7.13)$$

We now take care of I_2 by showing that

$$I_2 = \frac{c_1}{n} \int f * \varphi^{(4)} \log f * p_n = \frac{c_1}{n} \int f * \varphi^{(4)} \log f * \varphi + o(n^{-1}). \qquad (7.14)$$

To this end it suffices to show that $\int f * \varphi^{(4)} \log f * p_n \to \int f * \varphi^{(4)} \log f * \varphi$. As we already observed $f * p_n \to f * \varphi$ pointwise. Taking into account the bound (7.10), to find a majorant m_1 of $f * \varphi^{(4)} \log f * p_n$, it suffices to observe that

$|\varphi^{(4)}(t)| \le C_5 e^{-t^2/4}$ and thus

$$|f * \varphi^{(4)}|(x) \le 2(\varphi * |\varphi^{(4)}|)(x) \le 2C_5 \int e^{-(x-t)^2/2} e^{-t^2/4} dt \le 8C_5 e^{-x^2/6}.$$

One can then take $m_1(x) = 8C_5 C_2 e^{-x^2/6}(1 + x^2)$.

By Jensen's inequality,

$$I_1 = \int f * \varphi \log f * p_n \le \int f * \varphi \log f * \varphi = -h(X_0 + Z). \qquad (7.15)$$

Putting (7.15), (7.14) and (7.13) together we get

$$\int f * p_n \log f * p_n \le \int f * \varphi \log f * \varphi + \frac{c_1}{n} \int (f * \varphi)^{(4)} \log(f * \varphi) + o(n^{-1}).$$

This is

$$h(X_0 + Z) \le h(X_0 + Z_n) + \frac{1}{n} \cdot \frac{1}{4!} (\mathbb{E} X_0^4 - 3) \int (f * \varphi)^{(4)} \log(f * \varphi) + o(n^{-1}).$$

It is therefore enough to construct X_0 (satisfying all previous conditions) such that

$$(\mathbb{E} X_0^4 - 3) \int (f * \varphi)^{(4)} \log(f * \varphi) < 0.$$

It actually suffices to construct a smooth compactly supported even function g such that $\int g = \int g x^2 = \int g x^4 = 0$ and the function $f = \varphi + \varepsilon g$ satisfies

$$\int (f * \varphi)^{(4)} \log(f * \varphi) > 0$$

for some fixed small ε. We then perturb g a bit to get $\mathbb{E} X_0^4 < 3$ instead of $\mathbb{E} X_0^4 = 3$. This can be done without affecting log-concavity.

Let $\varphi_2(x) = (\varphi * \varphi)(x) = \frac{1}{2\sqrt{\pi}} e^{-x^2/4}$. Note that $\varphi_2^{(4)}(x) = \varphi_2(x)(\frac{3}{4} - \frac{3}{4} x^2 + \frac{1}{16} x^4)$. We have

$$\int (f * \varphi)^{(4)} \log(f * \varphi) = \int (\varphi_2 + \varepsilon \varphi * g)^{(4)} \log(\varphi_2 + \varepsilon \varphi * g)$$

$$= \int (\varphi_2 + \varepsilon \varphi * g)^{(4)} \left(\log(\varphi_2) + \varepsilon \frac{\varphi * g}{\varphi_2} \right.$$

$$\left. - \frac{1}{2} \varepsilon^2 \left(\frac{\varphi * g}{\varphi_2} \right)^2 + r_\varepsilon(x) \right) dx.$$

We shall show that $\int |(\varphi_2 + \varepsilon\varphi * g)^{(4)}||r_\varepsilon| \leq C_8|\varepsilon|^3$. To justify this we first observe that by Taylor's formula with the Lagrange reminder, we have

$$|\log(1+a) - a + a^2/2| \leq \frac{1}{3}\frac{|a|^3}{(1-|a|)^3} \qquad |a| < 1. \qquad (7.16)$$

Due to the fact that g is bounded and compactly supported, we have

$$|\varphi * g|(x) \leq C_6 \int_{-C_6}^{C_6} \varphi(x-t)dt \leq 2C_6^2\varphi((|x|-C_6)_+) \leq 2C_6^2 e^{-(|x|-C_6)_+^2/2}.$$

Thus

$$\frac{|\varphi * g|(x)}{\varphi_2(x)} \leq 4\sqrt{\pi}C_6^2 e^{x^2/4}e^{-(|x|-C_6)_+^2/2} \leq C_7.$$

Using (7.16) with $a = \varepsilon\frac{\varphi * g}{\varphi_2}$ and $|\varepsilon| < \frac{1}{2C_7}$ (in which case $|a| \leq 1/2$) we get

$$|r_\varepsilon(x)| = \left|\log\left(1 + \varepsilon\frac{\varphi * g}{\varphi_2}\right) - \varepsilon\frac{\varphi * g}{\varphi_2} + \frac{1}{2}\varepsilon^2\left(\frac{\varphi * g}{\varphi_2}\right)^2\right| \leq \frac{|\varepsilon|^3}{3}C_7^3\frac{1}{(1-\frac{1}{2})^3}.$$

Thus

$$\int |(\varphi_2 + \varepsilon\varphi * g)^{(4)}||r_\varepsilon| \leq \frac{8}{3}C_7^3|\varepsilon|^3\int\left(|\varphi_2^{(4)}| + \frac{1}{2C_7}\varphi * |g^{(4)}|\right) \leq C_8|\varepsilon|^3.$$

Therefore

$$\int (f * \varphi)^{(4)}\log(f * \varphi) = \int (\varphi_2 + \varepsilon\varphi * g)^{(4)}\left(\log(\varphi_2) + \varepsilon\frac{\varphi * g}{\varphi_2}\right.$$

$$\left. -\frac{1}{2}\varepsilon^2\left(\frac{\varphi * g}{\varphi_2}\right)^2\right) + o(\varepsilon^2).$$

Integrating by parts we see that the leading term in the above equation is

$$\int \varphi_2^{(4)}\log\varphi_2 = \int \varphi_2^{(4)}(x)\log\left(\frac{1}{2\sqrt{\pi}}e^{-x^2/4}\right)dx$$

$$= -\int \varphi_2^{(4)}(x)\left(\log(2\sqrt{\pi}) + \frac{1}{4}x^2\right)dx$$

$$= -\int \varphi_2(x)\left(\log(2\sqrt{\pi}) + \frac{1}{4}x^2\right)^{(4)}dx = 0.$$

The term in front of ε vanishes. Indeed, $\int \varphi_2^{(4)} \frac{\varphi * g}{\varphi_2} = \int (\frac{3}{4} - \frac{3}{4}x^2 + \frac{1}{16}x^4)(\varphi * g)$ which can be seen to vanish after using Fubini's theorem thanks to g being orthogonal to $1, x, \ldots, x^4$. Moreover, $\int (\varphi * g)^{(4)} \log(\varphi_2) = \int (\varphi * g)(\log \frac{1}{2\sqrt{\pi}} - \frac{x^2}{4})^{(4)} = 0$. The term in front of ε^2 is equal to

$$J = \int \frac{(\varphi * g)^{(4)}(\varphi * g)}{\varphi_2} - \frac{1}{2} \int \frac{\varphi_2^{(4)}(\varphi * g)^2}{\varphi_2^2} = J_1 - J_2.$$

The first integral is equal to

$$J_1 = \int \int \int 2\sqrt{\pi} e^{x^2/4} g^{(4)}(s) g(t) \frac{1}{2\pi} e^{-(x-s)^2/2} e^{-(x-t)^2/2} dx\,ds\,dt.$$

Now,

$$\int 2\sqrt{\pi} e^{x^2/4} \frac{1}{2\pi} e^{-(x-s)^2/2} e^{-(x-t)^2/2} dx = \frac{2e^{\frac{1}{6}(-s^2 + 4st - t^2)}}{\sqrt{3}}.$$

Therefore,

$$J_1 = \frac{2}{\sqrt{3}} \int \int e^{\frac{1}{6}(-s^2 + 4st - t^2)} g^{(4)}(s) g(t) ds\,dt.$$

If we integrate the first integral four times by parts we get

$$J_1 = \frac{2}{81\sqrt{3}} \int \int e^{\frac{1}{6}(-s^2 + 4st - t^2)} \Big[27 + s^4 - 8s^3 t - 72t^2$$

$$+ 16t^4 - 8st(-9 + 4t^2) + 6s^2(-3 + 4t^2) \Big] g(s) g(t) ds\,dt$$

Since $\varphi_2^{(4)}/\varphi_2^2 = \frac{\sqrt{\pi}}{8}(12 - 12x^2 + x^4) e^{x^2/4}$, we get

$$J_2 = \int \int \int \frac{\sqrt{\pi}}{16}(12 - 12x^2 + x^4) e^{x^2/4} g(s) g(t) \frac{1}{2\pi} e^{-(x-s)^2/2} e^{-(x-t)^2/2} dx\,ds\,dt.$$

Since

$$\int \frac{\sqrt{\pi}}{16}(12 - 12x^2 + x^4) e^{x^2/4} \frac{1}{2\pi} e^{-(x-s)^2/2} e^{-(x-t)^2/2} dx$$

$$= \frac{1}{81\sqrt{3}} e^{\frac{1}{6}(-s^2 + 4st - t^2)} \Big[27 + (s+t)^2(-18 + (s+t)^2) \Big],$$

we arrive at

$$J_2 = \int \int \frac{1}{81\sqrt{3}} e^{\frac{1}{6}(-s^2+4st-t^2)} \left[27 + (s+t)^2(-18+(s+t)^2)\right] g(s)g(t)\mathrm{d}s\mathrm{d}t.$$

Thus $J = J_1 - J_2$ becomes

$$J = J(g) = \frac{1}{81\sqrt{3}} \int \int e^{\frac{1}{6}(-s^2+4st-t^2)} \Big[27 + s^4 - 20s^3t - 126t^2 + 31t^4$$

$$+ 6s^2(-3+7t^2) + s(180t - 68t^3)\Big] g(s)g(t)\mathrm{d}s\mathrm{d}t.$$

The function

$$g(s) = \left(\frac{7280}{69}|s|^3 - \frac{11025}{23}s^2 + \frac{49000}{69}|s| - \frac{7875}{23}\right) \mathbf{1}_{[1,2]}(|s|)$$

is compactly supported and it satisfies $\int g = \int gx^2 = \int gx^4 = 0$. Numerical computations show that for this g we have $J(g) > 0.003$. However, this function is not smooth. To make it smooth it is enough to consider $g_\varepsilon = g * \frac{1}{\varepsilon}\psi(\cdot/\varepsilon)$ where ψ is smooth, compactly supported and integrates to 1. Then for any $\varepsilon > 0$ the function g_ε is smooth, compactly supported and satisfies $\int g_\varepsilon = \int g_\varepsilon x^2 = \int g_\varepsilon x^4 = 0$. To see this denote for simplicity $h = \frac{1}{\varepsilon}\psi(\cdot/\varepsilon)$ and observe that, e.g.,

$$\int g_\varepsilon(x)x^4\mathrm{d}x = \int g(t)h(s)(s+t)^4\mathrm{d}s\mathrm{d}t$$

$$= \int g(t)h(s)(s^4 + 4s^3t + 6s^2t^2 + 4st^3 + t^4)\mathrm{d}t\mathrm{d}s = 0,$$

since the integral with respect to t vanishes because of the properties of g. Taking $\varepsilon \to 0^+$, the corresponding functional $J(g_\varepsilon)$ converges to $J(g)$ due to the convergence of g_ε to g is L_1 and uniform boundedness of g_ε. As a consequence, for small $\varepsilon > 0$ we have $J(g_\varepsilon) > 0.001$. It suffices to pick one particular ε with this property.

$\square$

7.3 Entropy Power Inequalities Under Symmetries

The heart of the folklore proof of $h(S_2) \geq h(S_1)$ for symmetric distributions (see, e.g., [39]) is that for possibly dependent random variables X_1 and X_2, the $SL(n, \mathbb{R})$-

invariance of differential entropy combined with subadditivity imply that

$$h(X_1, X_2) = h\left(\frac{X_1 + X_2}{\sqrt{2}}, \frac{X_1 - X_2}{\sqrt{2}}\right)$$

$$\leq h\left(\frac{X_1 + X_2}{\sqrt{2}}\right) + h\left(\frac{X_1 - X_2}{\sqrt{2}}\right).$$

If the distribution of (X_1, X_2) is the same as that of $(X_1, -X_2)$, we deduce that

$$h\left(\frac{X_1 + X_2}{\sqrt{2}}\right) \geq \frac{h(X_1, X_2)}{2}. \tag{7.17}$$

If, furthermore, X_1 and X_2 are i.i.d., then $h(X_1, X_2) = 2h(X_1)$, yielding $h(S_2) \geq h(S_1)$. Note that under the i.i.d. assumption, the requirement that the distributions of (X_1, X_2) and $(X_1, -X_2)$ coincide is equivalent to the requirement that X_1 (or X_2) has a symmetric distribution.

Without assuming symmetry but assuming independence, we can use the fact from [23] that $h(X - Y) \leq 3h(X + Y) - h(X) - h(Y)$ for independent random variables X, Y to deduce $\frac{1}{2}[h(X_1) + h(X_2)] \leq h\left(\frac{X_1 + X_2}{\sqrt{2}}\right) + \frac{1}{4}\log 2$. In the i.i.d. case, the improved bound $h(X - Y) \leq 2h(X + Y) - h(X)$ holds [29], which implies $h(X_1) \leq h\left(\frac{X_1 + X_2}{\sqrt{2}}\right) + \frac{1}{6}\log 2$. These bounds are, however, not particularly interesting since they are weaker than the classical entropy power inequality; if they had recovered it, these ideas would have represented by far its most elementary proof.

Hao and Jog [20] generalized the inequality (7.17) to the case where one has n random variables, under a natural n-variable extension of the distributional requirement, namely unconditionality. However, they used a proof that goes through Fisher information inequalities, similar to the original Stam proof of the full entropy power inequality. The main observation of this section is simply that under certain circumstances, one can give a direct and simple proof of the Hao–Jog inequality, as well as others like it, akin to the 2-line proof of the inequality (7.17) given above. The "certain circumstances" have to do with the existence of appropriate linear transformations that respect certain symmetries– specifically Hadamard matrices.

Let us first outline how this works in the real case. Suppose n is a dimension for which there exists a *Hadamard matrix*– namely, a $n \times n$ matrix with all its entries being 1 or -1, and its rows forming an orthogonal set of vectors. Dividing each row by its length $\sqrt{n}$ results in an orthogonal matrix O, all of whose entries are $\pm\frac{1}{\sqrt{n}}$. By unconditionality, each coordinate of the vector OX has the same distribution as $\frac{X_1 + \cdots + X_n}{\sqrt{n}}$. Hence

$$h(X) = h(OX) \leq \sum_{j=1}^{n} h((OX)_j) = nh\left(\frac{X_1 + \cdots + X_n}{\sqrt{n}}\right),$$

where the inequality follows from subadditivity of entropy. This is exactly the Hao-Jog inequality for those dimensions where a Hadamard matrix exists. It would be interesting to find a way around the dimensional restriction, but we do not currently have a way of doing so.

As is well known, other than the dimensions 1 and 2, Hadamard matrices may only exist for dimensions that are multiples of 4. As of this date, Hadamard matrices are known to exist for all multiples of 4 up to 664 [22], and it is a major open problem whether they in fact exist for all multiples of 4. (Incidentally, we note that the question of existence of Hadamard matrices can actually be formulated in the entropy language. Indeed, Hadamard matrices are precisely those that saturate the obvious bound for the entropy of an orthogonal matrix [19].)

In contrast, complex Hadamard matrices exist in every dimension. A *complex Hadamard matrix* of order n is a $n \times n$ matrix with complex entries all of which have modulus 1, and whose rows form an orthogonal set of vectors in $\mathbb{C}^n$. To see that complex Hadamard matrices always exist, we merely exhibit the Fourier matrices, which are a well known example of them: these are defined by the entries $H_{j,k} = \exp\{\frac{2\pi i (j-1)(k-1)}{n}\}$, for $j, k = 1, \ldots, n$, and are related to the discrete Fourier transform (DFT) matrices. Complex Hadamard matrices play an important role in quantum information theory [37]. They also yield Theorem 7.2.

Proof of Theorem 7.2 Take any $n \times n$ unitary matrix U which all entries are complex numbers of the same modulus $\frac{1}{\sqrt{n}}$; such matrices are easily constructed by multiplying a complex Hadamard matrix by $n^{-1/2}$. (For instance, one could take $U = \frac{1}{\sqrt{n}}[e^{2\pi i kl/n}]_{k,l}$.) By complex-unconditionality, each coordinate of the vector UX has the same distribution, the same as $\frac{X_1 + \cdots + X_n}{\sqrt{n}}$. Therefore, by subadditivity,

$$h(X) = h(UX) \le \sum_{j=1}^{n} h\left((UX)_j\right) = nh\left(\frac{X_1 + \cdots + X_n}{\sqrt{n}}\right),$$

which finishes the proof. □

Let us mention that the invariance idea above also very simply yields the inequality

$$D(X) \le \frac{1}{2}|h(X_1 + X_2) - h(X_1 - X_2)|,$$

where $D(X)$ denotes the relative entropy of the distribution of X from the closest Gaussian (which is the one with matching mean and covariance matrix), and X_1, X_2 are independent copies of a random vector X in $\mathbb{R}^n$. First observed in [30, Theorem 10], this fact quantifies the distance from Gaussianity of a random vector in terms of how different the entropies of the sum and difference of i.i.d. copies of it are.

Finally, we mention that the idea of considering two i.i.d. copies and using invariance (sometimes called the "doubling trick") has been used in sophisticated ways as a key tool to study both functional inequalities [5, 12, 24] and problems in network information theory (see, e.g., [14, 18]).

Acknowledgements We learned about the Edgeworth expansion used in our proof of Theorem 7.1 from S. Bobkov during the AIM workshop *Entropy power inequalities*. We are immensely grateful to him as well as the AIM and the organisers of the workshop which was a valuable and unique research experience.

We would like to thank the anonymous referee for his or her careful reading of the manuscript and suggesting several clarifications and improvements.

References

1. Aimpl: entropy power inequalities (2017). http://aimpl.org/entropypower/
2. S. Artstein, K.M. Ball, F. Barthe, A. Naor, Solution of Shannon's problem on the monotonicity of entropy. J. Am. Math. Soc. **17**(4), 975–982 (2004)
3. K. Ball, P. Nayar, T. Tkocz, A reverse entropy power inequality for log-concave random vectors. Stud. Math. **235**(1), 17–30 (2016)
4. A.R. Barron, Entropy and the central limit theorem. Ann. Probab. **14**, 336–342 (1986)
5. F. Barthe, Optimal Young's inequality and its converse: a simple proof. Geom. Funct. Anal. **8**(2), 234–242 (1998)
6. S.G. Bobkov, G.P. Chistyakov, Entropy power inequality for the Rényi entropy. IEEE Trans. Inform. Theory **61**(2), 708–714 (2015)
7. S. Bobkov, M. Madiman, Dimensional behaviour of entropy and information. C. R. Acad. Sci. Paris Sér. I Math. **349**, 201–204 (2011)
8. S.G. Bobkov, M. Madiman, L. Wang, Fractional generalizations of Young and Brunn-Minkowski inequalities, in *Concentration, functional inequalities and isoperimetry*, ed. by C. Houdré, M. Ledoux, E. Milman, M. Milman. Contemporary Mathematics, vol. 545 (American Mathematical Society, Providence, 2011), pp. 35–53
9. S.G. Bobkov, G.P. Chistyakov, F. Götze, Rate of convergence and Edgeworth-type expansion in the entropic central limit theorem. Ann. Probab. **41**(4), 2479–2512 (2013)
10. S. Bobkov, G.P. Chistyakov, F. Götze, Rényi divergence and the central limit theorem. Ann. Probab. **47**(1), 270–323 (2019). arXiv:1608.01805
11. L. Boltzmann, *Lectures on gas theory* (Dover Publications, Mineola, 1995). Reprint of the 1896–1898 edition (Translated by S. G. Brush)
12. E.A. Carlen, Superadditivity of Fisher's information and logarithmic Sobolev inequalities. J. Funct. Anal. **101**(1), 194–211 (1991)
13. T.A. Courtade, Bounds on the Poincaré constant for convolution measures (preprint, 2018). arXiv:1807.00027
14. T.A. Courtade, A strong entropy power inequality. IEEE Trans. Inform. Theory **64**(4, part 1), 2173–2192 (2018)
15. A. Eskenazis, P. Nayar, T. Tkocz, Gaussian mixtures: entropy and geometric inequalities. Ann. Probab. **46**(5), 2908–2945 (2018)
16. M. Fradelizi, M. Madiman, A. Marsiglietti, A. Zvavitch, Do Minkowski averages get progressively more convex? C. R. Acad. Sci. Paris Sér. I Math. **354**(2), 185–189 (2016)
17. M. Fradelizi, M. Madiman, A. Marsiglietti, A. Zvavitch, The convexifying effect of Minkowski summation. EMS Surv. Math. Sci. **5**(1/2), 1–64 (2019). arXiv:1704.05486
18. Y. Geng, C. Nair, The capacity region of the two-receiver Gaussian vector broadcast channel with private and common messages. IEEE Trans. Inform. Theory **60**(4), 2087–2104 (2014)
19. H. Gopalkrishna Gadiyar, K.M. Sangeeta Maini, R. Padma, H.S. Sharatchandra, Entropy and Hadamard matrices. J. Phys. A **36**(7), L109–L112 (2003)

20. J. Hao, V. Jog, An entropy inequality for symmetric random variables, in Proceedings of IEEE International Symposium on Information Theory, IEEE (2018), pp. 1600–1604. arXiv:1801.03868.
21. O. Johnson, *Information theory and the central limit theorem* (Imperial College Press, London, 2004)
22. H. Kharaghani, B. Tayfeh-Rezaie, A Hadamard matrix of order 428. J. Combin. Des. **13**(6), 435–440 (2005)
23. I. Kontoyiannis, M. Madiman, Sumset and inverse sumset inequalities for differential entropy and mutual information. IEEE Trans. Inform. Theory **60**(8), 4503–4514 (2014)
24. E.H. Lieb, Gaussian kernels have only Gaussian maximizers. Invent. Math. **102**(1), 179–208 (1990)
25. M. Madiman, A primer on entropic limit theorems (preprint, 2017)
26. M. Madiman, A.R. Barron, The monotonicity of information in the central limit theorem and entropy power inequalities, in *Proceedings of IEEE International Symposium on Information Theory*, Seattle (2006), pp. 1021–1025
27. M. Madiman, A.R. Barron, Generalized entropy power inequalities and monotonicity properties of information. IEEE Trans. Inform. Theory **53**(7), 2317–2329 (2007)
28. M. Madiman, F. Ghassemi, Combinatorial entropy power inequalities: a preliminary study of the Stam region. IEEE Trans. Inform. Theory **65**(3), 1375–1386 (2019). arXiv:1704.01177
29. M. Madiman, I. Kontoyiannis, The entropies of the sum and the difference of two IID random variables are not too different, in *Proceedings of IEEE International Symposium on Information Theory*, Austin (2010)
30. M. Madiman, I. Kontoyiannis, Entropy bounds on Abelian groups and the Ruzsa divergence. IEEE Trans. Inform. Theory **64**(1), 77–92 (2018). arXiv:1508.04089
31. M. Madiman, J. Melbourne, P. Xu, Forward and reverse entropy power inequalities in convex geometry, in *Convexity and concentration*, ed. by E. Carlen, M. Madiman, E.M. Werner. IMA Volumes in Mathematics and its Applications, vol. 161 (Springer, Berlin, 2017), pp. 427–485. arXiv:1604.04225
32. M. Madiman, L. Wang, J.O. Woo, Majorization and Rényi entropy inequalities via Sperner theory Discrete Math. **342**(10), 2911–2923, (2019). arXiv:1712.00913
33. M. Madiman, L. Wang, J.O. Woo, Rényi entropy inequalities for sums in prime cyclic groups. (preprint, 2017). arXiv:1710.00812
34. C.E. Shannon, A mathematical theory of communication. Bell Syst. Tech. J. **27**, 379–423, 623–656 (1948)
35. D. Shlyakhtenko, A free analogue of Shannon's problem on monotonicity of entropy. Adv. Math. **208**(2), 824–833 (2007)
36. A.J. Stam, Some inequalities satisfied by the quantities of information of Fisher and Shannon. Inform. Control **2**, 101–112 (1959)
37. W. Tadej, K. Życzkowski, A concise guide to complex Hadamard matrices. Open Syst. Inf. Dyn. **13**(2), 133–177 (2006)
38. A.M. Tulino, S. Verdú, Monotonic decrease of the non-Gaussianness of the sum of independent random variables: a simple proof. IEEE Trans. Inform. Theory **52**(9), 4295–4297 (2006)
39. L. Wang, M. Madiman, Beyond the entropy power inequality, via rearrangements. IEEE Trans. Inform. Theory **60**(9), 5116–5137 (2014)
40. L. Wang, J.O. Woo, M. Madiman, A lower bound on the Rényi entropy of convolutions in the integers, in *IEEE International Symposium on Information Theory*, Honolulu (2014), pp. 2829–2833
41. Y. Yu, Letter to the editor: on an inequality of Karlin and Rinott concerning weighted sums of i.i.d. random variables. Adv. Appl. Probab. **40**(4), 1223–1226 (2008)

Chapter 8
On the Geometry of Random Polytopes

Shahar Mendelson

Abstract We present a simple proof to a fact recently established in Guédon et al. (Commun Contemp Math (to appear, 2018). arXiv:1811.12007): let ξ be a symmetric random variable that has variance 1, let $\Gamma = (\xi_{ij})$ be an $N \times n$ random matrix whose entries are independent copies of ξ, and set $X_1, \ldots, X_N$ to be the rows of Γ. Then under minimal assumptions on ξ and as long as $N \geq c_1 n$, with high probability

$$c_2\big(B_\infty^n \cap \sqrt{\log(eN/n)}B_2^n\big) \subset \mathrm{absconv}(X_1, \ldots, X_N).$$

8.1 Introduction

Let ξ be a symmetric random variable that has variance 1 and let $X = (\xi_1, \ldots, \xi_n)$ be the random vector whose coordinates are independent copies of ξ. Consider a random matrix Γ whose rows $X_1, \ldots, X_N$ are independent copies of X. In this note we explore the geometry of the random polytope

$$K = \mathrm{absconv}(X_1, \ldots, X_N) = \Gamma^* B_1^N ;$$

specifically, we study whether K is likely to contain a large canonical convex body.

One of the first results in this direction is from [4], where it is shown that if ξ is the standard Gaussian random variable, $0 < \alpha < 1$ and $N \geq c_0(\alpha)n$, then

$$c_1(\alpha)\sqrt{\log(eN/n)}B_2^n \subset \mathrm{absconv}(X_1, \ldots, X_N) \tag{8.1.1}$$

S. Mendelson (✉)
LPSM, Sorbonne University, Paris, France

Mathematical Sciences Institute, The Australian National University, Canberra, ACT, Australia
e-mail: shahar.mendelson@anu.edu.au

© Springer Nature Switzerland AG 2020   187
B. Klartag, E. Milman (eds.), *Geometric Aspects of Functional Analysis*,
Lecture Notes in Mathematics 2266,
https://doi.org/10.1007/978-3-030-46762-3_8

with probability at least $1 - 2 \exp(-c_2 N^{1-\alpha} n^\alpha)$. It should be noted that this estimate cannot be improved—up to the dependence of the constants on α (see, for example, the discussion in Section 4 of [9]).

The proof of (8.1.1) relies heavily on the tail behaviour of the Gaussian random variable. It is therefore natural to try and extend (8.1.1) beyond the Gaussian case, to random polytopes generated by more general random variables that still have 'well-behaved' tails. The problem was studied in [3] where X was assumed to be uniformly distributed in $\{-1, 1\}^n$ and it was shown that if $N \geq n \log^2 n$ then with high probability

$$c\left(B_\infty^n \cap \sqrt{\log(eN/n)} B_2^n\right) \subset \mathrm{absconv}(X_1, \ldots, X_N). \tag{8.1.2}$$

Remark 8.1.1 Note that here, the body that $\mathrm{absconv}(X_1, \ldots, X_N)$ contains is slightly smaller than in (8.1.1), as one has to intersect the Euclidean ball from (8.1.1) with the unit cube.

The optimal *sub-Gaussian estimate* was established in [9]:

Theorem 8.1.2 *Let ξ be a mean-zero random variable that has variance 1 and is L-sub-Gaussian.[1] Let $0 < \alpha < 1$ and set $N \geq c_0(\alpha)n$. Then with probability at least $1 - 2 \exp(-c_1 N^{1-\alpha} n^\alpha)$*

$$c_2(\alpha)\left(B_\infty^n \cap \sqrt{\log(eN/n)} B_2^n\right) \subset \mathrm{absconv}(X_1, \ldots, X_N), \tag{8.1.3}$$

where c_0 and c_2 are constants that depend on α and c_1 is an absolute constant.

While Theorem 8.1.2 resolves the problem when ξ is sub-Gaussian, the situation is less clear when ξ is heavy-tailed. That naturally leads to the following question:

Question 8.1.3 Under what conditions on ξ one still has that for $N \geq c_1 n$,

$$c_2(B_\infty^n \cap \sqrt{\log(eN/n)} B_2^n) \subset \mathrm{absconv}(X_1, \ldots, X_N) \tag{8.1.4}$$

with high probability?

Following the progress in [7], where Question 8.1.3 had been studied under milder moment assumptions on ξ than in Theorem 8.1.2, Question 8.1.3 was answered in [5] under a minimal small-ball condition on ξ.

Definition 8.1.4 A mean-zero random variable ξ satisfies a small-ball condition with constants κ and δ if

$$Pr(|\xi| \geq \kappa) \geq \delta. \tag{8.1.5}$$

---

[1] A centred random variable is L-sub-Gaussian if for every $p \geq 2$, $\|\xi\|_{L_p} \leq L\sqrt{p}\|\xi\|_{L_2}$.

Theorem 8.1.5 ([5]) *Let ξ be a symmetric, variance 1 random variable that satisfies (8.1.5) with constants κ and δ. For $0 < \alpha < 1$ there are constants c_1, c_2 and c_3 that depend on κ, δ and α for which the following holds. If $N \geq c_1 n$ then with probability at least $1 - 2\exp(-c_2 N^{1-\alpha} n^{\alpha})$,*

$$c_3(B_\infty^n \cap \sqrt{\log(eN/n)} B_2^n) \subset \mathrm{absconv}(X_1, \ldots, X_N).$$

Remark 8.1.6 In [5] the random variables (ξ_{ij}) are only assumed to be independent, symmetric and variance 1, with each one of the ξ_{ij}'s satisfying (8.1.5) with the same constants κ and δ. In what follows we consider only the case in which ξ_{ij} are independent copies of a single random variable ξ, though extending the presentation to the independent case is straightforward.

The original proof of Theorem 8.1.5 is based on the construction of a well-chosen net, and that construction is rather involved. Here we present a much simpler argument that is based on the *small-ball method* (see, e.g., [10–12]). As an added value, the method presented here gives more information than the assertion of Theorem 8.1.5, as is explained in what follows.

The starting point of the proof of Theorem 8.1.5 is straightforward: let

$$K = \mathrm{absconv}(X_1, \ldots, X_n) = \Gamma^* B_1^N$$

and set

$$L = B_\infty^n \cap \sqrt{\log(eN/n)} B_2^n.$$

By comparing the support functions of L and of K, one has to show that with the wanted probability, for every $z \in \mathbb{R}^n$, $h_L(z) \leq h_{cK}(z)$. And, since $h_{cK}(z) = c\|\Gamma z\|_\infty$, Theorem 8.1.5 can be established by showing that for suitable constants c_0 and c_1,

$$Pr(\exists z \in \partial L^\circ \quad \|\Gamma z\|_\infty \leq c_0) \leq 2\exp(-c_1 N^{1-\alpha} n^{\alpha}). \tag{8.1.6}$$

What we actually show is a stronger statement than (8.1.6): not only is there a high probability event on which

$$\inf_{z \in \partial L^\circ} \|\Gamma z\|_\infty \geq c_0,$$

but in fact, on that "good event", for each $z \in \partial L^\circ$, Γz has $\sim N^{1-\alpha} n^{\alpha}$ large coordinates, with each one of these coordinates satisfying that $|\langle z, X_i \rangle| \geq c_0$. Thus, the fact that $\|\Gamma z\|_\infty \geq c_0$ is exhibited by many coordinates and not just by a single one.

Proving that indeed, with high probability the smallest cardinality

$$\inf_{z \in \partial L^\circ} |\{i : |\langle z, X_i \rangle| \geq c_0\}|$$

is large is carried out in two steps:

Controlling a Single Point—See Corollary 8.2.4 For $0 < \alpha < 1$ and a well chosen $c_0 = c_0(\alpha)$ one establishes an *individual estimate*: that for every fixed $z \in \partial L^\circ$,

$$Pr(|\langle z, X \rangle| \geq 2c_0) \geq 4 \left(\frac{n}{N}\right)^\alpha.$$

In particular, if $X_1, \ldots, X_N$ are independent copies of X then with probability at least $1 - 2\exp(-c_2 N^{1-\alpha} n^\alpha)$,

$$\left|\{i : |\langle z, X_i \rangle| \geq 2c_0\}\right| \geq 2N^{1-\alpha} n^\alpha. \tag{8.1.7}$$

From a Single Function to Uniform Control Thanks to the high probability estimate with which (8.1.7) holds, it is possible to control uniformly any subset of ∂L° whose cardinality is at most $\exp(c_2 N^{1-\alpha} n^\alpha / 2)$. Let $\mathcal{T}$ be a minimal ρ-cover of ∂L° with respect to the ℓ_2 norm, and of the allowed cardinality. For every $z \in \partial L^\circ$, let $\pi z \in \mathcal{T}$ that satisfies $\|z - \pi z\|_2 \leq \rho$. The wanted uniform control is achieved by showing that

$$\sup_{z \in \partial L^\circ} \left|\{i : |\langle z - \pi z, X_i \rangle| \geq c_0\}\right| \leq N^{1-\alpha} n^\alpha$$

with probability at least $1 - 2\exp(-c_3(\alpha) N^{1-\alpha} n^\alpha)$.

Indeed, combining the two estimates it follows that with probability at least

$$1 - 2\exp(-c(\alpha) N^{1-\alpha} n^\alpha),$$

for every $z \in \partial L^\circ$, one has that

$$\left|\{i : |\langle \pi z, X_i \rangle| \geq 2c_0\}\right| \geq 2N^{1-\alpha} n^\alpha$$

and

$$\left|\{i : |\langle z - \pi z, X_i \rangle| \geq c_0\}\right| \leq N^{1-\alpha} n^\alpha.$$

Hence, on that event, for every $z \in \partial L^\circ$ there is $J_z \subset \{1, \ldots, n\}$ of cardinality at least $N^{1-\alpha} n^\alpha$, and for every $j \in J_z$,

$$|\langle z, X_i \rangle| \geq |\langle \pi z, X_i \rangle| - |\langle z - \pi z, X_i \rangle| \geq c_0,$$

implying that

$$\inf_{z \in \partial L^{\circ}} \left| \{ i : | \langle z, X_i \rangle | \ge c_0 \} \right| \ge N^{1-\alpha} n^{\alpha};$$

in particular, $\inf_{z \in \partial L^{\circ}} \|\Gamma z\|_{\infty} \ge c_0$ as required.

In the next section this line of reasoning is used to prove Theorem 8.1.5.

8.2   Proof of Theorem 8.1.5

Before we begin the proof, let us introduce some notation. Throughout, absolute constant are denoted by c, c_1, c' etc. Unless specified otherwise, the value of these constants may change from line to line. Constants that depend on some parameter α are denoted by $c(\alpha)$. We write $a \lesssim b$ if there is an absolute constant c such that $a \le cb$; $a \lesssim_{\alpha} b$ implies that $a \le c(\alpha)b$; and $a \sim b$ if both $a \lesssim b$ and $b \lesssim a$.

The required estimate for a single point follows very closely ideas from [13], which had been developed for obtaining lower estimates on the tails of marginals of the Rademacher vector $(\varepsilon_i)_{i=1}^{n}$, that is, on

$$Pr(| \sum_{i=1}^{n} \varepsilon_i z_i | > t)$$

as a function of the 'location' in $\mathbb{R}^n$ of $(z_i)_{i=1}^{n}$.

Fix $1 \le r \le n$ and consider the interpolation body $L_r = B_{\infty}^n \cap \sqrt{r} B_2^n$ and its dual $L_r^{\circ} = \text{conv}(B_1^n \cup (1/\sqrt{r}) B_2^n)$. The key estimate one needs to establish the wanted individual control is:

Theorem 8.2.1 *There exist constants c' and c'' that depend only on the small-ball constants of ξ (κ and δ) such that if $z \in \partial L_r^{\circ}$ then*

$$Pr\big(| \langle z, X \rangle | \ge c' \big) \ge 2 \exp(-c'' r).$$

Just as in [13], the proof of Theorem 8.2.1 is based on some well-known facts on the interpolation norm $\| \ \|_{L_r^{\circ}}$.

Lemma 8.2.2 *There exists an absolute constant c_0 such that for every $z \in \mathbb{R}^n$,*

$$\|z\|_{L_r^{\circ}} \le \sum_{i=1}^{r} z_i^* + \sqrt{r} \big(\sum_{i>r} (z_i^2)^* \big)^{1/2} \le c_0 \|z\|_{L_r^{\circ}},$$

where $(z_i^*)_{i=1}^{n}$ is the nonincreasing rearrangement of $(|z_i|)_{i=1}^{n}$.

Moreover, for very $z \in \mathbb{R}^n$ there is a partition of $\{1, \ldots, n\}$ to r disjoint blocks $I_1, \ldots, I_r$ such that

$$\frac{\|z\|_{L_r^\circ}}{\sqrt{2}} \leq \sum_{j=1}^{r} \left(\sum_{i \in I_j} z_i^2\right)^{1/2} \leq \|z\|_{L_r^\circ}.$$

The first part of Lemma 8.2.2 is due to Holmstedt (see Theorem 4.1 in [6]) and it gives useful intuition on the nature of the norm $\| \ \|_{L_r^\circ}$. The second part is Lemma 2 from [13] and it plays an essential role in what follows.

Before proving Theorem 8.2.1, we require an additional observation that is based on the small-ball condition satisfies by ξ.

Lemma 8.2.3 *Let $J \subset \{1, \ldots, n\}$ and set $Y = \sum_{j \in J} z_j \xi_j$ for $z = (z_i)_{i=1}^n$. Then*

$$\mathbb{E}|Y| \geq c(\kappa, \delta)\left(\sum_{j \in J} z_j^2\right)^{1/2},$$

where $c(\kappa, \delta) < 1$ is a constant the depends only on ξ's small-ball constants κ and δ.

Proof Let $(\varepsilon_j)_{j \in J}$ be independent, symmetric, $\{-1, 1\}$-valued random variables that are also independent of $(\xi_j)_{j \in J}$. Recall that ξ is symmetric and therefore $(\xi_j)_{j \in J}$ has the same distribution as $(\varepsilon_j \xi_j)_{j \in J}$. By Khintchine's inequality it is straightforward to verify that

$$\mathbb{E}|Y| = \mathbb{E}_\xi \mathbb{E}_\varepsilon \left|\sum_{j \in J} \varepsilon_j z_j \xi_j\right| \gtrsim \mathbb{E}_\xi \left(\sum_{j \in J} z_j^2 \xi_j^2\right)^{1/2}.$$

For $j \in J$ let $\eta_j = \mathbb{1}_{\{|\xi_j| \geq \kappa\}}$; thus, the η_j's are iid $\{0, 1\}$-valued random variables whose mean is at least δ, and point-wise

$$\left(\sum_{j \in J} z_j^2 \xi_j^2\right)^{1/2} \geq \kappa \left(\sum_{j \in J} \eta_j z_j^2\right)^{1/2}.$$

Hence, all that is left to complete the proof is to show that

$$\mathbb{E}\left(\sum_{j \in J} \eta_j z_j^2\right)^{1/2} \geq c(\delta)\left(\sum_{j \in J} z_j^2\right)^{1/2}.$$

Let $a_j = z_j^2/(\sum_{j \in J} z_j^2)$ and in particular, $\|(a_j)_{j \in J}\|_1 = 1$. Assume without loss of generality that $J = \{1, \ldots, \ell\}$ and that the a_j's are non-increasing, let $\gamma > 0$ be a parameter to be specified in what follows, and set $p = \mathbb{E}\eta_1 \geq \delta$.

Consider two cases:

- If $a_1 \geq \gamma p$ then with probability at least p, $\sum_{j=1}^{\ell} \eta_j a_j \geq a_1 \geq \gamma p$. In that case

$$\mathbb{E}\left(\sum_{j=1}^{\ell} \eta_j a_j\right)^{1/2} \geq \sqrt{\gamma} p^{3/2} \geq \sqrt{\gamma} \delta^{3/2}.$$

- Alternatively, $a_1 \leq \gamma p$, implying that

$$A = \sum_{j=1}^{\ell} a_j^2 \leq a_1 \sum_{j=1}^{\ell} a_j \leq \gamma p$$

because $\|(a_j)_{j=1}^{\ell}\|_1 = 1$.

By Bernstein's inequality,

$$Pr\left(\left|\sum_{j=1}^{\ell}(\eta_j - p)a_j\right| \geq \frac{p}{2}\right) \leq 2 \exp\left(-c_0 \min\left\{\frac{(p/2)^2}{pA}, \frac{p/2}{a_1}\right\}\right)$$

$$\leq 2 \exp(-c_1/\gamma) \leq \frac{1}{2}$$

provided that γ is a small-enough absolute constant. Using, once again, that $\|(a_j)_{j=1}^{\ell}\|_1 = 1$ it is evident that with probability at least $1/2$, $\sum_{j=1}^{\ell} \eta_j a_j \geq (1/2)p$ and therefore

$$\mathbb{E}\left(\sum_{j=1}^{\ell} \eta_j a_j\right)^{1/2} \geq \frac{\sqrt{p}}{4} \geq \frac{\sqrt{\delta}}{4}.$$

Thus, setting $c(\kappa, \delta) \sim \kappa \delta^{3/2}$ one has that

$$\mathbb{E}\left(\sum_{j=1}^{\ell} z_j^2 \xi_j^2\right)^{1/2} \geq c(\kappa, \delta)\left(\sum_{j=1}^{\ell} z_j^2\right)^{1/2},$$

as claimed. ∎

Proof of Theorem 8.2.1 Fix $z \in \partial L_r^{\circ}$ and recall that by Lemma 8.2.2 there is a decomposition of $\{1, \ldots, n\}$ to disjoint blocks $(I_j)_{j=1}^{r}$ such that

$$\sum_{j=1}^{r}\left(\sum_{i \in I_j} z_i^2\right)^{1/2} \geq \frac{1}{\sqrt{2}}. \tag{8.2.1}$$

Let $Y_j = \sum_{i \in I_j} z_i \xi_i$; observe that $Y_1, \ldots, Y_r$ are independent random variables and that by Lemma 8.2.3,

$$\mathbb{E}|Y_j| \geq c(\kappa, \delta) \Big(\sum_{i \in I_j} z_i^2 \Big)^{1/2}$$

for a constant $0 < c(\kappa, \delta) < 1$.

At the same time,

$$\mathbb{E}Y_j^2 = \sum_{i \in I_j} z_i^2 \mathbb{E}\xi_i^2 = \sum_{i \in I_j} z_i^2.$$

Therefore, by the Paley-Zygmund inequality (see, e.g., [2]), for any $0 < \theta < 1$,

$$Pr(|Y_j| \geq \theta \mathbb{E}|Y_j|) \geq (1 - \theta^2) \frac{(\mathbb{E}|Y_j|)^2}{\mathbb{E}Y_j^2}.$$

Setting $\theta = 1/2$,

$$Pr\Big(|Y_j| \geq \frac{1}{2}c(\kappa, \delta) \Big(\sum_{i \in I_j} z_i^2 \Big)^{1/2}\Big) \geq \frac{3}{4}c^2(\kappa, \delta),$$

and since Y_j is a symmetric random variable (because the ξ_i's are symmetric), it follows that

$$Pr\Big(Y_j \geq \frac{1}{2}c(\kappa, \delta) \Big(\sum_{i \in I_j} z_i^2 \Big)^{1/2}\Big) \geq \frac{3}{8}c^2(\kappa, \delta) \equiv c_1(\kappa, \delta).$$

For $1 \leq j \leq r$ let

$$\mathcal{B}_j = \Big\{ Y_j \geq \frac{1}{2}c(\kappa, \delta) \Big(\sum_{i \in I_j} z_i^2 \Big)^{1/2} \Big\}$$

which are independent events. Hence,

$$Pr\Big(\sum_{i=1}^n \xi_i z_i \geq \frac{1}{2}c(\kappa, \delta) \sum_{j=1}^r \Big(\sum_{i \in I_j} z_i^2 \Big)^{1/2}\Big) = Pr\Big(\sum_{j=1}^r Y_j \geq \frac{1}{2}c(\kappa, \delta) \sum_{j=1}^r \Big(\sum_{i \in I_j} z_i^2 \Big)^{1/2}\Big)$$

$$\geq \prod_{j=1}^r Pr(\mathcal{B}_j) \geq c_1^r(\kappa, \delta).$$

Thus, by (8.2.1), if $c' = \frac{1}{4}c(\kappa, \delta)$ and $c'' = \log(1/c_1(\kappa, \delta)) > 0$, one has

$$Pr\left(\sum_{i=1}^{n} \xi_i z_i \geq c'\right) \geq \exp(-c''r).$$

■

From here on, the constants c' and c'' denote the constants from Theorem 8.2.1.

Corollary 8.2.4 *For $0 < \alpha < 1$, κ and δ there are constants c_0 and c_1 that depend on α, κ and δ, and an absolute constant c_2 for which the following holds. If $N \geq c_0 n$, $r \leq c_1 \log(eN/n)$ and $z \in \partial L_r^\circ$ then with probability at least $1 - 2\exp(-c_2 N^{1-\alpha} n^\alpha)$,*

$$\left|\{i : |\langle z, X_i \rangle| \geq c'\}\right| \geq 2N^{1-\alpha} n^\alpha.$$

Proof Let $z \in \partial L_r^\circ$, and invoking Theorem 8.2.1,

$$Pr\left(|\langle z, X \rangle| \geq c'\right) \geq \exp(-c''r)$$

where c' and c'' depend only on κ and δ.

Set $r_0 = c_1 \log(eN/n)$ such that $\exp(-c''r_0) \geq 4(n/N)^\alpha$; thus, $c_1 = c_1(\alpha, \kappa, \delta)$. If $r \leq r_0$, $X_1, \ldots, X_N$ are independent copied of X and $\eta_i = \mathbb{1}_{\{|\langle z, X_i\rangle| \geq c'\}}$, then $\mathbb{E}\eta_i \geq 4(n/N)^\alpha$. Hence, by a standard concentration argument (e.g. Bernstein's inequality), with probability at least $1 - 2\exp(-c_2 N^{1-\alpha} n^\alpha)$,

$$\left|\{i : |\langle z, X_i \rangle| \geq c'\}\right| \geq 2N^{1-\alpha} n^\alpha,$$

where c_2 is an absolute constant.

■

Thanks to the high probability estimate with which Corollary 8.2.4 holds, one can control uniformly all the elements of a set $\mathcal{T} \subset \partial L_r^\circ$ as long as $|\mathcal{T}| \leq \exp(c_0 N^{1-\alpha} n^\alpha)$ for a suitable absolute constant c_0, and as long as $r \leq c(\alpha, \kappa, \delta) \log(eN/n)$. In that case, there is an event of probability at least $1 - 2\exp(-c_1 N^{1-\alpha} n^\alpha)$ such that for every $z \in \mathcal{T}$,

$$\left|\{i : |\langle z, X_i \rangle| \geq c'\}\right| \geq 2N^{1-\alpha} n^\alpha. \tag{8.2.2}$$

The natural choice of a set $\mathcal{T}$ is a minimal ρ-cover of ∂L_r° with respect to the ℓ_2 norm. Note that $L_r^\circ = \text{conv}(B_1^n \cup r^{-1/2} B_2^n) \subset B_2^n$, and so there is a ρ-cover of the allowed cardinality for

$$\rho \leq 5 \exp(-c_2(N/n)^{1-\alpha}),$$

where c_2 is an absolute constant.

Clearly, $\{z - \pi z : z \in \partial L_r^\circ\} \subset \rho B_2^n$, and as was explained in the introduction, to complete the proof of Theorem 8.1.5 it suffices to prove the following lemma:

Lemma 8.2.5 *Using the above notation, with probability at least* $1 - 2\exp(-c_3 N^{1-\alpha} n^\alpha)$, *one has*

$$Q = \sup_{u \in \rho B_2^n} \left|\{i : |\langle u, X_i\rangle| \geq c'/2\}\right| \leq N^{1-\alpha} n^\alpha. \tag{8.2.3}$$

Proof Observe that Q is the supremum of an empirical process indexed by a class of binary valued functions

$$F = \left\{f_z = \mathbb{1}_{\{|\langle z,\cdot\rangle| \geq c'/2\}} : z \in \rho B_2^n\right\};$$

in particular, for every $f_z \in F$,

$$\|f_z\|_{L_2} = Pr^{1/2}(|\langle z, X\rangle| \geq c'/2) \leq \frac{2\|\langle z, X\rangle\|_{L_2}}{c'} \leq \frac{2\rho}{c'}$$

$$= c_4(\kappa, \delta)\exp(-c_2(N/n)^{1-\alpha}).$$

By Talagrand's concentration inequality for bounded empirical processes ([14], see also [1, Chapter 12]), with probability at least $1 - 2\exp(-t)$,

$$Q \lesssim \mathbb{E}Q + \sqrt{t}\sqrt{N}\sup_{f_z \in F}\|f_z\|_{L_2} + t\sup_{f_z \in F}\|f_z\|_{L_\infty}$$

$$\lesssim \mathbb{E}Q + \sqrt{t}\sqrt{N}c_4(\kappa, \delta)\exp(-c_2(N/n)^{1-\alpha}) + t$$

$$= (1) + (2) + (3).$$

Let us show that for the right choice of t and N large enough, $Q \leq N^{1-\alpha}n^\alpha$. The required estimate on (2) and (3) clearly holds as long as

$$t \lesssim_{\kappa,\delta} N^{1-\alpha}n^\alpha \quad \text{and} \quad N \gtrsim_\alpha n.$$

As for $\mathbb{E}Q$, note that point-wise

$$\sup_{u \in \rho B_2^n}\left|\{i : |\langle u, X_i\rangle| \geq c'/2\}\right| \leq \frac{2}{c'}\sup_{u \in \rho B_2^n}\sum_{i=1}^{N}|\langle u, X_i\rangle|.$$

Let $(\varepsilon_i)_{i=1}^N$ be independent, symmetric, $\{-1, 1\}$-valued random variables that are independent of $(X_i)_{i=1}^N$. The Giné–Zinn symmetrization theorem (see, for example,

[15, Chapter 2.3]) implies that

$$\mathbb{E} \sup_{u \in \rho B_2^n} \left| \sum_{i=1}^{N} \left(|\langle u, X_i \rangle| - \mathbb{E} |\langle u, X_i \rangle| \right) \right| \leq 2\mathbb{E} \sup_{u \in \rho B_2^n} \left| \sum_{i=1}^{N} \varepsilon_i |\langle u, X_i \rangle| \right|,$$

and since $\phi(t) = |t|$ is a 1-Lipschitz function that satisfies $\phi(0) = 0$, it follows from the contraction inequality for Bernoulli processes (see, e.g., [8, Chapter 4]) that

$$\mathbb{E} \sup_{u \in \rho B_2^n} \left| \sum_{i=1}^{N} \varepsilon_i |\langle u, X_i \rangle| \right| \leq 2\mathbb{E} \sup_{u \in \rho B_2^n} \left| \sum_{i=1}^{N} \varepsilon_i \langle u, X_i \rangle \right|.$$

Therefore,

$$\mathbb{E}Q \leq \frac{2}{c'} \mathbb{E} \sup_{u \in \rho B_2^n} \sum_{i=1}^{N} |\langle u, X_i \rangle|$$

$$\leq \frac{4}{c'} \mathbb{E} \sup_{u \in \rho B_2^n} \left| \sum_{i=1}^{N} \varepsilon_i |\langle u, X_i \rangle| \right| + \frac{2N}{c'} \sup_{u \in \rho B_2^n} \mathbb{E} |\langle u, X_i \rangle|$$

$$\leq \frac{8}{c'} \mathbb{E} \sup_{u \in \rho B_2^n} \left\langle \sum_{i=1}^{N} \varepsilon_i X_i, u \right\rangle + \frac{2N}{c'} \rho$$

$$\leq \frac{8\rho}{c'} (\sqrt{Nn} + N) \lesssim_{\kappa, \delta} N \exp(-c_2 (N/n)^{1-\alpha}),$$

which is sufficiently small as long as $N \gtrsim_{\alpha, \kappa, \delta} n$.   ∎

8.3   Concluding Remarks

This proof of Theorem 8.1.5 is based on the small-ball method and follows an almost identical path to previous results that use the method: first, one obtains an individual estimate that implies that for each v in a fine-enough net, many of the values $(|\langle X_i, v \rangle|)_{i=1}^{N}$ are in the 'right range'; and then, that the 'oscillation vector' $(|\langle X_i, z - v \rangle|)_{i=1}^{N}$ does not spoil too many coordinates when v is 'close enough' to z. Thus, with high probability and uniformly in z, many of the values $(|\langle X_i, z \rangle|)_{i=1}^{N}$ are in the right range.

Having said that, there is one substantial difference between this proof and other instances in which the small-ball method had been used. Previously, individual estimates had been obtained in the small-ball regime; here the necessary regime is different: one requires a lower estimate on the tails of marginals of $X = (\xi_i)_{i=1}^{n}$. And indeed, the core of the proof is the individual estimate from Theorem 8.2.1, where one shows that if ξ satisfies a small-ball condition and X has iid coordinates distributed as ξ then its marginals exhibit a 'super-Gaussian' behaviour at the right level.

References

1. S. Boucheron, G. Lugosi, P. Massart, *Concentration inequalities* (Oxford University Press, Oxford, 2013). A nonasymptotic theory of independence
2. V.H. de la Peña, E. Giné, *Decoupling: from dependence to independence* (Springer, New York, 1999)
3. A. Giannopoulos, M. Hartzoulaki, Random spaces generated by vertices of the cube. Discrete Comput. Geom. **28**(2), 255–273 (2002)
4. E.D. Gluskin, Extremal properties of orthogonal parallelepipeds and their applications to the geometry of Banach spaces. Mat. Sb. **136**(1), 85–96 (1988)
5. O. Guédon, A.E. Litvak, K. Tatarko, Random polytopes obtained by matrices with heavy tailed entries. Commun. Contemp. Math. (to appear, 2018). arXiv:1811.12007
6. T. Holmstedt, Interpolation of quasi-normed spaces. Math. Scand. **26**, 177–199 (1970)
7. F. Krahmer, C. Kummerle, H. Rauhut, A quotient property for matrices with heavy-tailed entries and its application to noise-blind compressed sensing (2018). Manuscript, arXiv:1806.04261
8. M. Ledoux, M. Talagrand, *Probability in Banach spaces*. Classics in Mathematics (Springer, Berlin, 2011)
9. A.E. Litvak, A. Pajor, M. Rudelson, N. Tomczak-Jaegermann, Smallest singular value of random matrices and geometry of random polytopes. Adv. Math. **195**(2), 491–523 (2005)
10. S. Mendelson, A remark on the diameter of random sections of convex bodies, in *Geometric aspects of functional analysis*. Lecture Notes in Mathematics, vol. 2116 (Springer, Berlin, 2014), pp. 395–404
11. S. Mendelson, Learning without concentration. J. ACM **62**(3), Art. 21, 25 (2015)
12. S. Mendelson, Learning without concentration for general loss functions. Probab. Theory Relat. Fields **171**(1–2), 459–502 (2018)
13. S.J. Montgomery-Smith, The distribution of Rademacher sums. Proc. Am. Math. Soc. **109**(2), 517–522 (1990)
14. M. Talagrand, Sharper bounds for Gaussian and empirical processes. Ann. Probab. **22**(1), 28–76 (1994)
15. A.W. van der Vaart, J.A. Wellner, *Weak convergence and empirical processes*. Springer Series in Statistics (Springer, New York, 1996)

Chapter 9
Reciprocals and Flowers in Convexity

Emanuel Milman, Vitali Milman, and Liran Rotem

Abstract We study new classes of convex bodies and star bodies with unusual properties. First we define the class of reciprocal bodies, which may be viewed as convex bodies of the form "$1/K$". The map $K \mapsto K'$ sending a body to its reciprocal is a duality on the class of reciprocal bodies, and we study its properties.

To connect this new map with the classic polarity we use another construction, associating to each convex body K a star body which we call its flower and denote by $K^\clubsuit$. The mapping $K \mapsto K^\clubsuit$ is a bijection between the class $\mathcal{K}_0^n$ of convex bodies and the class $\mathcal{F}^n$ of flowers. Even though flowers are in general not convex, their study is very useful to the study of convex geometry. For example, we show that the polarity map $\circ : \mathcal{K}_0^n \to \mathcal{K}_0^n$ decomposes into two separate bijections: First our flower map $\clubsuit : \mathcal{K}_0^n \to \mathcal{F}^n$, followed by a slight modification Φ of the spherical inversion which maps $\mathcal{F}^n$ back to $\mathcal{K}_0^n$. Each of these maps has its own properties, which combine to create the various properties of the polarity map.

We study the various relations between the four maps $\prime$, $\circ$, $\clubsuit$ and Φ and use these relations to derive some of their properties. For example, we show that a convex body K is a reciprocal body if and only if its flower $K^\clubsuit$ is convex.

We show that the class $\mathcal{F}^n$ has a very rich structure, and is closed under many operations, including the Minkowski addition. This structure has corollaries for the other maps which we study. For example, we show that if K and T are reciprocal bodies so is their "harmonic sum" $(K^\circ + T^\circ)^\circ$. We also show that the volume $\left| \left(\sum_i \lambda_i K_i \right)^\clubsuit \right|$ is a homogeneous polynomial in the λ_i's, whose coefficients can be

E. Milman
Technion – Israel Institute of Technology, Haifa, Israel
e-mail: emilman@tx.technion.ac.il

V. Milman
Tel-Aviv University, Tel Aviv, Israel
e-mail: milman@tauex.tau.ac.il

L. Rotem (✉)
University of Minnesota, Minneapolis, MN, USA
e-mail: lrotem@umn.edu

© Springer Nature Switzerland AG 2020
B. Klartag, E. Milman (eds.), *Geometric Aspects of Functional Analysis*,
Lecture Notes in Mathematics 2266,
https://doi.org/10.1007/978-3-030-46762-3_9

called "♣-type mixed volumes". These mixed volumes satisfy natural geometric inequalities, such as an elliptic Alexandrov–Fenchel inequality. More geometric inequalities are also derived.

9.1   Introduction

In this paper we study new classes of convex bodies and star bodies in $\mathbb{R}^n$ with some unusual properties. We will provide precise definitions below, but let us first describe the general program of what will follow.

One of our new classes, "reciprocal" bodies, may be viewed as bodies of the form "$\frac{1}{K}$" for a convex body K. They appear as the image of a new "quasi-duality" operation on the class $\mathcal{K}_0^n$ of convex bodies. We denote this new map by $K \mapsto K'$. This operation reverses order (with respect to inclusions) and has the property $K''' = K'$. Hence the map $'$ is indeed a duality on its image.

This new operation is connected to the classical operation of polarity $\circ : K \mapsto K^\circ$ via another construction, which we call simply the "flower" of a body K and denote by $\clubsuit : K \mapsto K^\clubsuit$. We provide the definition of $K^\clubsuit$ in Definition 9.3 below, but an equivalent description which sheds light on the "flower" nomenclature is

$$K^\clubsuit = \bigcup \left\{ B\left(\frac{x}{2}, \frac{|x|}{2}\right) : x \in K \right\}$$

(see Proposition 9.19). Here $B(y, r)$ is the Euclidean ball with center $y \in \mathbb{R}^n$ and radius $r \geq 0$. In other words, $K^\clubsuit$ is the union of all balls having diameter $[0, x]$ with $x \in K$.

In general, $K^\clubsuit$ is a star body which is not necessarily convex. The flower of a convex body was previously studied for very different reasons in the field of stochastic geometry—see Remark 9.7. We show that our new map $'$ is precisely $K' = (K^\clubsuit)^\circ$. We also show that K belongs to the image of $'$, i.e. K is a reciprocal body, if and only if $K^\clubsuit$ is convex. This means that such reciprocal bodies are in some sense "more convex" than other convex bodies, and can also be called "doubly convex" bodies.

Interestingly, the flower map $\clubsuit$ is also connected to the n-dimensional spherical inversion Φ when applied to star bodies (Φ is defined by applying the pointwise map $\mathcal{I}(x) = \frac{x}{|x|^2}$ and taking set complement—see Definition 9.11). We describe the class of convex bodies on which Φ preserves convexity.

The method of study of these questions looks novel and some of the results are not intuitive. Just as an example, we show that if $\Phi(A)$ and $\Phi(B)$ are convex (for some star bodies A and B) then $\Phi(A + B)$ is convex as well, where $A + B$ is the Minkowski addition (see Corollary 9.37).

The family $\mathcal{F}^n$ of flowers should play a central role in the study of convexity. It has a very rich structure. For example, it is closed under the Minkowski addition, and is also preserved by orthogonal projections and sections. "Flower mixed volumes" also exist and, perhaps most interestingly, we have a decomposition of the classical polarity operation as

$$\mathcal{K}_0^n \xrightarrow{\clubsuit} \mathcal{F}^n \xrightarrow{\Phi} \mathcal{K}_0^n.$$

Here the maps $\clubsuit$ and Φ are 1–1 and onto, and we have $\circ = \Phi\clubsuit$ in the sense that $K^\circ = \Phi\left(K^{\clubsuit}\right)$ for all $K \in \mathcal{K}_0^n$.

The class of reciprocal bodies also looks interesting. No polytope belongs to this class, and no centrally symmetric ellipsoids (besides Euclidean balls centered at 0). At the same time this class is clearly important, as seen from its properties and the fact that it coincides with the "doubly convex" bodies. We provide several two-dimensional pictures to help create some intuition about this class of reciprocal bodies and about the class of flowers.

To make the above claims more precise, let us now give some basic definitions and fix our notation. The reader may consult [14] for more information. By a *convex body* in $\mathbb{R}^n$ we mean a set $K \subseteq \mathbb{R}^n$ which is closed and convex. We will always assume further that $0 \in K$, but we do not assume that K is compact or has non-empty interior. We denote the set of all such bodies by $\mathcal{K}_0^n$. The *support function* of K is the function $h_K : S^{n-1} \to [0, \infty]$ defined by $h_K(\theta) = \sup_{x \in K} \langle x, \theta \rangle$. Here $S^{n-1} = \{\theta \in \mathbb{R}^n : |\theta| = 1\}$ is the unit Euclidean sphere, and $\langle \cdot, \cdot \rangle$ is the standard scalar product on $\mathbb{R}^n$. The function h_K uniquely defines the body K.

The Minkowski sum of two convex bodies is defined by

$$K + T = \overline{\{x + y : x \in K, \ y \in T\}}$$

(the closure is not needed if K or T is compact). The homothety operation is defined by $\lambda K = \{\lambda x : x \in K\}$. These operations are related to the support function by the identity $h_{\lambda K + T} = \lambda h_K + h_T$.

We say that $A \subseteq \mathbb{R}^n$ is a star set if A is non-empty and $x \in A$ implies that $\lambda x \in A$ for all $0 \leq \lambda \leq 1$. The *radial function* $r_A : S^{n-1} \to [0, \infty]$ of A is defined by $r_A(\theta) = \sup\{\lambda \geq 0 : \lambda\theta \in A\}$. For us, a *star body* is simply a star set which is radially closed, in the sense that $r_A(\theta)\theta \in A$ for all directions $\theta \in S^{n-1}$ satisfying $r_A(\theta) < \infty$. For such bodies r_A uniquely defines A.

The *polarity map* $\circ : \mathcal{K}_0^n \to \mathcal{K}_0^n$ maps every body K to its polar

$$K^\circ = \left\{y \in \mathbb{R}^n : \langle x, y \rangle \leq 1 \text{ for all } x \in K \right\}. \tag{9.1.1}$$

It follows that $h_K = \frac{1}{r_{K^\circ}}$. The polarity map is a duality in the following sense:

- It is order reversing: If $K \subseteq T$ then $K^\circ \supseteq T^\circ$.
- It is an involution: $K^{\circ\circ} = K$ for all $K \in \mathcal{K}_0^n$ (if A is only a star body, then $A^{\circ\circ}$ is the closed convex hull of A).

In fact, it was proved in [1] that the polarity map is essentially the *only* duality on $\mathcal{K}_0^n$ (see also [15] which proves a strengthening of this fact). Similar results on different classes of convex bodies were proved earlier in [6] and [3].

The structure of a set equipped with a duality relation is common in mathematics. A basic example is the set $[0, \infty]$ equipped with the inversion $x \mapsto x^{-1}$ (we set of course $0^{-1} = \infty$ and $\infty^{-1} = 0$). Following this analogy, one may think of K° as a certain inverse "K^{-1}". This point of view can indeed be useful—see for example [12] and [9].

However, in recent works ([10, 11]), the authors discussed the application of functions such as $x \mapsto x^\alpha$ ($0 \le \alpha \le 1$) and $x \mapsto \log x$ to convex bodies. Applying the same idea to the inversion $x \mapsto \frac{1}{x}$, we obtain a new notion of the reciprocal body "K^{-1}". Recall that given a function $g : S^{n-1} \to [0, \infty]$, its *Alexandrov body*, or *Wulff shape*, is defined by

$$A[g] = \left\{ x \in \mathbb{R}^n : \langle x, \theta \rangle \le g(\theta) \text{ for all } \theta \in S^{n-1} \right\}.$$

In other words, $A[g]$ is the biggest convex body such that $h_{A[g]} \le g$. In particular, for every convex body K we have $K = A[h_K]$. We may now define:

Definition 9.1 Given $K \in \mathcal{K}_0^n$, the *reciprocal body* $K' \in \mathcal{K}_0^n$ is defined by $K' = A\left[\frac{1}{h_K}\right]$.

More explicitly, we have

$$K' = \bigcap_{\theta \in S^{n-1}} H^- \left(\theta, h_K(\theta)^{-1} \right),$$

where $H^-(\theta, c) = \{x \in \mathbb{R}^n : \langle x, \theta \rangle \le c\}$.

The idea of constructing new interesting convex bodies as Alexandrov bodies is not new. As one important recent example, Böröczky, Lutwak, Yang and Zhang consider in [4] the body $A\left[h_K^{1-\lambda} h_L^\lambda\right]$, which they call the λ-logarithmic mean of K and L.

Figure 9.1 depicts some simple convex bodies in $\mathbb{R}^2$ and their reciprocal. Some basic properties of the reciprocal map $K \mapsto K'$ are immediate from the definition:

Proposition 9.2 *For all $K, T \in \mathcal{K}_0^n$ we have:*

1. $K' \subseteq K^\circ$, *with an equality if and only if K is a Euclidean ball.*
2. *If $K \supseteq T$ then $K' \subseteq T'$.*
3. $K'' \supseteq K$.
4. $K''' = K'$.

Proof For (1), note that for every $\theta \in S^{n-1}$ we have $1 = \langle \theta, \theta \rangle \le h_K(\theta) h_{K^\circ}(\theta)$. Hence $K^\circ = A[h_{K^\circ}] \ge A\left[\frac{1}{h_K}\right] = K'$. An equality $K' = K^\circ$ implies that $h_{K^\circ} = \frac{1}{h_K}$, or equivalently $r_K = \frac{1}{h_{K^\circ}} = h_K$. This implies that K is a ball.

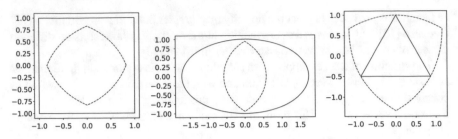

Fig. 9.1 Convex bodies (solid) and their reciprocals (dashed)

Property (2) is obvious from the definition.

For property (3), we know that $h_{K'} \leq \frac{1}{h_K}$ so $K'' = A\left[\frac{1}{h_{K'}}\right] \geq A[h_K] = K$.

Finally, (4) is a formal consequence of (2) and (3): We know that $K'' \supseteq K$, so $K''' \subseteq K'$. On the other hand applying (3) to K' gives $K''' \supseteq K'$.   □

Let us write

$$\mathcal{R}^n = \left\{ K' : \ K \in \mathcal{K}_0^n \right\}.$$

Note that properties (2) and (4) above imply that $'$ is a duality on the class $\mathcal{R}^n$. Also note that $K \in \mathcal{R}^n$ if and only if $K'' = K$.

Our next goal is to give an alternative description of the reciprocal body K'. Towards this goal we define:

Definition 9.3

1. For a convex body $K \in \mathcal{K}_0^n$ we denote by $K^{\clubsuit}$ the star body with radial function $r_{K^{\clubsuit}} = h_K$.
2. We say that a star body $A \subseteq \mathbb{R}^n$ is a *flower* if $A = \bigcup_{x \in C} B\left(\frac{x}{2}, \frac{|x|}{2}\right)$, where $C \subseteq \mathbb{R}^n$ is some closed set. The class of all flowers in $\mathbb{R}^n$ is denoted by $\mathcal{F}^n$.

The two parts of the definition are related by the following:

Theorem 9.4 *For every* $K \in \mathcal{K}_0^n$ *we have* $K^{\clubsuit} \in \mathcal{F}^n$. *Moreover, the map* $\clubsuit : \mathcal{K}_0^n \to \mathcal{F}^n$ *is a bijection. Equivalently, every flower A is of the form* $A = K^{\clubsuit}$ *for a unique* $K \in \mathcal{K}_0^n$; *We have* $A = \bigcup_{x \in K} B\left(\frac{x}{2}, \frac{|x|}{2}\right)$, *and we simply say that A is the flower of* K.

This theorem is a combination of Proposition 9.17(2), Proposition 9.19, and Remark 9.21.

As we will see flowers play an important role in connecting the reciprocity map to the polarity map. Note that in general $K^{\clubsuit}$ is not convex. Figure 9.2 depicts the flowers of some convex bodies in $\mathbb{R}^2$. Another example that will be important in the sequel is the following:

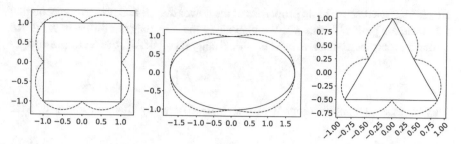

Fig. 9.2 Convex bodies (solid) and their flowers (dashed)

Example 9.5 For $x \in \mathbb{R}^n$ write $[0, x] = \{\lambda x : 0 \leq \lambda \leq 1\}$. Also denote the Euclidean ball with center x and radius $r > 0$ by $B(x, r)$, and write $B_x = B\left(\frac{x}{2}, \frac{|x|}{2}\right)$. Then $[0, x]^{\clubsuit} = B_x$. Indeed, a direct computation gives

$$h_{[0,x]}(\theta) = r_{B_x}(\theta) = \max\{\langle x, \theta \rangle, 0\}.$$

The identity $[0, x]^{\clubsuit} = B_x$ is also a classical theorem in geometry sometimes referred to as Thales's theorem: If an interval $[a, b] \subseteq \mathbb{R}^n$ is a diameter of a ball B, then ∂B is precisely the set of points y such that $\angle ayb = 90^\circ$.

The polarity map, the reciprocal map and the flower are all related via the following formula:

Proposition 9.6 *For every* $K \in \mathcal{K}_0^n$ *we have* $\left(K^{\clubsuit}\right)^\circ = K'$.

Note that even though in general $K^{\clubsuit} \notin \mathcal{K}_0^n$, we may still compute its polar using (9.1.1).

Proof By definition $x \in \left(K^{\clubsuit}\right)^\circ$ if and only if $\langle x, y \rangle \leq 1$ for all $y \in K^{\clubsuit}$. It is obviously enough to check this for $y \in \partial K^{\clubsuit}$, i.e. $y = r_{K^{\clubsuit}}(\theta)\theta = h_K(\theta)\theta$ for some $\theta \in S^{n-1}$.

Hence $x \in \left(K^{\clubsuit}\right)^\circ$ if and only if for all $\theta \in S^{n-1}$ we have $\langle x, h_K(\theta)\theta \rangle \leq 1$, or $\langle x, \theta \rangle \leq \frac{1}{h_K(\theta)}$. This means that $x \in A\left[\frac{1}{h_K}\right] = K'$. $\qquad \square$

Remark 9.7 The flower of a convex body was studied in stochastic geometry under the name "Voronoi Flower" (see e.g. [16]). The reason for the name is the following relation to Voronoi tessellations: For a discrete set of points $P \subseteq \mathbb{R}^n$, consider the (open) Voronoi cell

$$Z = \{x \in \mathbb{R}^n : |x - 0| < |x - y| \text{ for all } y \in P\}.$$

Then for any convex body K we have $Z \supseteq K$ if and only if $P \cap \left(2K^{\clubsuit}\right) = \emptyset$. It follows that if for example P is chosen according to a homogeneous Poisson point process, then the probability that $Z \supseteq K$ is computable from the volume of $K^{\clubsuit}$.

In Sect. 9.2 we discuss basic properties of the flower map ♣ and prove representation formulas for both $K^{\clubsuit}$ and K'. We also study the pre-images of a body $K \in \mathcal{R}^n$ under the reciprocity map. Since $\prime$ is not a duality on all of $\mathcal{K}_0^n$, the set of pre-images

$$\left\{ A \in \mathcal{K}_0^n : A' = K \right\}$$

may in general contain more than one body. We study this set, and prove the following results:

Theorem 9.8

1. *If $K \in \mathcal{R}^n$ is a* smooth *convex body then $K = A'$ for a unique $A \in \mathcal{K}_0^n$.*
2. *For a general $K \in \mathcal{R}^n$, the set $\left\{ A \in \mathcal{K}_0^n : A' = K \right\}$ is a convex subset of $\mathcal{K}_0^n$.*

The main goal of Sect. 9.3 is to prove the following theorem, characterizing the class $\mathcal{R}^n$ of reciprocal bodies:

Theorem 9.9 $K \in \mathcal{R}^n$ *if and only if $K^{\clubsuit}$ is convex.*

As a corollary we obtain:

Corollary 9.10 *For every $K \in \mathcal{R}^n$ and every subspace $E \subseteq \mathbb{R}^n$ one has $\left(\mathrm{Proj}_E K \right)' = \mathrm{Proj}_E K'$, where Proj_E denotes the orthogonal projection onto E.*

As we will see, this corollary is false without the assumption that $K \in \mathcal{R}^n$. We will prove Theorem 9.9 by connecting the various maps we constructed so far with another duality on the class of star-bodies:

Definition 9.11

1. Let $\mathcal{I} : \mathbb{R}^n \setminus \{0\} \to \mathbb{R}^n \setminus \{0\}$ denote the spherical inversion $\mathcal{I}(x) = \frac{x}{|x|^2}$.
2. Given a star body A, we denote by $\Phi(A)$ the star body with radial function $r_{\Phi(A)} = \frac{1}{r_A}$.

The map $A \mapsto \Phi(A)$ is obviously a duality on the class of star bodies. It is sometimes called star duality and denoted by A^* (see [13]), but we will prefer the notation $\Phi(A)$. Note that Φ is "essentially the same" as the pointwise map $\mathcal{I}$ in the sense that $\partial \Phi(A) = \mathcal{I}(\partial A)$, but $\mathcal{I}$ maps the interior of A to the exterior of $\Phi(A)$ and vice versa. Here by the boundary ∂A of a star body A we mean

$$\partial A = \left\{ r_A(\theta)\theta : \theta \in S^{n-1} \text{ such that } 0 < r_A(\theta) < \infty \right\}.$$

One interesting relation between Φ and our previous definitions is the following (see Propositions 9.28(2) and 9.33):

Theorem 9.12 Φ *is a bijection between* $\mathcal{K}_0^n$ *and* $\mathcal{F}^n$. *Moreover, the polarity map decomposes as*

$$\circ : \mathcal{K}_0^n \xrightarrow{\;\clubsuit\;} \mathcal{F}^n \xrightarrow{\;\Phi\;} \mathcal{K}_0^n,$$

in the sense that $\Phi\left(K^{\clubsuit}\right) = K^{\circ}$ *for all* $K \in \mathcal{K}_0^n$.

In Sect. 9.4 we use the results of Sect. 9.3 to further study the class of flowers, with applications to the study of reciprocity and the map Φ. First we understand when the map Φ preserves convexity. By Theorem 9.12, as Φ is an involution, we know that $\Phi(A)$ is convex if and only if A is a flower. When A is in addition convex, we have:

Theorem 9.13 *If* $K \in \mathcal{K}_0^n$ *then* $\Phi(K)$ *is convex if and only if* $K^{\circ} \in \mathcal{R}^n$.

(See Proposition 9.33). We then show that the class $\mathcal{F}^n$ has a lot of structure:

Theorem 9.14 *Fix* $A, B \in \mathcal{F}^n$ *and a linear subspace* $E \subseteq \mathbb{R}^n$. *Then* $A + B$ *and* conv A *are flowers in* $\mathbb{R}^n$, *and* $A \cap E$ *and* $\mathrm{Proj}_E A$ *are flowers in* E.

(See Propositions 9.35, 9.39 and 9.40). As corollaries we obtain:

Corollary 9.15

1. *If* $K, T \in \mathcal{R}^n$ *then* $(K^{\circ} + T^{\circ})^{\circ} \in \mathcal{R}^n$.
2. *If* K, T *are convex bodies then* $\Phi\left(\Phi(K) + \Phi(T)\right)$ *is also convex.*

As another corollary we construct a new addition $\oplus$ on $\mathcal{K}_0^n$ such that the class $\mathcal{R}^n$ is closed under $\oplus$. Moreover, when restricted to $\mathcal{R}^n$, this new addition has all properties one may expect: it is associative, commutative and monotone, it has $\{0\}$ as an identity element, and it satisfies $\lambda K \oplus \mu K = (\lambda + \mu) K$.

The final Sect. 9.5 is devoted to the study of inequalities. We begin by showing that the maps $\clubsuit$, Φ and $\prime$ are all convex in appropriate senses. We also study the functional $K \mapsto \left| K^{\clubsuit} \right|$, where $|\cdot|$ denotes the volume. We prove results that are analogous to Minkowski's theorem of polynomiality of volume and to the Alexandrov–Fenchel inequality:

Theorem 9.16 *Fix* $K_1, K_2, \ldots, K_m \in \mathcal{K}_0^n$. *Then*

$$\left| (\lambda_1 K_1 + \lambda_2 K_2 + \cdots + \lambda_m K_m)^{\clubsuit} \right| = \sum_{i_1, i_2, \ldots, i_n = 1}^{m} V^{\clubsuit}(K_{i_1}, K_{i_2}, \ldots, K_{i_n}) \cdot \lambda_{i_1} \lambda_{i_2} \cdots \lambda_{i_n},$$

where the coefficients are given by

$$V^{\clubsuit}(K_1, K_2, \ldots, K_n) = \left| B_2^n \right| \cdot \int_{S^{n-1}} h_{K_1}(\theta) h_{K_2}(\theta) \cdots h_{K_n}(\theta) \mathrm{d}\sigma(\theta)$$

(Here B_2^n denotes the unit Euclidean ball). Moreover, for every $K_1, K_2, \ldots, K_n \in \mathcal{K}_0^n$ we have

$$V^\clubsuit(K_1, K_2, K_3, \ldots, K_n)^2 \leq V^\clubsuit(K_1, K_1, K_3, \ldots K_n) \cdot V^\clubsuit(K_2, K_2, K_3, \ldots, K_n).$$

These results and their proofs are similar in spirit to the dual Brunn–Minkowski theory which was developed by Lutwak in [8]. We also prove a Kubota type formula for the new $\clubsuit$-quermassintegrals, and use it to compare them with the classical definition.

9.2 Properties of Reciprocity and Flowers

We begin this section with some basic properties of flowers:

Proposition 9.17

1. *For every $K \in \mathcal{K}_0^n$ we have $K^\clubsuit \supseteq K$, with equality if and only if K is an Euclidean ball.*
2. *If $K^\clubsuit = T^\clubsuit$ for $K, T \in \mathcal{K}_0^n$ then $K = T$.*
3. *Let $\{K_i\}_{i \in I}$ be a family of convex bodies. Then $\left(\text{conv}\left(\bigcup_{i \in I} K_i\right)\right)^\clubsuit = \bigcup_{i \in I} K_i^\clubsuit$.*
4. *For every $K \in \mathcal{K}_0^n$ and every subspace $E \subseteq \mathbb{R}^n$ we have $\left(\text{Proj}_E K\right)^\clubsuit = K^\clubsuit \cap E$ (where the $\clubsuit$ on the left hand side is taken inside the subspace E).*

Proof For (1) we have $r_{K^\clubsuit} = h_K \geq r_K$. The equality case is the same as in Proposition 9.2(1).

(2) is obvious since h_K uniquely defines K. For (3), write $A = \text{conv}\left(\bigcup_{i \in I} K_i\right)$ and $B = \bigcup_{i \in I} K_i^\clubsuit$. Then

$$r_{A^\clubsuit} = h_A = \max_{i \in I} h_{K_i} = \max_{i \in I} r_{K_i^\clubsuit} = r_B,$$

so $A^\clubsuit = B$.

Finally, for (4), since both bodies are inside E it is enough to check that their radial functions coincide in E. But if $\theta \in S^{n-1} \cap E$ then

$$r_{(\text{Proj}_E K)^\clubsuit}(\theta) = h_{\text{Proj}_E K}(\theta) = h_K(\theta) = r_{K^\clubsuit}(\theta) = r_{K^\clubsuit \cap E}(\theta),$$

proving the claim. $\qquad\qquad\qquad\qquad\qquad\qquad\qquad\qquad\qquad\qquad\qquad\qquad\qquad\qquad\square$

We will also need the following computation:

Lemma 9.18 *Let $B_x = B\left(\frac{x}{2}, \frac{|x|}{2}\right)$ be the ball with center $\frac{x}{2}$ and radius $\frac{|x|}{2}$. Let P_x be the paraboloid,*

$$P_x = \left\{ y \in \mathbb{R}^n : \langle y, x \rangle \leq 1 - \frac{1}{4}|x|^2 \left|Proj_{x^\perp} y\right|^2 \right\},$$

where $Proj_{x^\perp}$ denotes the orthogonal projection to the hyperplane orthogonal to x. Then $B_x^\circ = P_x$.

Proof It is enough to prove the result for $x = e_n = (0, 0, \ldots, 0, 1)$. Indeed, we can a write $x = \lambda \cdot u(e_n)$ for some orthogonal matrix u and some $\lambda > 0$, and then

$$(B_x)^\circ = \left(\lambda \cdot u\left(B_{e_n}\right)\right)^\circ = \frac{1}{\lambda} \cdot u\left(B_{e_n}^\circ\right) = \frac{1}{\lambda} \cdot u\left(P_{e_n}\right) = P_x.$$

Write a general point $y \in \mathbb{R}^n$ as $y = (z, t) \in \mathbb{R}^{n-1} \times \mathbb{R}$. Since $B_x = [0, x]^\clubsuit$ we know that

$$r_{B_{e_n}}(z, t) = h_{[0, e_n]}(z, t) = \max\{t, 0\}.$$

Hence we have

$$h_{B_{e_n}}(z, t) = \max_{\theta \in S^{n-1}} \langle (z, t), r_{B_{e_n}}(\theta)\theta \rangle = \max_{(u, s) \in S^{n-1}} \langle (z, t), (u, s) \rangle \max\{s, 0\}$$

$$= \max_{(u, s) \in \mathbb{R}^{n-1} \times \mathbb{R}} \left(\frac{\langle z, u \rangle + ts}{|u|^2 + s^2} \cdot \max\{s, 0\} \right).$$

It is obviously enough to maximize over $s > 0$, and by homogeneity we may take $s = 1$. It is also clear that the maximum is attained when $u = r \cdot \frac{z}{|z|}$ for some r. Therefore

$$h_{B_{e_n}}(z, t) = \max_r \left(\frac{r|z| + t}{r^2 + 1} \right).$$

We see that $(z, t) \in B_{e_n}^\circ$ if and only if for all r we have $\frac{r|z| + t}{r^2 + 1} \leq 1$, or $r^2 - |z| r + 1 - t \geq 0$. This happens exactly when $|z|^2 - 4(1 - t) \leq 0$, or $t \leq 1 - \frac{|z|^2}{4}$. Hence $B_{e_n}^\circ = P_{e_n}$ like we wanted. □

Hence we obtain the following descriptions of $K^\clubsuit$ and K':

Proposition 9.19 *For every $K \in \mathcal{K}_0^n$ we have $K^\clubsuit = \bigcup_{x \in K} B_x$, and $K' = \bigcap_{x \in K} P_x$.*

Proof Since $K = \text{conv}\left(\bigcup_{x \in K}[0, x]\right)$, Proposition 9.17(3) implies that $K^{\clubsuit} = \bigcup_{x \in K} B_x$. Hence

$$K' = \left(K^{\clubsuit}\right)^{\circ} = \bigcap_{x \in K} B_x^{\circ} = \bigcap_{x \in K} P_x.$$

$\square$

Remark 9.20 If K is compact, the same proof shows that it is enough to consider only $x \in \partial K$. In fact we can do a bit more: recall that $x \in \partial K$ is an extremal point for K if any representation $x = (1 - \lambda)y + \lambda z$ for $0 < \lambda < 1$ and $y, z \in K$ implies that $y = z = x$. Denote the set of extremal points by $\text{Ext}(K)$. By the Krein–Milman theorem[1] we have $K = \text{conv}\left(\bigcup_{x \in \text{Ext}(K)}[0, x]\right)$, so $K^{\clubsuit} = \bigcup_{x \in \text{Ext}(K)} B_x$ and $K' = \bigcap_{x \in \text{Ext}(K)} P_x$. In particular if K is a polytope then $K^{\clubsuit}$ is the union of finitely many balls and K' is the intersection of finitely many paraboloids.

Remark 9.21 The formulas of Proposition 9.19 can be used to define $K^{\clubsuit}$ and K' for non-convex sets (say compact). However, it turns out that under such definitions we have $K^{\clubsuit} = (\text{conv} K)^{\clubsuit}$ and $K' = (\text{conv } K)'$, so essentially nothing new is gained. To see that $K^{\clubsuit} = (\text{conv} K)^{\clubsuit}$ note that by the remark above

$$(\text{conv} K)^{\clubsuit} = \bigcup_{x \in \text{Ext}(\text{conv } K)} B_x \subseteq \bigcup_{x \in K} B_x = K^{\clubsuit}.$$

Let us now give one application of Proposition 9.19. We say that $K \in \mathcal{K}_0^n$ is *smooth* if K is compact, $0 \in \text{int } K$, and at every point $x \in \partial K$ there exists a unique supporting hyperplane to K. We say that $K \in \mathcal{K}_0^n$ is *strictly convex* if K is compact, $0 \in \text{int } K$ and $\text{Ext}(K) = \partial K$. It is a standard fact in convexity that K is smooth if and only if its polar K° is strictly convex (see, e.g. Proposition 1.e.2 of [7]).

Theorem 9.22 *Assume $K \in \mathcal{K}_0^n$ is compact and $0 \in \text{int } K$. Then K' is strictly convex.*

Ideologically, the theorem follows from the fact that for every $0 < r < R < \infty$ the family

$$\{P_x \cap B(0, R) : r < |x| < R\}$$

is "uniformly convex", i.e. has a uniform lower bound on its modulus of convexity. It then follows that an arbitrary intersection of such bodies will be strictly convex as well. In particular, since for $R > 0$ large enough we have $K' = \bigcap_{x \in \partial K} (P_x \cap B(0, R))$, it follows that K' is strictly convex. Since filling in the

---

[1] In the finite dimensional case the Krein–Milman theorem was first proved by Minkowski. See [14] and in particular the first note of Section 1.4.

computational details is tedious and not very illuminating, we will omit the formal proof.

Instead, let us now fix a reciprocal body $K \in \mathcal{R}^n$, and discuss the class of "pre-reciprocals" $\{A \in \mathcal{K}_0^n : A' = K\}$. It is obvious that such pre-reciprocals are in general not unique. For example, if $A \notin \mathcal{R}^n$ then A and A'' are two different pre-reciprocals of A'.

However, sometimes it is true that the pre-reciprocal is unique:

Proposition 9.23 *Let K be a smooth convex body. Then there exists at most one body A such that $A' = K$.*

Proof Assume $A' = B' = K$. Then $(A^\clubsuit)^\circ = (B^\clubsuit)^\circ = K$, which implies that $\mathrm{conv}\,(A^\clubsuit) = \mathrm{conv}\,(B^\clubsuit) = K^\circ$.

Since $\mathrm{conv}\,(A^\clubsuit) = K^\circ$ we have $A^\clubsuit \supseteq \mathrm{Ext}(K^\circ)$. Since K is smooth its polar is strictly convex, so $A^\clubsuit \supseteq \partial K^\circ$. But $A^\clubsuit$ is a star body, so we must have $A^\clubsuit = K^\circ$. Similarly $B^\clubsuit = K^\circ$, and since $A^\clubsuit = B^\clubsuit$ we conclude that $A = B$. $\qquad\square$

When K is not smooth it may have many pre-reciprocals, but something can still be said: The set $\mathcal{D}(K) = \{A \in \mathcal{K}_0^n : A' = K\}$ is a convex subset of $\mathcal{K}_0^n$.

Theorem 9.24

1. *Fix $K \in \mathcal{K}_0^n$ such that $0 \in \mathrm{int}\,K$. If $A, B \in \mathcal{D}(K)$ then $\lambda A + (1 - \lambda)B \in \mathcal{D}(K)$ for all $0 \leq \lambda \leq 1$.*
2. *If $K \in \mathcal{K}_0^n$ and $\mathcal{D}(K) \neq \emptyset$ then K' is the largest body in $\mathcal{D}(K)$.*

For the proof we need the following lemma:

Lemma 9.25 *Let $X, Y \subseteq \mathbb{R}^n$ be compact sets such that $\mathrm{conv}\,X = \mathrm{conv}\,Y = T$. Then $\mathrm{conv}\,(X \cap Y) = \mathrm{conv}\,(X \cup Y) = T$.*

Proof For the union this is trivial: On the one hand $\mathrm{conv}\,(X \cup Y) \supseteq \mathrm{conv}\,X = T$. On the other hand $X \cup Y \subseteq T$ and T is convex, so $\mathrm{conv}\,(X \cup Y) \subseteq T$.

For the intersection, the inclusion $\mathrm{conv}\,(X \cap Y) \subseteq T$ is again obvious. Conversely, since $\mathrm{conv}\,X = \mathrm{conv}\,Y = T$ it follows that $X, Y \supseteq \mathrm{Ext}(T)$, so $X \cap Y \supseteq \mathrm{Ext}\,T$. It follows from the Krein--Milman theorem that $\mathrm{conv}\,(X \cap Y) \supseteq \mathrm{conv}\,(\mathrm{Ext}\,T) = T$. $\qquad\square$

Proof of Theorem 9.24 For (1), fix $A, B \in \mathcal{D}(K)$. Since $A' = B' = K$ we have $\mathrm{conv}\,(A^\clubsuit) = \mathrm{conv}\,(B^\clubsuit) = K^\circ$. Since $0 \in \mathrm{int}\,K$ we know that K° is compact, and hence $A^\clubsuit$ and $B^\clubsuit$ are compact as well.

Write $C = \lambda A + (1 - \lambda)B$. We have

$$r_{C^\clubsuit} = h_C = \lambda h_A + (1 - \lambda)h_B \leq \max\{h_A, h_B\} = \max\{r_{A^\clubsuit}, r_{B^\clubsuit}\} = r_{A^\clubsuit \cup B^\clubsuit}.$$

Hence $C^{\clubsuit} \subseteq A^{\clubsuit} \cup B^{\clubsuit}$, and similarly $C^{\clubsuit} \supseteq A^{\clubsuit} \cap B^{\clubsuit}$. It follows that

$$K^{\circ} = \operatorname{conv}\left(A^{\clubsuit} \cap B^{\clubsuit}\right) \subseteq \operatorname{conv} C^{\clubsuit} \subseteq \operatorname{conv}\left(A^{\clubsuit} \cup B^{\clubsuit}\right) = K^{\circ},$$

so $C' = \left(C^{\clubsuit}\right)^{\circ} = K^{\circ\circ} = K$.

For (2), $\mathcal{D}(K) \neq \emptyset$ exactly means that $K \in \mathcal{R}^n$, so $K'' = K$ and $K' \in \mathcal{D}(K)$. For any other $A \in \mathcal{D}(K)$ we have $A \subseteq A'' = K'$ so K' is indeed the largest body in $\mathcal{D}(K)$. □

Note that Theorem 9.24 gives us a *partition* of the family of compact convex bodies in $\mathbb{R}^n$ into *convex* sub-families, where A and B belong to the same sub-family if and only if $A' = B'$.

We conclude this section by turning our attention to Theorem 9.9. For the full proof we will need some new ideas, presented in the next section. But the ideas we developed so far suffice to give a simple geometric proof of the theorem in some cases. We find it worthwhile, as the proof of Sect. 9.3 is not intuitive, and the following proof shows why convexity of $K^{\clubsuit}$ plays a role. Let us show the following:

Proposition 9.26 *Assume that $K \in \mathcal{R}^n$ is smooth. Then $K^{\clubsuit}$ is convex.*

Proof Assume by contradiction that $K^{\clubsuit}$ is not convex. Then we can choose a point $x \in \partial K^{\clubsuit} \cap \operatorname{int}\left(\operatorname{conv} K^{\clubsuit}\right)$. Write $\hat{x} = \frac{x}{|x|}$. Since

$$h_K\left(\hat{x}\right) = r_{K^{\clubsuit}}\left(\hat{x}\right) = |x|,$$

we conclude that the hyperplane $H_x = \{z : \langle z - x, x \rangle = 0\}$ is a supporting hyperplane for K. Fix a point $y \in \partial K \cap H_x$.

Since $y \in K$ we know that $[0, y] \subseteq K$, so $B_y = [0, y]^{\clubsuit} \subseteq K^{\clubsuit}$. We claim that $B_y \cap \partial K^{\clubsuit} = \{x\}$. Indeed, by elementary geometry (see Example 9.5) we know that $w \in \partial B_y$ if and only if $\angle 0wy = 90°$, i.e. $\langle w, y - w \rangle = 0$. This is also easy to check algebraically. Since $y \in H_x$ we know that $\langle y - x, x \rangle = 0$, so $x \in B_y$.

Conversely, if $w \in B_y \cap \partial K^{\clubsuit}$ then $y \in H_w = \{z : \langle z - w, w \rangle = 0\}$. Again since $w \in \partial K^{\clubsuit}$ we conclude that H_w is a supporting hyperplane for K. Since H_x and H_w are two supporting hyperplanes passing through y, and since K is smooth, we must have $H_x = H_w$, so $x = w$. This proves the claim.

It follows in particular that $B_y \subseteq \operatorname{int}\left(\operatorname{conv} K^{\clubsuit}\right)$. Since B_y is compact and $\operatorname{int}\left(\operatorname{conv} K^{\clubsuit}\right)$ is open, it follows that $B_z \subseteq \operatorname{int}\left(\operatorname{conv} K^{\clubsuit}\right)$ for all z close enough to y. In particular one may take $z = (1 + \varepsilon)y$ for a small enough $\varepsilon > 0$. Since $y \in \partial K, z \notin K$.

Define $P = \operatorname{conv}(K, z) = \operatorname{conv}(K \cup [0, z])$. Then

$$P^{\clubsuit} = K^{\clubsuit} \cup [0, z]^{\clubsuit} = K^{\clubsuit} \cup B_z \subseteq \operatorname{conv}\left(K^{\clubsuit}\right).$$

Hence $\operatorname{conv}\left(P^{\clubsuit}\right) = \operatorname{conv}\left(K^{\clubsuit}\right)$, so $P' = K'$. But then $K'' = P'' \supseteq P \supsetneq K$, so $K \notin \mathcal{R}^n$. □

9.3 The Spherical Inversion and a Proof of Theorem 9.9

The main goal of this section is to prove Theorem 9.9: $K \in \mathcal{R}^n$ if and only if $K^{\clubsuit}$ is convex. For the proof we will use the maps $\mathcal{I}$ and Φ from Definition 9.3. We will also use the following well-known property of $\mathcal{I}$ (see e.g. Theorem 5.2 of [2]):

Fact 9.27 Let $A \subseteq \mathbb{R}^n$ be a sphere or a hyperplane. Then $\mathcal{I}(A)$ is a hyperplane if $0 \in A$, and a sphere if $0 \notin A$.

It follows that if B is any ball such that $0 \in B$, then $\Phi(B)$ is either a ball (if $0 \in \operatorname{int} B$) or a half-space (if $0 \in \partial B$).

Since in this section we will compose many operations, it will be more convenient to write them in function notation, where composition is denoted by juxtaposition. For example, by $\circ \Phi \clubsuit K$ we mean $\left(\Phi\left(K^{\clubsuit}\right)\right)^{\circ}$. In particular $\circ\circ = \operatorname{conv}$, the (closed) convex hull operation. We have the following relations between the different maps:

Proposition 9.28 *If* $K \in \mathcal{K}_0^n$ *then*

1. $\circ \clubsuit K = K'$.
2. $\Phi \clubsuit K = \circ K$.
3. $\Phi \circ K = \clubsuit K$.
4. $\clubsuit \circ K = \Phi K$.
5. $(\circ K)' = \circ \Phi K$.

Proof Identity (1) is the same as Proposition 9.6.

For (2) we compare radial functions:

$$r_{\Phi \clubsuit K} = \frac{1}{r_{\clubsuit K}} = \frac{1}{h_K} = r_{\circ K}.$$

(3) follows from (2) by applying Φ to both sides.

For (4) we apply (3) to $\circ K$ instead of K and obtain

$$\clubsuit \circ K = \Phi \circ \circ K = \Phi K.$$

(5) is obtained from (4) by taking polar of both sides and applying (1). $\qquad\square$

Note that Proposition 9.28(2) provides a decomposition of the classical duality to a "global" part (the flower) and an "essentially pointwise" part (the map Φ). Also note that the identities (2) and (3) actually hold for all star bodies, since $\clubsuit A = \clubsuit \operatorname{conv} A$ and $\circ A = \circ \operatorname{conv} A$. The convexity of K is crucial however for identity (4), and for general star bodies we only have $\clubsuit \circ A = \Phi \operatorname{conv} A$.

We will also need to know the following construction and its properties, which may be of independent interest:

Definition 9.29 The *spherical inner hull* of a convex body K is defined by

$$\text{Inn}_S K = \bigcup \{B(x, |x|) : \ B(x, |x|) \subseteq K\}.$$

Proposition 9.30 *Fix* $K \in \mathcal{K}_0^n$. *Then*

1. We have the identity

$$\text{Inn}_S K = \Phi \operatorname{conv} \Phi K = \Phi \circ \circ \Phi K \tag{9.3.1}$$

2. $\text{Inn}_S K \in \mathcal{K}_0^n$. In other words, (9.3.1) always defines a convex subset of K.
3. $\text{Inn}_S K$ is the largest star body $A \subseteq K$ such that $\Phi(A)$ is convex. In particular $\text{Inn}_S K = K$ if and only if $\Phi(K)$ is convex.

Proof For (1) we should prove that $\Phi \operatorname{conv} \Phi K = \text{Inn}_S K$, or equivalently that $\operatorname{conv} \Phi K = \Phi \text{Inn}_S K$. Since Φ is a duality on star bodies we have

$$\Phi \text{Inn}_S K = \bigcap \{\Phi B(x, |x|) : \ \Phi B(x, |x|) \supseteq \Phi K\}.$$

Since $\{B(x, |x|) : x \in \mathbb{R}^n\}$ is exactly the family of all balls having 0 on their boundary, $\{\Phi B(x, |x|) : x \in \mathbb{R}^n\}$ is the family of all affine half-spaces with 0 in their interior. Hence

$$\Phi \text{Inn}_S K = \bigcap \left\{ H : \begin{array}{l} H \text{ is a half-space} \\ 0 \in \operatorname{int} H \text{ and } H \supseteq \Phi K \end{array} \right\} = \operatorname{conv} \Phi K$$

which is what we wanted to prove.

To show (2), fix $x, y \in \text{Inn}_S K$ and $0 < \lambda < 1$. We have $x \in B(a, |a|) \subseteq K$ and $y \in B(b, |b|) \subseteq K$ for some $a, b \in \mathbb{R}^n$. Hence

$$(1 - \lambda)x + \lambda y \in (1 - \lambda)B(a, |a|) + \lambda B(b, |b|)$$
$$= B((1 - \lambda)a + \lambda b, (1 - \lambda)|a| + \lambda|b|) \subseteq K.$$

Consider the ball $B = B((1 - \lambda)a + \lambda b, (1 - \lambda)|a| + \lambda|b|)$. Obviously $0 \in B$. We know that ΦB is either a ball or a half-space. In particular it is convex, so $\text{Inn}_S B = \Phi \operatorname{conv} \Phi B = \Phi \Phi B = B$. Hence $(1 - \lambda)x + \lambda y \in \text{Inn}_S B$ and we can find $c \in \mathbb{R}^n$ such that

$$(1 - \lambda)x + \lambda y \in B(c, |c|) \subseteq B \subseteq K.$$

It follows that $(1 - \lambda)x + \lambda y \in \text{Inn}_S K$ and the proof of (2) is complete.

Finally we prove (3). The inequality $\mathrm{Inn}_S\, K \subseteq K$ is obvious from the definition. Since

$$\Phi\,(\mathrm{Inn}_S\, K) = \Phi\Phi\,\mathrm{conv}\,\Phi K = \mathrm{conv}\,\Phi K,$$

we see that $\Phi\,(\mathrm{Inn}_S\, K)$ is convex. Next, we fix a star body $A \subseteq K$ such that $\Phi(A)$ is convex. Then $\Phi(A) \supseteq \Phi(K)$, and since $\Phi(A)$ is convex it follows that $\Phi\,(A) \supseteq \mathrm{conv}\,\Phi\,(K)$. Hence

$$A = \Phi\Phi A \subseteq \Phi\,\mathrm{conv}\,\Phi K = \mathrm{Inn}_S\, K,$$

which is what we wanted to prove. $\square$

Now we can finally prove Theorem 9.9:

Proof of Theorem 9.9 We start with the easy implication which does not require Proposition 9.30: Assume $\clubsuit K$ is convex. Then by Proposition 9.28(4) we have $\clubsuit \circ \clubsuit K = \Phi\clubsuit K$. Hence

$$K'' = \circ\clubsuit \circ \clubsuit K = \circ\Phi\clubsuit K = \circ\circ K = K,$$

so $K \in \mathcal{R}^n$.

Conversely, assume that $K \in \mathcal{R}^n$. Then $K'' = K$, meaning that $\circ\clubsuit \circ \clubsuit K = K$. As $\clubsuit = \Phi\circ$ we have $\circ\Phi \circ \circ\Phi \circ K = K$. Applying $\clubsuit$ to both sides we get $\clubsuit \circ \Phi \circ \circ\Phi \circ K = \clubsuit K$.

Since $\circ K \in \mathcal{K}_0^n$, Proposition 9.30 implies that $\Phi \circ \circ\Phi \circ K \in \mathcal{K}_0^n$. Hence by Proposition 9.28(4) we have

$$\clubsuit \circ \Phi \circ \circ\Phi \circ K = \Phi\Phi \circ \circ\Phi \circ K = \circ\circ \Phi \circ K = \circ\circ \clubsuit K.$$

We showed that $\clubsuit K = \circ\circ \clubsuit K = \mathrm{conv}\,(\clubsuit K)$, so $\clubsuit K$ is convex. $\square$

As a corollary of the theorem we have the following result about projections:

Proposition 9.31 *Fix* $K \in \mathcal{R}^n$ *and a subspace* $E \subseteq \mathbb{R}^n$. *Then* $\left(\mathrm{Proj}_E\, K\right)' = \mathrm{Proj}_E\, K'$.

The reciprocity on the left hand side is taken of course inside the subspace E. This identity should be compared with the standard identity

$$\mathrm{Proj}_E\, K^\circ = (K \cap E)^\circ \tag{9.3.2}$$

which holds for the polarity map.

Proof Since $K \in \mathcal{R}^n$ we know that $K^{\clubsuit}$ is convex. By Proposition 9.17(4) and (9.3.2) we have

$$\left(\text{Proj}_E K\right)' = \left(\left(\text{Proj}_E K\right)^{\clubsuit}\right)^{\circ} = \left(K^{\clubsuit} \cap E\right)^{\circ} = \text{Proj}_E \left(K^{\clubsuit}\right)^{\circ} = \text{Proj}_E K'.$$

$\square$

Remark 9.32 Note that we only claimed the identity for reciprocal bodies. In fact, if $\left(\text{Proj}_E K\right)' = \text{Proj}_E K'$ for all one-dimensional subspaces E, then $K \in \mathcal{R}^n$. To see this, note $K'' \in \mathcal{R}^n$ and $K' = K'''$, so by Proposition 9.31 we have

$$\left(\text{Proj}_E K\right)' = \text{Proj}_E K' = \text{Proj}_E K''' = \left(\text{Proj}_E K''\right)'.$$

Since every one-dimensional convex body is a reciprocal body we deduce that $\text{Proj}_E K = \text{Proj}_E K''$ for all one-dimensional subspaces E, so $K = K'' \in \mathcal{R}^n$.

9.4 Structures on the Class of Flowers and Applications

In general, the map Φ does not preserve convexity. We begin this section by understanding when $\Phi(A)$ is convex:

Proposition 9.33 Let A be a star body. Then $\Phi(A)$ is convex if and only if A is a flower.

Furthermore, the following are equivalent for a convex body $K \in \mathcal{K}_0^n$:

1. $\Phi(K)$ is convex.
2. $K^{\circ} \in \mathcal{R}^n$.
3. $\text{Inn}_S K = K$.

Proof For the first statement, note that if $A = T^{\clubsuit}$ is a flower then $\Phi(A) = \Phi\left(T^{\clubsuit}\right) = T^{\circ}$ is convex (see Proposition 9.28(2)). Conversely, Assume $\Phi(A) = T$ is convex. Then $\Phi(A) = T = \Phi\left((T^{\circ})^{\clubsuit}\right)$, so $A = (T^{\circ})^{\clubsuit}$ is a flower.

For the second statement, the equivalence between (1) and (2) is exactly Theorem 9.9: $K^{\circ} \in \mathcal{R}^n$ if and only if $(K^{\circ})^{\clubsuit} = \Phi(K)$ is convex. The equivalence between (1) and (3) was part of Proposition 9.30. $\square$

Of course, since Φ is an involution, the first statement of Proposition 9.33 means that the image $\Phi\left(\mathcal{K}_0^n\right)$ is exactly the class of flowers. As for the second statement, we remark that there are convex bodies $K \in \mathcal{R}^n$ such that $K^{\circ} \notin \mathcal{R}^n$, so these are indeed different classes of convex bodies. For example, take any compact convex body with $0 \in \text{int } T$ and $T \notin \mathcal{R}^n$, and take $K = T' \in \mathcal{R}^n$. Since $K = T' = (T'')'$ but $T \neq T''$ Proposition 9.23 implies that K is not smooth. Hence K° is not strictly convex, so by Theorem 9.22 we have $K^{\circ} \notin \mathcal{R}^n$.

We will now use Proposition 9.33 to study some structures on the class of flowers. Recall that the radial sum $A \widetilde{+} B$ of two star bodies A and B is given by

$r_{A \widetilde{+} B} = r_A + r_B$. It is immediate that if A and B are flowers then so is $A \widetilde{+} B$, and in fact

$$K^{\clubsuit} \widetilde{+} T^{\clubsuit} = (K + T)^{\clubsuit}. \tag{9.4.1}$$

It is less obvious that the class of flowers is also closed under the Minkowski addition. To prove this fact we will first need:

Proposition 9.34 *Let B be any Euclidean ball with $0 \in B$. Then B is a flower.*

Proof Fact 9.27 implies that $\Phi(B)$ is always convex. Proposition 9.33 finishes the proof. □

Theorem 9.35 *Assume A and B are two flowers (which are not necessarily convex). Then $A + B$ is also a flower, where $+$ is the usual Minkowski sum.*

Proof Write $A = K^{\clubsuit}$ and $B = T^{\clubsuit}$ for $K, T \in \mathcal{K}_0^n$. By Proposition 9.19 we have

$$A = \bigcup_{x \in K} B_x \quad \text{and} \quad B = \bigcup_{y \in T} B_y.$$

Hence

$$A + B = \bigcup_{\substack{x \in K \\ y \in T}} (B_x + B_y) = \bigcup_{\substack{x \in K \\ y \in T}} B\left(\frac{x+y}{2}, \frac{|x| + |y|}{2}\right).$$

Since $0 \in B\left(\frac{x+y}{2}, \frac{|x|+|y|}{2}\right)$ the previous proposition implies that every such ball is a flower. Since $A + B$ is a union of such balls, the claim follows (see Proposition 9.17(3)). □

Remark 9.36 Equation (9.4.1) shows that the radial sum of flowers corresponds to the Minkowski sum of convex bodies. Similarly, Theorem 9.35 implies that the Minkowski sum of flowers corresponds to an addition of convex bodies, defined implicitly by

$$K^{\clubsuit} + T^{\clubsuit} = (K \oplus T)^{\clubsuit}. \tag{9.4.2}$$

The addition $\oplus$ is associative, commutative, monotone and has $\{0\}$ as its identity element. However, in general it does not satisfy $K \oplus K = 2K$, and in fact $K \oplus K$ is usually not homothetic to K. The identity $K \oplus K = 2K$ does hold if K is a reciprocal body. Moreover, if $K, T \in \mathcal{R}^n$ then by Theorem 9.9 $K^{\clubsuit}$ and $T^{\clubsuit}$ are convex, so $(K \oplus T)^{\clubsuit}$ is convex and $K \oplus T \in \mathcal{R}^n$ as well. In other words, $\mathcal{R}^n$ is closed under $\oplus$.

Theorem 9.35 can be equivalently stated in the language of the map Φ:

Corollary 9.37 *Let A and B be star bodies such that $\Phi(A)$, $\Phi(B)$ are convex. Then $\Phi(A + B)$ is convex as well.*

There is also a similar statement for reciprocal bodies:

Proposition 9.38 *If $K, T \in \mathcal{R}^n$ then $(K^\circ + T^\circ)^\circ \in \mathcal{R}^n$.*

Proof Write $A = K'$ and $B = T'$. Then $K = K'' = A' = (A^\clubsuit)^\circ$. Since A is a reciprocal body $A^\clubsuit$ is convex, so $K^\circ = (A^\clubsuit)^{\circ\circ} = A^\clubsuit$. In the same way we have $T^\circ = B^\clubsuit$. Hence K° and T° are both flowers, so by the previous Proposition $K^\circ + T^\circ$ is a flower. If we write $K^\circ + T^\circ = C^\clubsuit$ then $(K^\circ + T^\circ)^\circ = C' \in \mathcal{R}^n$. $\square$

A similar phenomenon holds regarding sections and projections. If $A \subseteq \mathbb{R}^n$ is a flower and E is a subspace of $\mathbb{R}^n$ then we already saw in Proposition 9.17(4) that $A \cap E$ is a flower in E, and in fact $\left(\text{Proj}_E K\right)^\clubsuit = K^\clubsuit \cap E$. It is less clear, but still true, that $\text{Proj}_E A$ is a flower as well:

Proposition 9.39 *If $A \subseteq \mathbb{R}^n$ is a flower and E is a subspace of $\mathbb{R}^n$, then $\text{Proj}_E A$ is a flower in E.*

Proof If $A = K^\clubsuit$ then $A = \bigcup_{x \in K} B_x$, and then

$$\text{Proj}_E A = \bigcup_{x \in K} \text{Proj}_E B_x.$$

Each projection $\text{Proj}_E B_x$ is a Euclidean ball in E that contains the origin, so by Proposition 9.34 is a flower. It follows that $\text{Proj}_E A$ is a flower as well. $\square$

The last operation we would like to mention which preserves the class of flowers is the convex hull:

Proposition 9.40 *If $A \subseteq \mathbb{R}^n$ is a flower so is $\text{conv}(A)$, and in fact $\text{conv}\left(K^\clubsuit\right) = (K'')^\clubsuit$.*

Proof Using the notation of Sect. 9.3 we have $(K'')^\clubsuit = \clubsuit \circ \clubsuit \circ \clubsuit K$. Since $\circ \clubsuit K = K'$ is obviously a reciprocal body, Theorem 9.9 implies that $\clubsuit \circ \clubsuit K$ is convex. Hence by Proposition 9.28 parts (4) and (2) we have

$$(K'')^\clubsuit = \clubsuit \circ (\clubsuit \circ \clubsuit K) = \Phi\clubsuit \circ \clubsuit K = \circ \circ \clubsuit K = \text{conv}\left(K^\clubsuit\right).$$

$\square$

More structure on the class of flowers can be obtained by transferring known results about the class $\mathcal{K}_0^n$ of convex bodies. First let us define the "inverse flower" operation:

Definition 9.41 The *core* of a flower A is defined by

$$A^{-\clubsuit} = \left\{ x \in \mathbb{R}^n : \ B_x \subseteq A \right\}.$$

In a recent paper ([17]) Zong defined the core of a convex body T to be the Alexandrov body $A[r_T]$. This is equivalent to our definition, though we apply it to flowers and not to convex bodies. The core operation $-\clubsuit$ is indeed the inverse operation to $\clubsuit$: For every $K \in \mathcal{K}_0^n$ we have

$$\left(K^{\clubsuit} \right)^{-\clubsuit} = \left\{ x \in \mathbb{R}^n : \ [0, x]^{\clubsuit} \subseteq K^{\clubsuit} \right\} = \left\{ x \in \mathbb{R}^n : \ [0, x] \subseteq K \right\} = K.$$

Equivalently, for every flower A the set $K = A^{-\clubsuit}$ is a convex body and $K^{\clubsuit} = A$.

We already referred in the introduction to a characterization of the polarity from [1] and [15]. Essentially the same result can also be formulated in terms of order-preserving transformations. We say that a map $T : \mathcal{K}_0^n \to \mathcal{K}_0^n$ is order-preserving if $A \subseteq B$ if and only if $T(A) \subseteq T(B)$. Then the theorem states that the only order-preserving bijections $T : \mathcal{K}_0^n \to \mathcal{K}_0^n$ are the (pointwise) linear maps. From here we deduce:

Proposition 9.42 *Let $T : \mathcal{F}^n \to \mathcal{F}^n$ be an order-preserving bijection on the class of flowers. Then there exists an invertible linear map $u : \mathbb{R}^n \to \mathbb{R}^n$ such that $T(A) = \left(u A^{-\clubsuit} \right)^{\clubsuit}$.*

Proof Define $S : \mathcal{K}_0^n \to \mathcal{K}_0^n$ by $S(K) = \left(T \left(K^{\clubsuit} \right) \right)^{-\clubsuit}$. Then S is easily seen to be an order preserving bijection on the class $\mathcal{K}_0^n$. Hence by the above-mentioned result from [1] there exists a linear map $u : \mathbb{R}^n \to \mathbb{R}^n$ such that $S(K) = uK$. It follows that $T(A) = \left(u A^{-\clubsuit} \right)^{\clubsuit}$ like we wanted. $\qquad\square$

Note that even though S in the proof above is linear, the map T is in general not even a pointwise map. In fact, it can be quite complicated—it does not preserve convexity for example.

With the same proof one may also characterize all dualities on flowers, i.e. all order-reversing involutions:

Proposition 9.43 *Let $T : \mathcal{F}^n \to \mathcal{F}^n$ be an order-reversing involution on the class of flowers. Then there exists an invertible symmetric linear map $u : \mathbb{R}^n \to \mathbb{R}^n$ such that $T(A) = \left(\left(u A^{-\clubsuit} \right)^{\circ} \right)^{\clubsuit}$.*

We conclude this section with a nice example. Let B be any Euclidean ball with $0 \in B$. By Proposition 9.34 we know that $B = K^{\clubsuit}$ for some body K. What is K? It turns out that K is an ellipsoid. As $(uK)^{\clubsuit} = u \left(K^{\clubsuit} \right)$ for every orthogonal matrix u, the body K is clearly a body of revolution. Hence the problem is actually two-dimensional and we may assume that $n = 2$.

Up to rotation, every ellipse has the form

$$E = \left\{ \frac{(x - x_0)^2}{a^2} + \frac{(y - y_0)^2}{b^2} \leq 1 \right\} \subseteq \mathbb{R}^2$$

for $a > b > 0$. Recall that (x_0, y_0) is the center of the ellipse. If we write $c = \sqrt{a^2 - b^2}$ then $p_1 = (x_0 + c, y_0)$ and $p_2 = (x_0 - c, y_0)$ are the foci of E, and

$$E = \left\{ q \in \mathbb{R}^2 : |q - p_1| + |q - p_2| = 2a \right\}.$$

The number $e = \sqrt{1 - \frac{b^2}{a^2}}$ is the eccentricity of E. Obviously every ellipse in $\mathbb{R}^2$ is uniquely determined by its center, its eccentricity and one of its focus points. We then have:

Proposition 9.44 *Let $E \subseteq \mathbb{R}^2$ be an ellipse with center at $p \in \mathbb{R}^2$, one focus point at 0 and eccentricity e. Then:*

1. *$E^{\clubsuit}$ is a ball with center p and radius $\frac{|p|}{e}$.*
2. *E' is an ellipse with center $\widetilde{p} = -\frac{e^2}{1-e^2} \cdot \frac{p}{|p|^2}$, a focus point at 0 and eccentricity e.*

Proof By rotating and scaling it is enough to assume that the center of the ellipse is at $p = (1, 0)$. We then have

$$E = \left\{ (x, y) : \frac{(x - 1)^2}{a^2} + \frac{y^2}{a^2 - 1} \leq 1 \right\},$$

where $a = \frac{1}{e} > 1$. To prove (1), consider the centered ellipse $\widetilde{E} = E - p$. For such ellipses it is well-known that $h_{\widetilde{E}}(x, y) = \sqrt{a^2 x^2 + (a^2 - 1)y^2}$, and then

$$h_E(x, y) = h_{\widetilde{E}}(x, y) + h_{\{(1,0)\}}(x, y) = \sqrt{a^2 x^2 + (a^2 - 1)y^2} + x$$

(note that we consider $h_{\widetilde{E}}$ and h_E not as functions on S^{n-1}, but as 1-homogeneous functions defined on all of $\mathbb{R}^n$). Therefore

$$E^{\clubsuit} = \left\{ (x, y) : |(x, y)| \leq r_{E^{\clubsuit}} \left(\frac{(x, y)}{|(x, y)|} \right) \right\} = \left\{ (x, y) : h_E(x, y) \geq |(x, y)|^2 \right\}$$

$$= \left\{ (x, y) : \sqrt{a^2 x^2 + (a^2 - 1)y^2} + x \geq x^2 + y^2 \right\}$$

$$= \left\{ (x, y) : (x - 1)^2 + y^2 \leq a^2 \right\},$$

where the last equality follows from simple algebraic manipulations. We see that $E^\clubsuit$ is indeed a ball with center $p = (1, 0)$ and radius $\frac{|p|}{e} = a$.

To prove (2), recall that $E' = (E^\clubsuit)^\circ$. Like before, if $\tilde{B} = B((0, 0), a)$ is the centered ball then

$$h_{E^\clubsuit}(x, y) = h_{\tilde{B}}(x, y) + h_{\{(1,0)\}}(x, y) = a\sqrt{x^2 + y^2} + x.$$

Hence

$$E' = \left(E^\clubsuit\right)^\circ = \left\{(x, y) : h_{E^\clubsuit}(x, y) \leq 1\right\}$$

$$= \left\{(x, y) : a\sqrt{x^2 + y^2} + x \leq 1\right\}.$$

Again, some algebraic manipulations will give us the (unpleasant) canonical form

$$E' = \left\{(x, y) : \frac{\left(x + \frac{1}{a^2 - 1}\right)^2}{\left(\frac{a}{a^2 - 1}\right)^2} + \frac{y^2}{\frac{1}{a^2 - 1}} \leq 1\right\}.$$

Hence the center of E' is indeed at $\left(-\frac{1}{a^2 - 1}, 0\right) = \left(-\frac{e^2}{1 - e^2}, 0\right) = -\frac{e^2}{1 - e^2} \cdot \frac{p}{|p|^2}$. The distance from the center to the foci is

$$\sqrt{\left(\frac{a}{a^2 - 1}\right)^2 - \frac{1}{a^2 - 1}} = \frac{1}{a^2 - 1} = \frac{e^2}{1 - e^2},$$

so one of the focus points is indeed the origin. Finally, the eccentricity of E' is indeed

$$\sqrt{1 - \frac{\frac{1}{a^2 - 1}}{\left(\frac{a}{a^2 - 1}\right)^2}} = \frac{1}{a} = e.$$

$\square$

This proposition also gives a nice example of the addition $\oplus$ defined in (9.4.2): For every $x_1, x_2, \ldots, x_m \in \mathbb{R}^n$ the body $\bigoplus_{i=1}^m [0, x_i]$ is an ellipsoid. Indeed, we have

$$\left(\bigoplus_{i=1}^m [0, x_i]\right)^\clubsuit = \sum_{i=1}^m [0, x_i]^\clubsuit = \sum_{i=1}^m B_{x_i}$$

which is a Euclidean ball, so by the last computation $\bigoplus_{i=1}^m [0, x_i]$ is an ellipsoid of revolution with one focus point at 0.

9.5 Geometric Inequalities

In this final section we discuss several inequalities involving flowers and reciprocal bodies. We begin by showing that the various operations constructed in this paper are convex maps. A theorem of Firey from [5] implies that the polarity map $\circ$: $\mathcal{K}_0^n \to \mathcal{K}_0^n$ is convex with respect to the Minkowski addition: For every $K, T \in \mathcal{K}_0^n$ and every $0 \le \lambda \le 1$ one has

$$((1 - \lambda)K + \lambda T)^\circ \subseteq (1 - \lambda)K^\circ + \lambda T^\circ.$$

The maps $\clubsuit$ and Φ are also convex with respect to the Minkowski addition on their appropriate domains:

Theorem 9.45 *The map* $\clubsuit$: $\mathcal{K}_0^n \to \mathcal{F}^n$ *is convex. The map* Φ *is convex when applied to arbitrary star bodies.*

Proof For any two star bodies A and B we have $r_{A+B} \ge r_A + r_B$. Hence for $K, T \in \mathcal{K}_0^n$ and $0 \le \lambda \le 1$ we have

$$r_{((1-\lambda)K+\lambda T)^\clubsuit} = h_{(1-\lambda)K+\lambda T} = (1 - \lambda)h_K + \lambda h_T$$

$$= (1 - \lambda)r_{K^\clubsuit} + \lambda r_{T^\clubsuit} \le r_{(1-\lambda)K^\clubsuit + \lambda T^\clubsuit}.$$

It follows that $((1 - \lambda)K + \lambda T)^\clubsuit \subseteq (1 - \lambda)K^\clubsuit + \lambda T^\clubsuit$ so $\clubsuit$ is convex.

For the convexity of Φ fix star bodies A and B and $0 \le \lambda \le 1$, and note that

$$r_{\Phi((1-\lambda)A+\lambda B)} = \frac{1}{r_{(1-\lambda)A+\lambda B}} \le \frac{1}{(1 - \lambda)r_A + \lambda r_B} \overset{(*)}{\le} \frac{1-\lambda}{r_A} + \frac{\lambda}{r_B}$$

$$= (1 - \lambda)r_{\Phi(A)} + \lambda r_{\Phi(B)} \le r_{(1-\lambda)\Phi(A)+\lambda\Phi(B)},$$

where the inequality $(*)$ is the convexity of the map $x \mapsto \frac{1}{x}$ on $(0, \infty)$. $\qquad \square$

Convexity of the reciprocal map is more delicate. For general convex bodies $K, T \in \mathcal{K}_0^n$ the inequality

$$((1 - \lambda)K + \lambda T)' \subseteq (1 - \lambda)K' + \lambda T'$$

is *false*. It becomes true if we further assume that K and T are reciprocal bodies: If $K \in \mathcal{R}^n$ then $K^\clubsuit$ is convex, which means that $\frac{1}{r_{K^\clubsuit}} = \frac{1}{h_K}$ is the support function of a convex body. Hence $h_{K'} = h_{A[1/h_K]} = \frac{1}{h_K}$ and similarly $h_{T'} = \frac{1}{h_T}$. Therefore we indeed have

$$h_{((1-\lambda)K+\lambda T)'} \le \frac{1}{h_{(1-\lambda)K+\lambda T}} = \frac{1}{(1 - \lambda)h_K + \lambda h_T} \le \frac{1-\lambda}{h_K} + \frac{\lambda}{h_T}$$

$$= (1 - \lambda)h_{K'} + \lambda h_{T'} = h_{(1-\lambda)K'+\lambda T'}.$$

However, one cannot really say that $\prime$ is a convex map on $\mathcal{R}^n$ in the standard sense, since the class $\mathcal{R}^n$ is not closed with respect to the Minkowski addition. In Eq. (9.4.2) of the previous section we defined a new addition $\oplus$ which does preserve the class $\mathcal{R}^n$, and the following holds:

Proposition 9.46 *The reciprocal map* $\prime : \mathcal{R}^n \to \mathcal{R}^n$ *is convex with respect to the addition* $\oplus$.

Proof For every $K, T \in \mathcal{R}^n$ we have

$$h_{K \oplus T} = r_{(K \oplus T)^\clubsuit} = r_{K^\clubsuit + T^\clubsuit} \geq r_{K^\clubsuit} + r_{T^\clubsuit} = h_K + h_T = h_{K+T},$$

so $K \oplus T \supseteq K + T$. Hence by the convexity of $\circ$ we have

$$
((1-\lambda)K \oplus \lambda T)' = \left[((1-\lambda)K \oplus \lambda T)^\clubsuit \right]^\circ = \left((1-\lambda)K^\clubsuit + \lambda T^\clubsuit \right)^\circ
$$
$$
\subseteq (1-\lambda)\left(K^\clubsuit \right)^\circ + \lambda \left(T^\clubsuit \right)^\circ \subseteq (1-\lambda)K' \oplus \lambda T'.
$$

$\square$

We now turn our attention to numerical inequalities involving flowers. To each body K we can associate a new numerical parameter which is $\left| K^\clubsuit \right|$, the volume of the flower of K. It was explained in Remark 9.7 why this volume is important in stochastic geometry. We then have the following reverse Brunn-Minkowski inequality:

Proposition 9.47 *For every* $K, T \in \mathcal{K}_0^n$ *one has* $\left| (K+T)^\clubsuit \right|^{\frac{1}{n}} \leq \left| K^\clubsuit \right|^{\frac{1}{n}} + \left| T^\clubsuit \right|^{\frac{1}{n}}$.

Proof Recall that for every star body A in $\mathbb{R}^n$ we may integrate by polar coordinates and deduce that $|A| = \left| B_2^n \right| \cdot \int_{S^{n-1}} r_A(\theta)^n \mathrm{d}\sigma(\theta)$. Here σ denotes the uniform probability measure on the sphere. It follows that for every $K \in \mathcal{K}_0^n$ we have

$$
\left| K^\clubsuit \right| = \left| B_2^n \right| \cdot \int_{S^{n-1}} h_K(\theta)^n \mathrm{d}\sigma(\theta). \tag{9.5.1}
$$

In other words, $\left| K^\clubsuit \right|^{\frac{1}{n}}$ is proportional to $\| h_K \|_{L^n(S^{n-1})}$, where $L^n(S^{n-1})$ is the relevant L^p space. Therefore the required inequality is nothing more than Minkowski's inequality (the triangle inequality for L^p-norms, in our case for $p = n$). $\square$

Similarly, we have an analogue of Minkowski's theorem on the polynomiality of volume. Recall that for every fixed convex bodies $K_1, K_2, \ldots, K_m$ we have

$$
|\lambda_1 K_1 + \lambda_2 K_2 + \cdots + \lambda_m K_m| = \sum_{i_1, i_2, \ldots, i_n = 1}^{m} V(K_{i_1}, K_{i_2}, \ldots, K_{i_n}) \cdot \lambda_{i_1} \lambda_{i_2} \cdots \lambda_{i_n},
$$

Where we take the coefficients $V(K_{i_1}, K_{i_2}, \ldots, K_{i_n})$ to be symmetric with respect to a permutation of the arguments. The number $V(K_1, K_2, \ldots, K_n)$ is called the mixed volume of $K_1, K_2, \ldots, K_n$ and is fundamental to convex geometry. We then have:

Proposition 9.48 *Fix $K_1, K_2, \ldots, K_m \in \mathcal{K}_0^n$. Then for every $\lambda_1, \lambda_2, \ldots, \lambda_m \geq 0$ one has*

$$\left| (\lambda_1 K_1 + \lambda_2 K_2 + \cdots + \lambda_m K_m)^{\clubsuit} \right| = \sum_{i_1, i_2, \ldots, i_n = 1}^{m} V^{\clubsuit}(K_{i_1}, K_{i_2}, \ldots, K_{i_n}) \cdot \lambda_{i_1} \lambda_{i_2} \cdots \lambda_{i_n},$$

where the coefficients are given by

$$V^{\clubsuit}(K_1, K_2, \ldots, K_n) = \left| B_2^n \right| \cdot \int_{S^{n-1}} h_{K_1}(\theta) h_{K_2}(\theta) \cdots h_{K_n}(\theta) \mathrm{d}\sigma(\theta). \qquad (9.5.2)$$

The proof is immediate from formula (9.5.1). Moreover, the new $\clubsuit$-mixed volumes satisfy a reverse (elliptic) Alexandrov-Fenchel type inequality:

Proposition 9.49 *For every $K_1, K_2, \ldots, K_n \in \mathcal{K}_0^n$ we have*

$$V^{\clubsuit}(K_1, K_2, K_3, \ldots, K_n)^2 \leq V^{\clubsuit}(K_1, K_1, K_3, \ldots K_n) \cdot V^{\clubsuit}(K_2, K_2, K_3, \ldots, K_n), \qquad (9.5.3)$$

as well as

$$V^{\clubsuit}(K_1, K_2, \ldots, K_n) \leq \left(\prod_{i=1}^{n} \left| K_i^{\clubsuit} \right| \right)^{\frac{1}{n}}. \qquad (9.5.4)$$

Moreover, assume that the bodies K_i are all compact and $K_i \neq \{0\}$ for all $1 \leq i \leq n$. Then equality occurs in (9.5.3) if and only if K_1 and K_2 are homothetic, and in (9.5.4) if and only if $K_1, K_2, \ldots, K_n$ are all homothetic.

Proof Apply Hölder's inequality to formula (9.5.2). □

These results and their proofs are very closely related to the dual Brunn–Minkowski theory which was developed by Lutwak in [8].

Next we would like to compare the $\clubsuit$-mixed volume $V^{\clubsuit}(K_1, K_2, \ldots, K_n)$ with the classical mixed volume $V(K_1, K_2, \ldots, K_n)$. Since $\left| T^{\clubsuit} \right| \geq |T|$ for every $T \in \mathcal{K}_0^n$, one may conjecture that $V^{\clubsuit}(K_1, K_2, \ldots, K_n) \geq V(K_1, K_2, \ldots, K_n)$. This is not true however, as the next example shows:

Example 9.50 Let $\{e_1, e_2\}$ be the standard basis of $\mathbb{R}^2$. Define $K = [-e_1, e_1]$ and $T = [-e_2, e_2]$. Then $|\lambda K + \mu T| = 4\lambda\mu$ which implies that $V(K, T) = 2$.

On the other hand by Formula (9.5.2) we have

$$V^\clubsuit(K, T) = \left|B_2^2\right| \cdot \int_{S^1} h_K(\theta) h_T(\theta) d\sigma(\theta) = \pi \cdot \frac{1}{2\pi} \int_0^{2\pi} |\cos\theta| \, |\sin\theta| \, d\theta = 1,$$

so $V(K, T) > V^\clubsuit(K, T)$.

However, in one case we can compare the $\clubsuit$-mixed volume with the classical one. Recall that for $K \in \mathcal{K}_0^n$ and $0 \le i \le n$ the i'th quermassintegral of K is defined by

$$W_i(K) = V\left(\underbrace{K, K, \ldots, K}_{n-i \text{ times}}, \underbrace{B_2^n, B_2^n, \ldots, B_2^n}_{i \text{ times}}\right).$$

Kubota's formula then states that

$$W_{n-i}(K) = \frac{|B_2^n|}{|B_2^i|} \cdot \int_{G(n,i)} \left|\operatorname{Proj}_E K\right| d\mu(E),$$

where $G(n, i)$ is the set of all i-dimensional linear subspaces of $\mathbb{R}^n$, and μ is the Haar probability measure on $G(n, i)$.

We define the $\clubsuit$-quermassintegrals in the obvious way as $W_i^\clubsuit(K) = V^\clubsuit(\underbrace{K, \ldots, K}_{n-i}, \underbrace{B_2^n, \ldots, B_2^n}_{i})$. We then have a Kubota–type formula:

Theorem 9.51 *For every $K \in \mathcal{K}_0^n$ and every $0 \le i \le n$ we have*

$$W_{n-i}^\clubsuit(K) = \frac{|B_2^n|}{|B_2^i|} \cdot \int_{G(n,i)} \left|(\operatorname{Proj}_E K)^\clubsuit\right| d\mu(E),$$

where μ is the Haar probability measure on $G(n, i)$ and the flower map $\clubsuit$ on the right hand side is taken inside the subspace E.

Proof If $T \subseteq \mathbb{R}^m$ then integrating in polar coordinates we have $|T| = |B_2^m| \cdot \int_{S^{m-1}} r_T(\theta)^m d\sigma_m(\theta)$, where σ_m denotes the Haar probability measure on S^{m-1}. Therefore

$$\int_{G(n,i)} \left|(\operatorname{Proj}_E K)^\clubsuit\right| d\mu(E) = \int_{G(n,i)} \left|K^\clubsuit \cap E\right| d\mu(E)$$

$$= \left|B_2^i\right| \int_{G(n,i)} \int_{S_E} r_{K^\clubsuit}(\theta)^i d\sigma_E(\theta) d\mu(E)$$

$$= \left|B_2^i\right| \int_{S^{n-1}} r_{K^\clubsuit}(\theta)^i d\sigma_n(\theta)$$

$$= \left| B_2^i \right| \int_{S^{n-1}} h_K(\theta)^i \, \mathrm{d}\sigma_n(\theta)$$

$$= \frac{\left| B_2^i \right|}{\left| B_2^n \right|} W_{n-i}^{\clubsuit}(K).$$

□

And as a corollary we obtain:

Corollary 9.52 *For every* $K \in \mathcal{K}_0^n$ *and* $0 \leq i \leq n$ *we have* $W_i^{\clubsuit}(K) \geq W_i(K)$.

Proof We have

$$W_{n-i}(K) = \frac{\left| B_2^n \right|}{\left| B_2^i \right|} \cdot \int_{G(n,i)} \left| \mathrm{Proj}_E K \right| \mathrm{d}\mu(E)$$

$$\leq \frac{\left| B_2^n \right|}{\left| B_2^i \right|} \cdot \int_{G(n,i)} \left| (\mathrm{Proj}_E K)^{\clubsuit} \right| \mathrm{d}\mu(E) = W_{n-i}^{\clubsuit}(K).$$

□

It is well known that $W_{n-1}(K)$ is (up to normalization) the mean width of K. Hence from formula (9.5.2) we immediately have $W_{n-1}^{\clubsuit}(K) = W_{n-1}(K)$. The Alexandrov-Fenchel inequality and its flower version from Proposition 9.49 then imply that

$$\left(\frac{|K|}{\left| B_2^n \right|} \right)^{\frac{1}{n}} \leq \left(\frac{W_1(K)}{\left| B_2^n \right|} \right)^{\frac{1}{n-1}} \leq \cdots \leq \left(\frac{W_{n-2}(K)}{\left| B_2^n \right|} \right)^{\frac{1}{2}} \leq \frac{W_{n-1}(K)}{\left| B_2^n \right|}$$

$$= \frac{W_{n-1}^{\clubsuit}(K)}{\left| B_2^n \right|} \leq \left(\frac{W_{n-2}^{\clubsuit}(K)}{\left| B_2^n \right|} \right)^{\frac{1}{2}} \leq \cdots \leq \left(\frac{W_1^{\clubsuit}(K)}{\left| B_2^n \right|} \right)^{\frac{1}{n-1}} \leq \left(\frac{|K^{\clubsuit}|}{\left| B_2^n \right|} \right)^{\frac{1}{n}}$$

which gives another proof of the relation $W_i^{\clubsuit}(K) \geq W_i(K)$.

We conclude this paper with a remark regarding the distance of flowers and reciprocal bodies to the Euclidean ball. We restrict ourselves to bodies which are compact and contain 0 at their interior. The *geometric distance* between such bodies K and T is

$$d(K, T) = \inf \left\{ \frac{b}{a} : aK \subseteq T \subseteq bK \right\}.$$

Recall that a body K is centrally symmetric if $K = -K$.

Proposition 9.53

1. If a flower A is centrally symmetric and convex, then $d\left(A, B_2^n\right) \leq 2$.
2. If $K \in \mathcal{R}^n$ is centrally symmetric, then $d(K, B_2^n) \leq 2$.

Proof To prove the first assertion, write $A = K^{\clubsuit}$ and let $R = \max_{x \in K} |x|$. Since $K \subseteq R \cdot B_2^n$ we have $A \subseteq R \cdot B_2^n$.

On the other hand, fix $x \in K$ with $|x| = R$ and note that $B_x = [0, x]^{\clubsuit} \subseteq K^{\clubsuit} = A$. Since K is centrally symmetric we also have $-x \in K$, so $B_{-x} \subseteq A$. Hence

$$\frac{R}{2} \cdot B_2^n \subseteq \operatorname{conv}\left(B_x \cup B_{-x}\right) \subseteq A,$$

so $d\left(A, B_2^n\right) \leq 2$.

For the second assertion, fix a centrally symmetric reciprocal body K and define $T = K'$. Then $K = T' = \left(T^{\clubsuit}\right)^{\circ}$. Since T is a reciprocal body $T^{\clubsuit}$ is convex, so $d\left(T^{\clubsuit}, B_2^n\right) = d(K^{\circ}, B_2^n) \leq 2$. Since polarity preserves the geometric distance we also have $d\left(K, B_2^n\right) \leq 2$. $\qquad\square$

Note that this result is false if K is not centrally symmetric. For example, we already saw in Proposition 9.44 that if $E \subseteq \mathbb{R}^2$ is any ellipse with a focus point at 0 then $E \in \mathcal{R}^2$, and the distance $d(E, B_2^2)$ is obviously unbounded on this class of ellipses. In fact, even the modified distance $\widetilde{d}(E, B_2^2) = \inf_{x \in \mathbb{R}^2} d\left(E + x, B_2^2\right)$ is unbounded on this class.

Acknowledgements The authors would like to thank M. Gromov and R. Schneider for a useful exchange of messages regarding this paper. They would also like to thank R. Gardner and D. Hug for introducing them to the useful references [16] and [17], and the anonymous referee for their detailed comments. The second and third named authors were jointly funded by BSF grant 2016050. The second named author was also funded by ISF grant 519/17.

References

1. S. Artstein-Avidan, V. Milman, The concept of duality for measure projections of convex bodies. J. Funct. Anal. **254**(10), 2648–2666 (2008)
2. D. Blair, *Inversion theory and conformal mapping*. Student Mathematical Library (American Mathematical Society, Providence, 2000)
3. K.J. Böröczky, R. Schneider, A characterization of the duality mapping for convex bodies. Geom. Funct. Anal. **18**(3), 657–667 (2008)
4. K.J. Böröczky, E. Lutwak, D. Yang, G. Zhang, The log-Brunn-Minkowski inequality. Adv. Math. **231**(3–4), 1974–1997 (2012)
5. W.J. Firey, Polar means of convex bodies and a dual to the Brunn-Minkowski theorem. Can. J. Math. **13**, 444–453 (1961)
6. P.M. Gruber, The endomorphisms of the lattice of norms in finite dimensions. Abhandlungen aus dem Mathematischen Seminar der Universität Hamburg **62**(1), 179–189 (1992)
7. J. Lindenstrauss, L. Tzafriri, *Classical Banach spaces II*. Ergebnisse der Mathematik und ihrer Grenzgebiete, vol. 97 (Springer, Berlin, 1979)

8. E. Lutwak, Dual mixed volumes. Pac. J. Math. **58**(2), 531–538 (1975)
9. V. Milman, L. Rotem, Non-standard constructions in convex geometry; geometric means of convex bodies, in *Convexity and concentration*, ed. by E. Carlen, M. Madiman, E. Werner. The IMA Volumes in Mathematics and its Applications, vol. 161 (Springer, New York, 2017), pp. 361–390
10. V. Milman, L. Rotem, Powers and logarithms of convex bodies. C.R. Math. **355**(9), 981–986 (2017)
11. V. Milman, L. Rotem, Weighted geometric means of convex bodies, in *Selim Krein Centennial*, ed. by P. Kuchment, E. Semenov. Contemporary Mathematics (AMS, Providence, 2019)
12. I. Molchanov, Continued fractions built from convex sets and convex functions. Commun. Contemp. Math. **17**(5), 1550003 (2015)
13. M. Moszyńska, Quotient star bodies, intersection bodies, and star duality. J. Math. Anal. Appl. **232**(1), 45–60 (1999)
14. R. Schneider, *Convex bodies: the Brunn-Minkowski theory*. Encyclopedia of Mathematics and its Applications, 2nd edn. (Cambridge University Press, Cambridge, 2014)
15. B. Slomka, On duality and endomorphisms of lattices of closed convex sets. Adv. Geom. **11**(2), 225–239 (2011)
16. E. Spodarev (ed.), *Stochastic geometry, spatial statistics and random fields*. Lecture Notes in Mathematics, vol. 2068 (Springer, Berlin, 2013)
17. C. Zong, A computer approach to determine the densest translative tetrahedron packings. Exp. Math. (2018). https://doi.org/10.1080/10586458.2019.1582378

Chapter 10
Moments of the Distance Between Independent Random Vectors

Assaf Naor and Krzysztof Oleszkiewicz

Abstract We derive various sharp bounds on moments of the distance between two independent random vectors taking values in a Banach space.

10.1 Introduction

Throughout what follows, all Banach spaces are tacitly assumed to be separable. This assumption removes the need to discuss measurability side-issues; alternatively one could consider throughout only the special case of finitely-supported random variables, which captures all of the key ideas. We will also tacitly assume that all Banach spaces are over the complex scalars $\mathbb{C}$. This assumption is convenient for the ensuing proofs, but the main statements (namely, those that do not mention complex scalars explicitly) hold over the real scalars as well, through a standard complexification procedure. All the notation and terminology from Banach space theory that occurs below is basic and standard, as in e.g. [15].

Our starting point is the following question. What is the smallest $C > 0$ such that for every Banach space $(F, \| \cdot \|_F)$ and every two independent F-valued integrable

A.N. was supported by the Packard Foundation and the Simons Foundation. The research that is presented here was conducted under the auspices of the Simons Algorithms and Geometry (A&G) Think Tank. K.O. was partially supported by the National Science Centre, Poland, project number 2012/05/B/ST1/00412.

A. Naor
Department of Mathematics, Princeton University, Princeton, NJ, USA
e-mail: naor@math.princeton.edu

K. Oleszkiewicz (✉)
Institute of Mathematics, University of Warsaw, Warszawa, Poland

Institute of Mathematics, Polish Academy of Sciences, Warszawa, Poland
e-mail: koles@mimuw.edu.pl

© Springer Nature Switzerland AG 2020
B. Klartag, E. Milman (eds.), *Geometric Aspects of Functional Analysis*,
Lecture Notes in Mathematics 2266,
https://doi.org/10.1007/978-3-030-46762-3_10

random vectors $X, Y \in L_1(F)$ we have

$$\inf_{z \in F} \mathbb{E}\left[\|X - z\|_F + \|Y - z\|_F\right] \leqslant C\mathbb{E}\left[\|X - Y\|_F\right]? \tag{10.1}$$

We will reason that (10.1) holds with $C = 3$, and that $C = 3$ is the sharp constant here. More generally, we have the following theorem.

Theorem 10.1.1 *Suppose that $p \geqslant 1$ and $(F, \| \cdot \|_F)$ is a Banach space. Let $X, Y \in L_p(F)$ be two independent F-valued p-integrable random vectors. Then*

$$\inf_{z \in F} \mathbb{E}\left[\|X - z\|_F^p + \|Y - z\|_F^p\right] \leqslant \frac{3^p}{2^{p-1}}\mathbb{E}\left[\|X - Y\|_F^p\right]. \tag{10.2}$$

The constant $\frac{3^p}{2^{p-1}}$ in (10.2) cannot be improved.

The Banach space F that exhibits this sharpness of (10.2) is, of course, a subspace of ℓ_∞, but we do not know what is the optimal constant in (10.2) when $F = \ell_\infty$ itself. More generally, understanding the meaning of the optimal constant in (10.2) for specific Banach spaces is an interesting question, which we investigate in the rest of the present work for certain special classes of Banach spaces but do not fully resolve.

10.1.1 Geometric Motivation

Our interest in (10.1) arose from investigations of [1] in the context of Riemannian/Alexandrov geometry. It is well established throughout an extensive geometric literature that a range of useful quadratic distance inequalities for a metric space $(\mathcal{M}, d_{\mathcal{M}})$ arise if one imposes bounds on its curvature in the sense of Alexandrov. The term "quadratic" here indicates that these inequalities involve squares of distances between finite point configurations in $\mathcal{M}$. A phenomenon that was established in [1] is that any such quadratic metric inequality that holds for every Alexandrov space of nonnegative curvature becomes valid in any metric space whatsoever if one removes the squaring of the distances, i.e., in essence upon "linearization" of the inequality; see [1] for a precise formulation. This led naturally to the question whether the same phenomenon holds for Hadamard spaces (complete simply connected spaces whose Alexandrov curvature in nonpositive); see [1] for an extensive discussion as well as the recent negative resolution of this question in [11]. In the context of a Hadamard space $(\mathcal{M}, d_{\mathcal{M}})$, the analogue of (10.1) is that independent finitely-supported $\mathcal{M}$-valued random variables X, Y satisfy

$$\inf_{z \in \mathcal{M}} \mathbb{E}\left[d_{\mathcal{M}}(X, z)^2 + d_{\mathcal{M}}(Y, z)^2\right] \leqslant \mathbb{E}\left[d_{\mathcal{M}}(X, Y)^2\right]. \tag{10.3}$$

See [1] for a standard derivation of (10.3), where $z \in m$ is an appropriate "geometric barycenter," namely it is obtained as the minimizer of the expected squared distance from X to z. As explained in [1], by using (10.3) iteratively one can obtain quadratic metric inequalities that hold in any Hadamard space and serve as obstructions for certain geometric embeddings. The "linearized" version of (10.3), in the case of Banach spaces and allowing for a loss of a factor C, is precisely (10.1). So, in the spirit of [1] it is natural to ask what is the smallest C for which it holds. This is what we address here, leading to analytic questions about Banach spaces that are interesting in their own right from the probabilistic and geometric perspective. We note that there are questions along these lines that [1] raises and remain open; see e.g. [1, Question 32].

10.1.2 Probabilistic Discussion

The inequality which reverses (10.1) holds trivially as a consequence of the triangle inequality, even when X and Y are not necessarily independent. Namely, any $X, Y \in L_1(F)$ satisfy

$$\mathbb{E}\left[\|X - Y\|_F\right] \leqslant \inf_{z \in F} \mathbb{E}\left[\|X - z\|_F + \|Y - z\|_F\right].$$

So, the above discussion is about the extent to which this use of the triangle inequality can be reversed.

Since the upper bound that we seek is in terms of the distance in $L_p(F)$ between independent copies of X and Y, this can be further used to control from above expressions such as $\mathbb{E}[\|X - Y\|_F^p]$ for X and Y not necessarily independent in terms of $\mathbb{E}[\|X' - Y'\|_F^p]$, where X' and Y' are independent, X' has the same distribution as X, and Y' has the same distribution as Y.

In order to analyse the inequality (10.2) in a specific Banach space $(F, \|\cdot\|_F)$, we consider the following geometric moduli. Given $p \geqslant 1$ let $b_p(F, \|\cdot\|_F)$, or simply $b_p(F)$ if the norm is clear from the context, be the infimum over those $b > 0$ such that every independent F-valued random variables $X, Y \in L_p(F)$ satisfy

$$\inf_{z \in F} \mathbb{E}\left[\|X - z\|_F^p + \|Y - z\|_F^p\right] \leqslant b\mathbb{E}\left[\|X - Y\|_F^p\right]. \tag{10.4}$$

Thus, $b_p(F)$ is precisely the best possible constant in the $L_p(F)$-analogue of the aforementioned barycentric inequality (10.3). The use of the letter "b" in this notation is in reference to the word "barycentric." Theorem 10.1.1 asserts that $b_p(F) \leqslant 3^p/2^{p-1}$, and that this bound cannot be improved in general.

Let $m_p(F, \|\cdot\|_F) > 0$, or simply $m_p(F)$ if the norm is clear from the context, be the infimum over those $m > 0$ such that every independent F-valued random variables $X, Y \in L_p(F)$ satisfy

$$\mathbb{E}\left[\left\|X - \frac{1}{2}\mathbb{E}[X] - \frac{1}{2}\mathbb{E}[Y]\right\|_F^p + \left\|Y - \frac{1}{2}\mathbb{E}[X] - \frac{1}{2}\mathbb{E}[Y]\right\|_F^p\right] \leqslant m\mathbb{E}\left[\|X - Y\|_F^p\right].$$

(10.5)

The use of the letter "m" in this notation is in reference to the word "mixture," since the left-hand side of (10.5) is equal to $2\mathbb{E}[\|Z - \mathbb{E}[Z]\|_F^p]$, where $Z \in L_p(F)$ is distributed according to the mixture of the laws of X and Y, namely Z is the F-valued random vector such that for every Borel set $A \subseteq F$,

$$\mathbb{P}[Z \in A] = \frac{1}{2}\mathbb{P}[X \in A] + \frac{1}{2}\mathbb{P}[Y \in A]$$

(10.6)

Obviously $\mathfrak{b}_p(F) \leqslant m_p(F)$, because (10.5) corresponds to choosing $z = \frac{1}{2}\mathbb{E}[X] + \frac{1}{2}\mathbb{E}[Y] \in F$ in (10.4).

While we sometimes bound $m_p(F)$ directly, it is beneficial to refine the considerations through the study of two further moduli that are natural in their own right and, as we shall see later, their use can lead to better bounds. Firstly, let $\mathfrak{r}_p(F, \| \cdot \|_F)$, or simply $\mathfrak{r}_p(F)$ if the norm is clear from the context, be the infimum over those $\mathfrak{r} > 0$ such that every independent F-valued random variables $X, Y \in L_p(F)$ satisfy

$$\mathbb{E}\left[\|X - X'\|_F^p\right] + \mathbb{E}\left[\|Y - Y'\|_F^p\right] \leqslant \mathfrak{r}\mathbb{E}\left[\|X - Y\|_F^p\right],$$

(10.7)

where X', Y' are independent copies of X and Y, respectively. The use of the letter "$\mathfrak{r}$" in this notation is in reference to the word "roundness," as we shall next explain.

Observe also that (10.7) is a purely metric condition, i.e., it involves only distances between points. So, it makes sense to investigate (10.7) in any metric space $(\mathcal{M}, d_\mathcal{M})$, namely to study the inequality

$$\mathbb{E}\left[d_\mathcal{M}(X, X')^p\right] + \mathbb{E}\left[d_\mathcal{M}(Y, Y')^p\right] \leqslant \mathfrak{r}\mathbb{E}\left[d_\mathcal{M}(X, Y)^p\right].$$

(10.8)

One requires (10.8) to hold for $\mathcal{M}$-valued independent random variables X, X', Y, Y' (say, finitely-supported, to avoid measurability assumptions) such that each of the pairs X, X' and Y, Y' is identically distributed.

To the best of our knowledge, condition (10.8) was first studied systematically by Enflo [10], who defined a metric space $(\mathcal{M}, d_\mathcal{M})$ to have *generalized roundness* p it satisfies (10.8) with $\mathfrak{r} = 2$. He proved that L_p has generalized roundness p for $p \in [1, 2]$, and ingeniously used this notion to answer an old question of Smirnov. See [9] for a relatively recent example of substantial impact of Enflo's approach. By combining [14] with [19], a metric space $(\mathcal{M}, d_\mathcal{M})$ has generalized roundness p if and only if $(\mathcal{M}, d_\mathcal{M}^{p/2})$ embeds isometrically into a Hilbert space. The case $\mathfrak{r} > 2$ of (10.8) arose in [2] in the context of metric embeddings.

The final geometric modulus that we consider here is a quantity $j_p(F, \|\cdot\|_F)$, or simply $j_p(F)$ if the norm is clear from the context, that is defined to be the supremum over those $j \geqslant 1$ such that every independent and identically distributed F-valued random variables $Z, Z' \in L_p(F)$ satisfy

$$j\mathbb{E}\big[\|Z - \mathbb{E}[Z]\|_F^p\big] \leqslant \mathbb{E}\big[\|Z - Z'\|_F^p\big], \tag{10.9}$$

Note that (10.9) holds with $j = 1$ by Jensen's inequality, so we are asking here for an improvement of (this use of) Jensen's inequality by a definite factor; the letter "j" in this notation is in reference to "Jensen."

We have the following general bounds, which hold for every Banach space $(F, \|\cdot\|_F)$ and every $p \geqslant 1$.

$$\beta_p(F) \leqslant m_p(F) \leqslant \frac{2 + \tau_p(F)}{2j_p(F)}. \tag{10.10}$$

Indeed, we already observed the first inequality in (10.10), and the second inequality in (10.10) is justified by taking independent random variables $X, Y \in L_p(F)$, considering their mixture $Z \in L_p(F)$ as defined in (10.6), letting X', Y', Z' be independent copies of X, Y, Z, respectively, and proceeding as follows.

$$\mathbb{E}\bigg[\bigg\|X - \frac{1}{2}\mathbb{E}[X] - \frac{1}{2}\mathbb{E}[Y]\bigg\|_F^p + \bigg\|Y - \frac{1}{2}\mathbb{E}[X] - \frac{1}{2}\mathbb{E}[Y]\bigg\|_F^p\bigg]$$

$$\overset{(10.6)}{=} 2\mathbb{E}\big[\|Z - \mathbb{E}[Z]\|_F^p\big] \overset{(10.9)}{\leqslant} \frac{2}{j_p(F)}\mathbb{E}\big[\|Z - Z'\|_F^p\big]$$

$$\overset{(10.6)}{=} \frac{2}{j_p(F)}\bigg(\frac{1}{2}\mathbb{E}\big[\|X - Y\|_F^p\big] + \frac{1}{4}\mathbb{E}\big[\|X - X'\|_F^p\big] + \frac{1}{4}\mathbb{E}\big[\|Y - Y'\|_F^p\big]\bigg)$$

$$\leqslant \frac{2}{j_p(F)}\bigg(\frac{1}{2} + \frac{1}{4}\tau_p(F)\bigg)\mathbb{E}\big[\|X - Y\|_F^p\big].$$

Recalling the definition (10.5) of $m_p(F)$, this implies (10.10).

Here we prove the following bounds on $\beta_p(L_q), m_p(L_q), \tau_p(L_q), j_p(L_q)$ for $p, q \in [1, \infty)$.

Theorem 10.1.2 *For every $p, q \in [1, \infty)$ we have $j_p(L_q) = 2^{c(p,q)}$, where*

$$c(p, q) \overset{\text{def}}{=} \min\bigg\{1, p-1, \frac{p}{q}, \frac{p(q-1)}{q}\bigg\}$$

$$= \begin{cases} p - 1 & \text{if } 1 \leqslant p \leqslant q \leqslant 2 \text{ or } 1 \leqslant p \leqslant \frac{q}{q-1} \leqslant 2, \\ \frac{p(q-1)}{q} & \text{if } q \leqslant p \leqslant \frac{q}{q-1}, \\ \frac{p}{q} & \text{if } \frac{q}{q-1} \leqslant p \leqslant q, \\ 1 & \text{if } p \geqslant \frac{q}{q-1} \geqslant 2 \text{ or } p \geqslant q \geqslant 2. \end{cases} \tag{10.11}$$

We also have $\tau_p(L_q) \leqslant 2^{C(p,q)}$, where

$$
C(p,q) \overset{\text{def}}{=} \begin{cases} p-1 & \text{if } \frac{p}{p-1} \leqslant q \leqslant p, \\ \frac{p(q-2)}{q} + 1 & \text{if } \frac{q}{q-1} \leqslant p \leqslant q, \\ 2 - \frac{p}{q} & \text{if } q \geqslant 2 \text{ and } 1 \leqslant p \leqslant \frac{q}{q-1}, \\ \frac{p}{q} & \text{if } q \leqslant 2 \text{ and } q \leqslant p \leqslant \frac{q}{q-1}, \\ 1 & \text{if } 1 \leqslant p \leqslant q \leqslant 2. \end{cases} \tag{10.12}
$$

In fact, if $\frac{p}{p-1} \leqslant q \leqslant p$, then $\tau_p(L_q) = 2^{p-1}$, if $\frac{q}{q-1} \leqslant p \leqslant q$, then $\tau_p(L_q) = 2^{\frac{p(q-2)}{q}+1}$, and $\tau_p(L_q) = 2$ if $1 \leqslant p \leqslant q \leqslant 2$. Namely, the above bound on $\tau_p(L_q)$ is sharp in the first, second and fifth ranges in (10.12).

Furthermore, $\theta_p(L_q) = m_p(L_q) = 2^{2-p}$ if $p \leqslant q \leqslant 2$. More generally, we have the bound

$$
\theta_p(L_q) \leqslant m_p(L_q) \leqslant \min \left\{ \frac{3^p}{2^{p-1}} \left(\frac{\sqrt{2}}{3} \right)^{2c(p,q)}, \frac{2^{C(p,q)}+2}{2^{c(p,q)+1}} \right\}. \tag{10.13}
$$

The upper bound on $\theta_p(L_q)$ in (10.13) improves over (10.2) when $F = L_q$ for all values of $p, q \in [1, \infty)$. It would be interesting to find the exact value of $\theta_p(L_q)$ in the entire range $p, q \in [1, \infty)$. Note that the second quantity in the minimum in the right hand side of (10.13) corresponds to using (10.10) together with the bounds on $j_p(L_q)$ and $\tau_p(L_q)$ that Theorem 10.1.2 provides; when, say, $p = q$, this quantity is smaller than the first quantity in the minimum in the right hand side of (10.13) if and only if $1 \leqslant p < 3$.

Theorem 10.1.2 states that the constant $C(p,q)$ is sharp in the first, second and fifth ranges in (10.12). The following conjecture formulates what we expect to be the sharp values of $\tau_p(L_q)$ for all $p, q \in [1, \infty)$.

Conjecture 10.1.3 For all $p, q \in [1, \infty)$ we have $\tau_p(L_q) = 2^{C_{\text{opt}}(p,q)}$, where

$$
C_{\text{opt}}(p,q) \overset{\text{def}}{=} \max \left\{ 1, p-1, \frac{p(q-2)}{q} + 1 \right\}
$$

$$
= \begin{cases} p-1 & \text{if } p \geqslant 2 \text{ and } 1 \leqslant q \leqslant p, \\ \frac{p(q-2)}{q} + 1 & \text{if } q \geqslant 2 \text{ and } 1 \leqslant p \leqslant q, \\ 1 & \text{if } p, q \in [1, 2]. \end{cases} \tag{10.14}
$$

We will prove later that $\tau_p(L_q) \geqslant 2^{C_{\text{opt}}(p,q)}$, so Conjecture (10.1.3) is about improving our upper bounds on $\tau_p(L_q)$ in the remaining third and fourth ranges that appear in (10.12).

Question 10.1.4 Below we will obtain improvements over (10.2) for other spaces besides $\{L_q : q \in [1, \infty)\}$, including e.g. the Schatten–von Neumann trace classes (see e.g. [20]) $\{S_q \; : \; q \in (1, \infty)\}$. However, parts of Theorem 10.1.2 rely on "commutative" properties of L_q which are not valid for S_q, thus leading to even better bounds in the commutative setting. It would be especially interesting to obtain sharp bounds in noncommutative probabilistic inequalities such as the roundness inequality (10.7) when $F = S_q$. In particular, we ask what is the value of $z_1(S_1)$? At present, we know (as was already shown by Enflo [10]) that $z_1(L_1) = 2$ while the only bound that we have for S_1 is $z_1(S_1) \leq 4$. Note that 4 is a trivial upper bound here, which holds for every Banach space. Interestingly, it follows from [7] that $z_1(S_1) \geq 2\sqrt{2}$, as explained in Remark 10.3.1 below. So, there is a genuine difference between the commutative and noncommutative settings of L_1 and S_1, respectively. As a more modest question, is $z_1(S_1)$ strictly less than 4?

10.1.3   Complex Interpolation

We will use basic terminology, notation and results of complex interpolation of Banach spaces; the relevant background appears in [4, 8]. Theorem 10.1.2 is a special case of the following more general result about interpolation spaces. As such, it applies also to random variables that take values in certain spaces other than L_q, including, for examples, Schatten–von Neumann trace classes (see e.g. [20]) and, by an extrapolation theorem of Pisier[18], Banach lattices of nontrivial type.

Theorem 10.1.5 *Fix $\theta \in [0, 1]$ and $\frac{2}{2-\theta} \leq p \leq \frac{2}{\theta}$. Let $(F, \| \cdot \|_F), (H, \| \cdot \|_H)$ be a compatible pair of Banach spaces such that $(H, \| \cdot \|_H)$ is a Hilbert space. Then the following estimates hold true.*

$$z_p([F, H]_\theta) \leq 2^{1+(1-\theta)p} \qquad \text{and} \qquad j_p([F, H]_\theta) \geq 2^{\frac{\theta p}{2}}. \qquad (10.15)$$

Additionally, we have

$$\mathit{b}_p([F, H]_\theta) \leq m_p([F, H]_\theta) \leq \min \left\{ \frac{3^p}{2^{p-1}} \left(\frac{\sqrt{2}}{3} \right)^{p\theta}, \frac{1 + 2^{(1-\theta)p}}{2^{\frac{\theta p}{2}}} \right\}$$

$$= \begin{cases} \frac{3^p}{2^{p-1}} \left(\frac{\sqrt{2}}{3} \right)^{p\theta} & \text{if } \frac{1}{1-\theta} \leq p \leq \frac{2}{\theta}, \\ \frac{1 + 2^{(1-\theta)p}}{2^{\frac{\theta p}{2}}} & \text{if } \frac{2}{2-\theta} \leq p \leq \frac{1}{1-\theta}. \end{cases} \qquad (10.16)$$

(Note that if the first range of values of p in the right hand side of (10.16) is nonempty, then necessarily $\theta \leq \frac{2}{3}$.)

The deduction of Theorem 10.1.2 from Theorem 10.1.5 appears in Sect. 10.3 below; in most cases this deduction is nothing more than a direct substitution into

Theorem 10.1.5, but in some cases a further argument is needed. Theorem 10.1.5 itself is a special case of the following theorem.

Theorem 10.1.6 *Fix* $\theta \in [0, 1]$ *and* $p \in [1, \infty]$ *that satisfy* $\frac{2}{2-\theta} \leqslant p \leqslant \frac{2}{\theta}$. *Let* $(F, \|\cdot\|_F)$, $(H, \|\cdot\|_H)$ *be a compatible pair of Banach spaces such that* $(H, \|\cdot\|_H)$ *is a Hilbert space. Suppose that* $(\mathcal{X}, \mu)$ *and* $(\mathcal{Y}, \nu)$ *are probability spaces. Then, for every* $f \in L_p(\mu \times \nu; [F, H]_\theta)$ *we have*

$$2^{1+(1-\theta)p} \iint_{\mathcal{X} \times \mathcal{Y}} \|f(x, y)\|^p_{[F,H]_\theta} d\mu(x) d\nu(y)$$

$$\geqslant \iint_{\mathcal{X} \times \mathcal{X}} \left\| \int_{\mathcal{Y}} (f(x, y) - f(\chi, y)) d\nu(y) \right\|^p_{[F,H]_\theta} d\mu(x) d\mu(\chi)$$

$$+ \iint_{\mathcal{Y} \times \mathcal{Y}} \left\| \int_{\mathcal{X}} (f(x, y) - f(x, \upsilon)) d\mu(x) \right\|^p_{[F,H]_\theta} d\nu(y) d\nu(\upsilon), \qquad (10.17)$$

and

$$\frac{3^p}{2^{p-1}} \left(\frac{\sqrt{2}}{3} \right)^{p\theta} \iint_{\mathcal{X} \times \mathcal{Y}} \|f(x, y)\|^p_{[F,H]_\theta} d\mu(x) d\nu(y)$$

$$\geqslant \int_{\mathcal{X}} \left\| \int_{\mathcal{Y}} f(x, y) d\nu(y) - \frac{1}{2} \iint_{\mathcal{X} \times \mathcal{Y}} f(\chi, \upsilon) d\mu(\chi) d\nu(\upsilon) \right\|^p_{[F,H]_\theta} d\mu(x)$$

$$+ \int_{\mathcal{Y}} \left\| \int_{\mathcal{X}} f(x, y) d\mu(x) - \frac{1}{2} \iint_{\mathcal{X} \times \mathcal{Y}} f(\chi, \upsilon) d\mu(\chi) d\nu(\upsilon) \right\|^p_{[F,H]_\theta} d\nu(y).$$

$$(10.18)$$

Furthermore, if $g \in L_p(\mu \times \mu; [F, H]_\theta)$, *then*

$$2^{\left(1 - \frac{\theta}{2}\right)p} \iint_{\mathcal{X} \times \mathcal{X}} \|g(x, \chi)\|^p_{[F,H]_\theta} d\mu(x) d\mu(\chi)$$

$$\geqslant \int_{\mathcal{X}} \left\| \int_{\mathcal{X}} (g(x, \chi) - g(\chi, x)) d\mu(x) \right\|^p_{[F,H]_\theta} d\mu(\chi). \qquad (10.19)$$

Proof of Theorem 10.1.5 assuming Theorem 10.1.6 Let X and Y be independent p-integrable $[F, H]_\theta$-valued random vectors. Due to the independence assumption, without loss of generality there are probability spaces $(\mathcal{X}, \mu)$ and $(\mathcal{Y}, \nu)$ such that X and Y are elements of $L_p(\mu \times \nu; [F, H]_\theta)$ that depend only on the first variable and second variable, respectively. Then (10.17) and (10.18) applied to $f = X - Y$ become

$$\mathbb{E}\left[\|X - X'\|^p_{[F,H]_\theta} \right] + \mathbb{E}\left[\|Y - Y'\|^p_{[F,H]_\theta} \right] \leqslant 2^{1+(1-\theta)p} \mathbb{E}\left[\|X - Y\|^p_{[F,H]_\theta} \right],$$

and

$$\mathbb{E}\left[\left\|X - \frac{1}{2}\mathbb{E}[X] - \frac{1}{2}\mathbb{E}[Y]\right\|_{[F,H]_\theta}^p\right] + \mathbb{E}\left[\left\|Y - \frac{1}{2}\mathbb{E}[X] - \frac{1}{2}\mathbb{E}[Y]\right\|_{[F,H]_\theta}^p\right]$$

$$\leqslant \frac{3^p}{2^{p-1}}\left(\frac{\sqrt{2}}{3}\right)^{p\theta} \mathbb{E}\left[\|X - Y\|_{[F,H]_\theta}^p\right].$$

We therefore established the first inequality in (10.15) as well as the upper bound on $m_p([F, H]_\theta)$ that corresponds to the first term in the minimum that appears in (10.16).

Similarly, due to the fact that X and X' are i.i.d., without loss of generality there is a probability space $(\mathcal{X}, \mu)$ such that X and X' are elements of $L_p(\mu \times \mu; [F, H]_\theta)$ that depend only on the first variable and second variable, respectively. Then, (10.19) applied to $g = X - X'$ simplifies to give

$$\mathbb{E}\left[\|X - X'\|_{[F,H]_\theta}^p\right] \geqslant 2^{\frac{\theta p}{2}}\mathbb{E}\left[\|X - \mathbb{E}[X]\|_{[F,H]_\theta}^p\right].$$

This establishes the second inequality in (10.15), as well as the upper bound on $m_p([F, H]_\theta)$ that corresponds to the second term in the minimum that appears in (10.16), due to (10.10). $\qquad\square$

The first and third inequalities of Theorem 10.1.6 are generalizations of results that appeared in the literature. Specifically, (10.17) generalizes Lemma 6 of [2], and (10.19) generalizes Lemma 5 of [17], which is itself inspired by a step within the proof of Theorem 2 of [21]. The proof of Theorem 10.1.6, which appears in Sect. 10.3 below, differs from the proofs of [2, 17, 21], but relies on the same ideas.

10.2   Proof of Theorem 10.1.1

Let $(F, \|\cdot\|_F)$ be a Banach space. Fix $p \geqslant 1$. Theorem 10.1.1 asserts that $\mathfrak{b}_p(F) \leqslant \frac{3^p}{2^{p-1}}$. In fact, $m_p(F) \leqslant \frac{3^p}{2^{p-1}}$, which is stronger by (10.10). To see this, let $X, Y \in L_p(F)$ be independent random vectors and observe that

$$\mathbb{E}\left[\left\|X - \frac{1}{2}\mathbb{E}[X] - \frac{1}{2}\mathbb{E}[Y]\right\|_F^p\right] = \frac{3^p}{2^p}\mathbb{E}\left[\left\|\frac{2}{3}(X - \mathbb{E}[Y]) + \frac{1}{3}(\mathbb{E}[Y] - \mathbb{E}[X])\right\|_F^p\right]$$

$$\leqslant \frac{3^p}{2^p}\left(\frac{2}{3}\mathbb{E}\left[\|X - \mathbb{E}[Y]\|_F^p\right] + \frac{1}{3}\mathbb{E}\left[\|\mathbb{E}[Y] - \mathbb{E}[X]\|_F^p\right]\right)$$

$$\leqslant \frac{3^p}{2^p}\mathbb{E}\left[\|X - Y\|_F^p\right],$$

where the penultimate step holds due to the convexity of $\|\cdot\|_F^p$ and the final step holds because, by Jensen's inequality, both $\mathbb{E}\left[\|X - \mathbb{E}[Y]\|_F^p\right] = \mathbb{E}\left[\|\mathbb{E}_Y[X - Y]\|_F^p\right]$ and $\mathbb{E}\left[\|\mathbb{E}[Y - X]\|_F^p\right]$ are at most $\mathbb{E}\left[\|X - Y\|_F^p\right]$. The symmetric reasoning with X replaced by Y now gives

$$\mathbb{E}\left[\left\|X - \frac{1}{2}\mathbb{E}[X] - \frac{1}{2}\mathbb{E}[Y]\right\|_F^p\right] + \mathbb{E}\left[\left\|Y - \frac{1}{2}\mathbb{E}[X] - \frac{1}{2}\mathbb{E}[Y]\right\|_F^p\right]$$

$$\leqslant 2\max\left\{\mathbb{E}\left[\left\|X - \frac{1}{2}\mathbb{E}[X] - \frac{1}{2}\mathbb{E}[Y]\right\|_F^p\right], \mathbb{E}\left[\left\|Y - \frac{1}{2}\mathbb{E}[X] - \frac{1}{2}\mathbb{E}[Y]\right\|_F^p\right]\right\}$$

$$\leqslant \frac{3^p}{2^{p-1}}\mathbb{E}\left[\|X - Y\|_F^p\right].$$

This shows that $m_p(F) \leqslant \frac{3^p}{2^{p-1}}$. It remains to prove that the bound $\mathfrak{b}_p(F) \leqslant \frac{3^p}{2^{p-1}}$ is optimal for general F.

Fix an integer $n \geqslant 2$ and consider

$$F_n \stackrel{\text{def}}{=} \left\{x \in \mathbb{C}^{2n} : \sum_{k=1}^{2n} x_k = 0\right\},$$

equipped with supremum norm inherited from ℓ_∞^{2n}. We will prove that

$$\mathfrak{b}_p(F_n) \geqslant 2\left(\frac{3}{2} - \frac{1}{n}\right)^p \xrightarrow{n \to \infty} \frac{3^p}{2^{p-1}}. \tag{10.20}$$

Denote by $\{e_k\}_{k=1}^{2n}$ the standard coordinate basis of ℓ_∞^{2n}. Define two n-element sets $A_n, B_n \subseteq F_n$ by

$$A_n \stackrel{\text{def}}{=} \left\{(3n-2)e_j - (n+2)\sum_{k\in\{1,\ldots,n\}\smallsetminus\{j\}} e_k + (n-2)\sum_{k=n+1}^{2n} e_k : j \in \{1,\ldots,n\}\right\},$$

and

$$B_n \stackrel{\text{def}}{=} \left\{(n-2)\sum_{k=1}^{n} e_k + (3n-2)e_j - (n+2)\sum_{k\in\{n+1,\ldots,2n\}\smallsetminus\{j\}} e_k : j \in \{n+1,\ldots,2n\}\right\}.$$

Note that A_n and B_n are indeed subsets of F_n because $3n-2-(n-1)(n+2)+n(n-2) = 0$. Let X, Y be independent and uniformly distributed on A_n, B_n, respectively. One checks that $\|a - b\|_\infty = 2n$ for any $a \in A_n$ and $b \in B_n$. So, $\mathbb{E}\left[\|X - Y\|_\infty^p\right] = (2n)^p$. The desired bound (10.20) will follow if we demonstrate that

$$\forall z \in F_n, \qquad \mathbb{E}\left[\|X - z\|_\infty^p + \|Y - z\|_\infty^p\right] \geqslant 2(3n-2)^p. \tag{10.21}$$

The proof of (10.21) proceeds via symmetrization. For permutations $\sigma, \rho \in S_n$, define $T_{\sigma,\rho} : F_n \to F_n$ by

$$\forall x = (x_1, \ldots, x_{2n}) \in F_n,$$

$$T_{\sigma,\rho}(x) \overset{\text{def}}{=} \left(x_{\sigma(1)}, x_{\sigma(2)}, \ldots, x_{\sigma(n)}, x_{n+\rho(1)}, x_{n+\rho(2)}, \ldots, x_{n+\rho(n)} \right).$$

$T_{\sigma,\rho}$ is a linear isometry of F_n and the sets A_n and B_n are $T_{\sigma,\rho}$-invariant. Hence, for any $z \in F_n$,

$$\mathbb{E}\left[\|X - z\|_\infty^p \right] = \frac{1}{(n!)^2} \sum_{\sigma,\rho \in S_n} \mathbb{E}\left[\|T_{\sigma,\rho}(X) - z\|_\infty^p \right]$$

$$= \frac{1}{(n!)^2} \sum_{\sigma,\rho \in S_n} \mathbb{E}\left[\|X - T_{\sigma^{-1},\rho^{-1}}(z)\|_\infty^p \right]$$

$$\geqslant \mathbb{E}\left[\left\| X - \frac{1}{(n!)^2} \sum_{\sigma,\rho \in S_n} T_{\sigma^{-1},\rho^{-1}}(z) \right\|_\infty^p \right]$$

$$= \mathbb{E}\left[\left\| X - \frac{z_1 + \ldots + z_n}{n} \sum_{k=1}^{n} e_k + \frac{z_1 + \ldots + z_n}{n} \sum_{k=n+1}^{2n} e_k \right\|_\infty^p \right].$$

$$(10.22)$$

Denoting $u = (z_1 + \ldots + z_n)/n$, it follows from (10.22) that $\mathbb{E}\left[\|X - z\|_\infty^p \right] \geqslant |3n - 2 - u|^p$, because one of the first n coordinates of any member of the support of X equals $3n - 2$. The same argument with X replaced by Y gives that $\mathbb{E}\left[\|Y - z\|_\infty^p \right] \geqslant |3n - 2 + u|^p$, because now one of the last n coordinates of any member of the support of Y equals $3n - 2$. We conclude with the following application of the convexity of $|\cdot|^p$.

$$\mathbb{E}\left[\|X - z\|_\infty^p + \|Y - z\|_\infty^p \right] \geqslant |3n - 2 - u|^p + |3n - 2 + u|^p \geqslant 2(3n - 2)^p. \quad \square$$

Remark 10.2.1 It is worthwhile to examine what the above argument gives if we take the norm on F_n to be the norm inherited from ℓ_q^{2n}. One computes that $\|a - b\|_q = (2n)^{1+1/q}$ for every $a \in A_n$ and $b \in B_n$. So,

$$\mathbb{E}\left[\|X - Y\|_q^p \right] = (2n)^{\frac{p(q+1)}{q}}.$$

$$(10.23)$$

Also, it follows from the same reasoning that led to (10.22) that for every $z \in F_n$,

$$\mathbb{E}\left[\|X - z\|_q^p\right] \geq \mathbb{E}\left[\left\|X - u\sum_{k=1}^{n} e_k + u \sum_{k=n+1}^{n} e_k\right\|_q^p\right]$$

$$= \left(|3n - 2 - u|^q + (n-1)|n + 2 + u|^q + n|n - 2 + u|^q\right)^{\frac{p}{q}},$$

and

$$\mathbb{E}\left[\|Y - z\|_q^p\right] \geq \mathbb{E}\left[\left\|Y - u\sum_{k=1}^{n} e_k + u \sum_{k=n+1}^{n} e_k\right\|_q^p\right]$$

$$= \left(|3n - 2 + u|^q + (n-1)|n + 2 - u|^q + n|n - 2 - u|^q\right)^{\frac{p}{q}}.$$

Hence, using the convexity of the p'th power of the ℓ_q norm on $\mathbb{R}^3$, we see that

$$\mathbb{E}\left[\|X - z\|_q^p\right] + \mathbb{E}\left[\|Y - z\|_q^p\right] \geq 2\left((3n - 2)^q + (n-1)(n + 2)^q + n(n - 2)^q\right)^{\frac{p}{q}}. \tag{10.24}$$

By contrasting (10.23) with (10.24) we conclude that

$$\mathfrak{b}_p(F_n, \|\cdot\|_q) \geq 2\left((3n - 2)^q + (n-1)(n + 2)^q + n(n - 2)^q\right)^{\frac{p}{q}} (2n)^{-\frac{p(q+1)}{q}}.$$

In particular, if we take $p = q \geq 2$ and $n = \lceil q \rceil$, then we conclude that $\mathfrak{b}_q(F_n, \|\cdot\|_q) \geq \frac{c}{q}\left(\frac{3}{2}\right)^q$ for some universal constant $c > 0$. So, there is very little potential asymptotic gain (as $q \to \infty$) if we know that the Banach space of Theorem 10.1.1 admits an isometric embedding into L_q.

Above, and in what follows, we stated that a normed space admits an isometric embedding into L_q without specifying whether the embedding is linear or not. Later we will need such embeddings to be linear, so we recall that for any $q \geq 1$, by a classical differentiation argument (see [3, Chapter 7] for a thorough treatment of such reductions to the linear setting), a normed space embeds isometrically into L_q as a metric space if and only if it admits a linear isometric embedding into L_q.

Note that the phenomenon of Remark 10.2.1 is special to random variables that have different expectations. Namely, if $\mathbb{E}[X] = \mathbb{E}[Y]$, then by Jensen's inequality the ratio that defines $\mathfrak{b}_q(F)$ is at most 2 rather than the aforementioned exponential growth as $q \to \infty$. The following proposition shows that if F is a subspace of L_q for $q \geq 3$, then when $\mathbb{E}[X] = \mathbb{E}[Y]$ this ratio is at most 1, which is easily seen to be best possible (consider any nontrivial symmetric random variable X, and take Y to be identically 0).

Proposition 10.2.2 *Let $(F, \|\cdot\|_F)$ be a Banach space that admits an isometric embedding into L_q for some $q \in [3, \infty)$. Then, for any pair of independent F-*

valued random vectors $X, Y \in L_q(F)$ *with* $\mathbb{E}[X] = \mathbb{E}[Y]$,

$$\inf_{z \in F} \mathbb{E}\left[\|X - z\|_F^q + \|Y - z\|_F^q\right] \leqslant \mathbb{E}\left[\|X - \mathbb{E}[X]\|_F^q + \|Y - \mathbb{E}[X]\|_F^q\right]$$

$$\leqslant \mathbb{E}\left[\|X - Y\|_F^q\right]. \tag{10.25}$$

Proof L_q over $\mathbb{C}$ embeds isometrically into L_q over $\mathbb{R}$ (indeed, complex L_q is, as a real Banach space, the same as $L_q(\ell_2^2)$, so this follows from the fact that Hilbert space is isometric to a subspace of L_q). So, in Proposition 10.2.2 we may assume that F embeds isometrically into L_q over $\mathbb{R}$, and therefore by integration/Fubini it suffices to prove (10.25) for real-valued random variables. So, our goal is to show that if X, Y are independent mean-zero real random variables with $\mathbb{E}[|X|^q]$, $\mathbb{E}[|Y|^q] < \infty$, then

$$\mathbb{E}\left[|X + Y|^q\right] \geqslant \mathbb{E}\left[|X|^q\right] + \mathbb{E}\left[|Y|^q\right]. \tag{10.26}$$

The bound (10.25) would then follow by applying (10.26) to the mean-zero variables $X - \mathbb{E}[X]$ and $\mathbb{E}[X] - Y$.

Note in passing that the assumption $q \geqslant 3$ is crucial here, i.e. (10.26) fails if $q \in (0, 3) \smallsetminus \{2\}$. Indeed, if $\beta \in (0, \frac{1}{2})$ and $\mathbb{P}[X = 1 - \beta] = \mathbb{P}[Y = 1 - \beta] = \beta$ and $\mathbb{P}[X = -\beta] = \mathbb{P}[Y = -\beta] = 1 - \beta$, then $\mathbb{E}[X] = \mathbb{E}[Y] = 0$ but

$$\frac{\mathbb{E}[|X + Y|^q]}{\mathbb{E}[|X|^q] + \mathbb{E}[|Y|^q]} = \frac{\beta^2 2^q (1 - \beta)^q + (1 - \beta)^2 2^q \beta^q + 2\beta(1 - \beta)(1 - 2\beta)^q}{2\beta(1 - \beta)^q + 2(1 - \beta)\beta^q}. \tag{10.27}$$

If $q \in (0, 2)$, then the right hand side of (10.27) equals $2^{q-2} < 1$ for $\beta = \frac{1}{2}$. If $q \in (2, 3)$, then the right hand side of (10.27) equals $1 + (2^{q-1} - q - 1)\beta + o(\beta)$, which is less than 1 for small β since $2^{q-1} - q - 1 < 0$ for $q \in (2, 3)$.

To prove (10.26), for every $s > 0$ and $x \in \mathbb{R}$, denote $\phi_s(x) = \text{sign}(x) \cdot |x|^s$. Observe that

$$\forall x, y \in \mathbb{R}, \quad \alpha(x, y) \overset{\text{def}}{=} |x + y|^q - |x|^q - |y|^q - q\phi_{q-1}(x)y - qx\phi_{q-1}(y) \geqslant 0. \tag{10.28}$$

Once (10.28) is proved, (10.26) would follow because

$$0 \leqslant \mathbb{E}[\alpha(X, Y)] = \mathbb{E}\left[|X + Y|^q\right] - \mathbb{E}\left[|X|^q\right] - \mathbb{E}\left[|Y|^q\right] - q\mathbb{E}[Y]\mathbb{E}\left[\phi_{q-1}(X)\right]$$

$$- q\mathbb{E}[X]\mathbb{E}\left[\phi_{q-1}(Y)\right]$$

$$= \mathbb{E}\left[|X + Y|^q\right] - \mathbb{E}\left[|X|^q\right] - \mathbb{E}\left[|Y|^q\right],$$

where the penultimate step uses the independence of X, Y and the last step uses $\mathbb{E}[X] = \mathbb{E}[Y] = 0$.

It suffices to prove (10.28) when $q > 3$; the case $q = 3$ follows by passing to the limit. Once checks that

$$\frac{\partial^3 \alpha}{\partial x^2 \partial y} = p(p-1)(p-2)\big(\phi_{p-3}(x+y) - \phi_{p-3}(x)\big)$$

$$\implies \text{sign}\left(\frac{\partial^3 \alpha}{\partial x^2 \partial y}(x, y)\right) = \text{sign}(y),$$

where the last step holds because ϕ_{p-3} is increasing. Hence, $y \mapsto \frac{\partial^2 \alpha}{\partial x^2}(x, y)$ is decreasing for $y < 0$ and increasing for $y > 0$. One checks that $\frac{\partial^2 \alpha}{\partial x^2}(x, 0) = 0$ for all $x \in \mathbb{R}$, so $\frac{\partial^2 \alpha}{\partial x^2}(x, y) \geqslant 0$. Thus $x \mapsto \alpha(x, y)$ is convex for every fixed $y \in \mathbb{R}$. But $\alpha(0, y) = \frac{\partial \alpha}{\partial x}(0, y) = 0$ for any $y \in \mathbb{R}$, i.e. the tangent to the graph of $x \mapsto \alpha(x, y)$ at $x = 0$ is the x-axis. Convexity implies that the graph of $x \mapsto \alpha(x, y)$ lies above the x-axis, as required. $\square$

We end this section with the following simpler metric space counterpart of Theorem 10.1.1.

Proposition 10.2.3 *Fix $p \geqslant 1$ and let X and Y be independent finitely supported random variables taking values in a metric space $(\mathcal{M}, d_{\mathcal{M}})$. Then*

$$\inf_{z \in \mathcal{M}} \mathbb{E}\big[d_{\mathcal{M}}(X, z)^p + d_{\mathcal{M}}(Y, z)^p\big] \leqslant (2^p + 1)\, \mathbb{E}\big[d_{\mathcal{M}}(X, Y)^p\big]. \tag{10.29}$$

The constant $2^p + 1$ in (10.29) is optimal.

Proof Let X' have the same distribution as X and be independent of X and Y. The point-wise inequality

$$d_{\mathcal{M}}(X, X')^p \leqslant \big(d_{\mathcal{M}}(X, Y) + d_{\mathcal{M}}(Y, X')\big)^p \leqslant 2^{p-1}\big(d_{\mathcal{M}}(X, Y)^p + d_{\mathcal{M}}(X', Y)^p\big)$$

is a consequence of the triangle inequality and the convexity of $(u \geq 0) \mapsto u^p$. By taking expectations, we obtain $\mathbb{E}\big[d_{\mathcal{M}}(X, X')^p\big] \leqslant 2^p \mathbb{E}[d_{\mathcal{M}}(X, Y)^p]$, so that

$$\inf_{z \in \mathcal{M}} \mathbb{E}\big[d_{\mathcal{M}}(X, z)^p + d_{\mathcal{M}}(Y, z)^p\big] \leqslant \mathbb{E}\big[d_{\mathcal{M}}(X, X')^p + d_{\mathcal{M}}(Y, X')^p\big]$$

$$\leqslant (2^p + 1)\, \mathbb{E}\big[d_{\mathcal{M}}(X, Y)^p\big].$$

To see that the constant $2^p + 1$ is optimal, fix $n \in \mathbb{N}$ and let $\mathcal{M}$ be the complete bipartite graph $\mathsf{K}_{n,n}$, equipped with its shortest-path metric. Equivalently, $\mathcal{M}$ can be partitioned into two n-point subsets L, R, and for distinct $x, y \in \mathcal{M}$ we have $d_{\mathcal{M}}(x, y) = 2$ if $\{x, y\} \subseteq L$ or $\{x, y\} \subseteq R$, while $d_{\mathcal{M}}(x, y) = 1$ otherwise. Let X be uniformly distributed over L and Y be uniformly distributed over R. Then $d_{\mathcal{M}}(X, Y) = 1$ point-wise. If $z \in L$, then $d_{\mathcal{M}}(Y, z) = 1$ point-wise, while

$\mathbb{P}[d_m(X, z) = 2] = \frac{n-1}{n}$ and $\mathbb{P}[d_m(X, z) = 0] = \frac{1}{n}$. Consequently,

$$\frac{\mathbb{E}[d_m(X, z)^p + d_m(Y, z)^p]}{\mathbb{E}[d_m(X, Y)^p]} = \frac{n-1}{n} 2^p + 1 \xrightarrow[n \to \infty]{} 2^p + 1.$$

By symmetry, the same holds if $z \in R$. $\square$

10.3   Proof of Theorem 10.1.6 and Its Consequences

Here we prove Theorem 10.1.6 and deduce Theorem 10.1.2.

Proof of Theorem 10.1.6 The assumption $\frac{2}{2-\theta} \leqslant p \leqslant \frac{2}{\theta}$ implies that $\frac{1}{p} = \frac{1-\theta}{q} + \frac{\theta}{2}$ for some (unique) $q \in [1, \infty]$. We will fix this value of q for the rest of the proof of Theorem 10.1.6. All of the desired bounds (10.17)–(10.19) hold true when $\theta = 0$, namely for every Banach space $(F, \|\cdot\|_F)$ and every $f \in L_q(\mu \times \nu; F)$ we have

$$2^{q+1} \iint_{\mathcal{X} \times \mathcal{Y}} \|f(x, y)\|_F^q d\mu(x) d\nu(y)$$

$$\geqslant \iint_{\mathcal{X} \times \mathcal{X}} \left\| \int_{\mathcal{Y}} (f(x, y) - f(\chi, y)) d\nu(y) \right\|_F^q d\mu(x) d\mu(\chi) \qquad (10.30)$$

$$+ \iint_{\mathcal{Y} \times \mathcal{Y}} \left\| \int_{\mathcal{X}} (f(x, y) - f(x, \upsilon)) d\mu(x) \right\|_F^q d\nu(y) d\nu(\upsilon),$$

and

$$\frac{3^q}{2^{q-1}} \iint_{\mathcal{X} \times \mathcal{Y}} \|f(x, y)\|_F^q d\mu(x) d\nu(y)$$

$$\geqslant \int_{\mathcal{X}} \left\| \int_{\mathcal{Y}} f(x, y) d\nu(y) - \frac{1}{2} \iint_{\mathcal{X} \times \mathcal{Y}} f(\chi, \upsilon) d\mu(\chi) d\nu(\upsilon) \right\|_F^q d\mu(x)$$

$$+ \int_{\mathcal{Y}} \left\| \int_{\mathcal{X}} f(x, y) d\mu(x) - \frac{1}{2} \iint_{\mathcal{X} \times \mathcal{Y}} f(\chi, \upsilon) d\mu(\chi) d\nu(\upsilon) \right\|_F^q d\nu(y).$$

$$(10.31)$$

Furthermore, if $g \in L_q(\mu \times \mu; F)$, then

$$2^q \iint_{\mathcal{X} \times \mathcal{X}} \|g(x, \chi)\|_F^q d\mu(x) d\mu(\chi) \geqslant \int_{\mathcal{X}} \left\| \int_{\mathcal{X}} (g(x, \chi) - g(\chi, x)) d\mu(x) \right\|_F^q d\mu(\chi).$$

$$(10.32)$$

Indeed, (10.30)–(10.32) are direct consequences of the triangle inequality in $L_q(\mu \times \nu; F)$ and $L_q(\mu \times \mu; F)$ and Jensen's inequality, with the appropriate interpretation when $q = \infty$.

By complex interpolation theory (specifically, by combining [4, Theorem 4.1.2] and [4, Theorem 5.1.2]), Theorem 10.1.6 will follow if we prove the $\theta = 1$ case of (10.17)–(10.19). To this end, as H is a Hilbert space and the inequalities in question are quadratic, it suffices to prove them coordinate-wise (with respect to any orthonormal basis of H), i.e., it suffices to show that for every ($\mathbb{C}$-valued) $f \in L_2(\mu \times \nu)$ and $g \in L_2(\mu \times \mu)$,

$$
2 \iint_{\mathcal{X} \times \mathcal{Y}} |f(x, y)|^2 d\mu(x) d\nu(y)
$$

$$
\geqslant \iint_{\mathcal{X} \times \mathcal{X}} \left| \int_{\mathcal{Y}} (f(x, y) - f(\chi, y)) d\nu(y) \right|^2 d\mu(x) d\mu(\chi) \tag{10.33}
$$

$$
+ \iint_{\mathcal{Y} \times \mathcal{Y}} \left| \int_{\mathcal{X}} (f(x, y) - f(x, \upsilon)) d\mu(x) \right|^2 d\nu(y) d\nu(\upsilon),
$$

and

$$
\iint_{\mathcal{X} \times \mathcal{Y}} |f(x, y)|^2 d\mu(x) d\nu(y)
$$

$$
\geqslant \int_{\mathcal{X}} \left| \int_{\mathcal{Y}} f(x, y) d\nu(y) - \frac{1}{2} \iint_{\mathcal{X} \times \mathcal{Y}} f(\chi, \upsilon) d\mu(\chi) d\nu(\upsilon) \right|^2 d\mu(x)
$$

$$
+ \int_{\mathcal{Y}} \left| \int_{\mathcal{X}} f(x, y) d\mu(x) - \frac{1}{2} \iint_{\mathcal{X} \times \mathcal{Y}} f(\chi, \upsilon) d\mu(\chi) d\nu(\upsilon) \right|^2 d\nu(y),
$$

$$\tag{10.34}$$

and

$$
2 \iint_{\mathcal{X} \times \mathcal{X}} |g(x, \chi)|^2 d\mu(x) d\mu(\chi) \geqslant \int_{\mathcal{X}} \left| \int_{\mathcal{X}} (g(x, \chi) - g(\chi, x)) d\mu(x) \right|^2 d\mu(\chi).
$$

$$\tag{10.35}$$

The following derivation of the quadratic scalar inequalities (10.33)–(10.35) is an exercise in linear algebra.

Let $\{\varphi_j\}_{j=0}^{\infty} \subseteq L_2(\mu)$ and $\{\psi_k\}_{k=0}^{\infty} \subseteq L_2(\nu)$ be any orthonormal bases of $L_2(\mu)$ and $L_2(\nu)$, respectively, for which $\varphi_0 = 1_{\mathcal{X}}$ and $\psi_0 = 1_{\mathcal{Y}}$. Then $\{\varphi_j \otimes \psi_k\}_{j,k=0}^{\infty}, \{\varphi_j \otimes \varphi_k\}_{j,k=0}^{\infty}$ and $\{\psi_j \otimes \psi_k\}_{j,k=0}^{\infty}$ are orthonormal bases of $L_2(\mu \times \nu), L_2(\mu \times \mu)$ and $L_2(\nu \times \nu)$, respectively, where for $\varphi \in L_2(\mu)$ and $\psi \in L_2(\nu)$ one defines (as usual) $\varphi \otimes \psi : \mathcal{X} \times \mathcal{Y} \to \mathbb{C}$ by setting $\varphi \otimes \psi(x, y) = \varphi(x)\psi(y)$ for $(x, y) \in \mathcal{X} \times \mathcal{Y}$. We therefore have the following expansions, in the sense of convergence in $L_2(\mu \times \nu)$ and $L_2(\mu \times \mu)$, respectively.

$$f = \sum_{j=0}^{\infty}\sum_{k=0}^{\infty}\langle \xi, \varphi_j \otimes \psi_k\rangle_{L_2(\mu\times\nu)}\varphi_j \otimes \psi_k \text{ and } g = \sum_{j=0}^{\infty}\sum_{k=0}^{\infty}\langle \zeta, \varphi_j \otimes \psi_k\rangle_{L_2(\mu\times\nu)}\varphi_j \otimes \psi_k.$$

In particular, by Parseval we have

$$\|f\|_{L_2(\mu\times\nu)}^2 = \sum_{j=0}^{\infty}\sum_{k=0}^{\infty}\left|\langle f, \varphi_j \otimes \psi_k\rangle_{L_2(\mu\times\nu)}\right|^2 \text{ and } \|g\|_{L_2(\mu\times\mu)}^2$$

$$= \sum_{j=0}^{\infty}\sum_{k=0}^{\infty}\left|\langle g, \varphi_j \otimes \varphi_k\rangle_{L_2(\mu\times\mu)}\right|^2. \tag{10.36}$$

Define $R_\chi f \in L_2(\mu \times \mu)$ by

$$R_\chi f \stackrel{\text{def}}{=} \sum_{j=1}^{\infty}\langle f, \varphi_j \otimes \psi_0\rangle_{L_2(\mu\times\nu)}\varphi_j \otimes \varphi_0 - \sum_{j=1}^{\infty}\langle f, \varphi_j \otimes \psi_0\rangle_{L_2(\mu\times\nu)}\varphi_0 \otimes \varphi_j.$$

So, $(\mu \times \mu)$-almost surely $R_\chi f(x, \chi) = \int_y \big(f(x, y) - f(\chi, y)\big)d\nu(y)$. Also, define $R_y f \in L_2(\nu \times \nu)$ by

$$R_y f \stackrel{\text{def}}{=} \sum_{j=1}^{\infty}\langle f, \varphi_0 \otimes \psi_j\rangle_{L_2(\mu\times\nu)}\psi_j \otimes \psi_0 - \sum_{j=1}^{\infty}\langle f, \varphi_0 \otimes \psi_j\rangle_{L_2(\mu\times\nu)}\psi_0 \otimes \psi_j.$$

So, $(\nu \times \nu)$-almost surely $R_y f(y, \upsilon) = \int_\chi \big(f(x, y) - f(x, \upsilon)\big)d\nu(x)$. By Parseval in $L_2(\mu \times \mu)$, $L_2(\nu \times \nu)$, $L_2(\mu \times \nu)$,

$$\|R_\chi f\|_{L_2(\mu\times\mu)}^2 + \|R_y f\|_{L_2(\nu\times\nu)}^2$$

$$= 2\sum_{j=1}^{\infty}\left(\left|\langle f, \varphi_j \otimes \psi_0\rangle_{L_2(\mu\times\nu)}\right|^2 + \left|\langle f, \varphi_0 \otimes \psi_j\rangle_{L_2(\mu\times\nu)}\right|^2\right) \stackrel{(10.36)}{\leqslant} 2\|f\|_{L_2(\mu\times\nu)}^2.$$

This is precisely (10.33).

Next, for every $\alpha, \beta \in \mathbb{C}$ define $S_\chi^\alpha f \in L_2(\mu)$ and $S_y^\beta f \in L_2(\nu)$ by

$$S_\chi^\alpha f \stackrel{\text{def}}{=} (1 - \alpha)\langle f, \varphi_0 \otimes \psi_0\rangle_{L_2(\mu\times\nu)}\varphi_0 + \sum_{j=1}^{\infty}\langle f, \varphi_j \otimes \psi_0\rangle_{L_2(\mu\times\nu)}\varphi_j,$$

and

$$S_{y}^{\beta} f \overset{\text{def}}{=} (1 - \beta)\langle f, \varphi_0 \otimes \psi_0 \rangle_{L_2(\mu \times v)} \varphi_0 + \sum_{j=1}^{\infty} \langle f, \varphi_0 \otimes \psi_j \rangle_{L_2(\mu \times v)} \psi_j.$$

In other words, we have the following identities μ-almost surely and v-almost surely, respectively.

$$S_{\chi}^{\alpha} f(x) = \int_{y} f(x, y) dv(y) - \alpha \iint_{\chi \times y} f(\chi, v) d\mu(\chi) dv(v),$$

and

$$S_{y}^{\beta} f(y) = \int_{\chi} f(x, y) d\mu(x) - \beta \iint_{\chi \times y} f(\chi, v) d\mu(\chi) dv(v),$$

By Parseval in $L_2(\mu)$, $L_2(v)$, $L_2(\mu \times v)$,

$$\left\| S_{\chi}^{\alpha} f \right\|_{L_2(\mu)}^2 + \left\| S_{y}^{\beta} f \right\|_{L_2(v)}^2$$

$$= \left(|1 - \alpha|^2 + |1 - \beta|^2 \right) \left| \langle f, \varphi_0 \otimes \psi_0 \rangle_{L_2(\mu \times v)} \right|^2$$

$$+ \sum_{j=1}^{\infty} \left(\left| \langle f, \varphi_j \otimes \psi_0 \rangle_{L_2(\mu \times v)} \right|^2 + \left| \langle f, \varphi_0 \otimes \psi_j \rangle_{L_2(\mu \times v)} \right|^2 \right)$$

$$\overset{(10.36)}{\leqslant} \max \left\{ |1 - \alpha|^2 + |1 - \beta|^2, 1 \right\} \| f \|_{L_2(\mu \times v)}^2.$$

The case $\alpha = \beta = \frac{1}{2}$ of this inequality is precisely (10.34). It is worthwhile to note in passing that this reasoning (substituted into the above interpolation argument) yields the following generalization of (10.18).

$$\max \left\{ \left(|1 - \alpha|^2 + |1 - \beta|^2 \right)^{p\theta}, 1 \right\} \left((1 + |\alpha|)^{\frac{2p(1-\theta)}{2-\theta p}} + (1 + |\beta|)^{\frac{2p(1-\theta)}{2-\theta p}} \right)^{1 - \frac{\theta p}{2}}$$

$$\times \iint_{\chi \times y} \| f(x, y) \|_{[F,H]_\theta}^p d\mu(x) dv(y)$$

$$\geqslant \int_{\chi} \left\| \int_{y} f(x, y) dv(y) - \alpha \iint_{\chi \times y} f(\chi, v) d\mu(\chi) dv(v) \right\|_{[F,H]_\theta}^p d\mu(x)$$

$$+ \int_{y} \left\| \int_{\chi} f(x, y) d\mu(x) - \beta \iint_{\chi \times y} f(\chi, v) d\mu(\chi) dv(v) \right\|_{[F,H]_\theta}^p dv(y).$$

$$(10.37)$$

For the justification of the remaining inequality (10.35), define $Tg \in L_2(\mu)$ by

$$Tg \stackrel{\text{def}}{=} \sum_{j=1}^{\infty} \left(\langle g, \varphi_0 \otimes \varphi_j \rangle_{L_2(\mu \times \mu)} - \langle g, \varphi_j \otimes \varphi_0 \rangle_{L_2(\mu \times \mu)} \right) \varphi_j.$$

In other words, μ-almost surely $Tg(\chi) = \int_{\mathcal{X}} (g(x, \chi) - g(\chi, x)) d\mu(x)$. By Parseval in $L_2(\mu)$, $L_2(\mu \times \nu)$,

$$\|Tg\|_{L_2(\mu)}^2 = \sum_{j=1}^{\infty} \left| \langle g, \varphi_0 \otimes \varphi_j \rangle_{L_2(\mu \times \mu)} - \langle g, \varphi_j \otimes \varphi_0 \rangle_{L_2(\mu \times \mu)} \right|^2$$

$$\leqslant \sum_{j=1}^{\infty} 2 \left(\left| \langle g, \varphi_0 \otimes \varphi_j \rangle_{L_2(\mu \times \mu)} \right|^2 + \left| \langle g, \varphi_j \otimes \varphi_0 \rangle_{L_2(\mu \times \mu)} \right|^2 \right)$$

$$\stackrel{(10.36)}{\leqslant} 2\|g\|_{L_2(\mu \times \mu)}^2,$$

where in the penultimate step we used the convexity of $(\zeta \in \mathbb{C}) \mapsto |\zeta|^2$. This is precisely (10.35). □

We will next deduce Theorem 10.1.2 from the special case of Theorem 10.1.6 that we stated as Theorem 10.1.5.

Proof of Theorem 10.1.2 The largest $\theta \in [0, 1]$ for which $\frac{2}{2-\theta} \leqslant p \leqslant \frac{2}{\theta}$ and also $\frac{1}{q} = \frac{1-\theta}{r} + \frac{\theta}{2}$ for some $r \geqslant 1$ is

$$\theta_{\max} = \theta_{\max}(p, q) \stackrel{\text{def}}{=} 2 \min \left\{ \frac{1}{p}, 1 - \frac{1}{p}, \frac{1}{q}, 1 - \frac{1}{q} \right\}.$$

We then have $L_q = [L_r, L_2]_{\theta_{\max}}$. Note that the quantity $c(p, q)$ that is defined in (10.11) is equal to $\frac{p}{2}\theta_{\max}$.

By (10.15) with $\theta = \theta_{\max}$ and $F = L_r$ we have $j_p(L_q) \geqslant 2^{c(p,q)}$. The matching upper bound $j_p(L_q) \leqslant 2^{c(p,q)}$ holds due to the following quick examples. If X is uniformly distributed on $\{-1, 1\}$, then $\mathbb{E}[|X - \mathbb{E}[X]|^p]$ and $\mathbb{E}|X - X'|^p = 2^{p-1}$. So, $j_p(\mathbb{R}) \leqslant 2^{p-1}$. If $\varepsilon \in (0, 1)$ and $\mathbb{P}[X_\varepsilon = 0] = 1 - \varepsilon$ and $\mathbb{P}[X_\varepsilon = 1] = \varepsilon$, then for $p > 1$,

$$j_p(\mathbb{R}) \leqslant \frac{\mathbb{E}\left[\|X_\varepsilon - X'_\varepsilon\|_q^p\right]}{\mathbb{E}\left[\|X_\varepsilon - \mathbb{E}[X_\varepsilon]\|_q^p\right]} = \frac{2\varepsilon(1 - \varepsilon)}{(1 - \varepsilon)\varepsilon^p + \varepsilon(1 - \varepsilon)^p} \xrightarrow[\varepsilon \to 0^+]{} 2.$$

If $n \in \mathbb{N}$ and X_n is uniformly distributed over $\{\pm e_1, \ldots, \pm e_n\}$, where $\{e_j\}_{j=1}^\infty$ is the standard basis of ℓ_p, then

$$
j_p(L_q) \leqslant j_p(\ell_q^n) \leqslant \frac{\mathbb{E}\left[\|X_n - X_n'\|_q^p\right]}{\mathbb{E}\left[\|X_n - \mathbb{E}[X_n]\|_q^p\right]} = \frac{n-1}{n} 2^{\frac{p}{q}} + \frac{1}{2n} 2^p \xrightarrow[n\to\infty]{} 2^{\frac{p}{q}}.
$$

If $r_1, \ldots, r_n$ are i.i.d. symmetric Bernoulli random variables viewed as elements of L_q, e.g. they can be the coordinate functions in $L_q(\{-1, 1\}^n)$, then let R_n be uniformly distributed over $\{\pm r_1, \ldots, \pm r_n\}$. Then,

$$
j_p(L_q) \leqslant \frac{\mathbb{E}\left[\|R_n - R_n'\|_q^p\right]}{\mathbb{E}\left[\|R_n - \mathbb{E}[R_n]\|_q^p\right]} = \frac{n-1}{n} 2^{\frac{p(q-1)}{q}} + \frac{1}{2n} 2^p \xrightarrow[n\to\infty]{} 2^{\frac{p(q-1)}{q}}.
$$

This completes the proof that $j_p(L_q) = 2^{c(p,q)}$.

Next, an application of (10.15) with $\theta = \theta_{\max}$ and $F = L_r$ gives $\iota_p(L_q) \leqslant 2^{1+(1-\theta_{\max})p}$. In other words,

$$
\mathbb{E}\left[\|X - X'\|_q^p\right] + \mathbb{E}\left[\|Y - Y'\|_q^p\right] \leqslant 2^{\max\left\{p-1, 3-p, 1+\frac{p(q-2)}{q}, 1+\frac{p(2-q)}{q}\right\}} \mathbb{E}\left[\|X - Y\|_q^p\right],
$$
$$
\tag{10.38}
$$

for every p-integrable independent L_q-valued random variables X, X', Y, Y' such that (X, Y) and (X', Y') are identically distributed. The bound (10.38) coincides with (10.15), where $C(p, q)$ is as in (10.12), only in the first two ranges that appear in (10.12), namely when $\frac{p}{p-1} \leqslant q \leqslant p$ or when $\frac{q}{q-1} \leqslant p \leqslant q$. For the remaining ranges that appear in (10.12), the bound (10.38) is inferior to (10.15), so we reason as follows.

For every $q, Q \in [1, \infty]$ satisfying $Q \geqslant q$, by [16, Remark 5.10] (the case $Q \in [1, 2]$ is an older result [6]) there exists an embedding $\mathfrak{s} = \mathfrak{s}_{q,Q} : L_q \to L_Q$ (given by an explicit formula) such that

$$
\forall x, y \in L_q, \qquad \|\mathfrak{s}(x) - \mathfrak{s}(y)\|_Q = \|x - y\|_q^{\frac{q}{Q}}. \tag{10.39}
$$

Apply (10.38) to the L_Q-valued random vectors $\mathfrak{s}(X), \mathfrak{s}(X'), \mathfrak{s}(Y), \mathfrak{s}(Y')$ with q replaced by Q and p replaced with $\frac{pQ}{q}$. The resulting estimate is

$$
\mathbb{E}\left[\|X - X'\|_q^p\right] + \mathbb{E}\left[\|Y - Y'\|_q^p\right]
$$
$$
\stackrel{(10.39)}{=} \mathbb{E}\left[\|\mathfrak{s}(X) - \mathfrak{s}(X')\|_Q^{\frac{pQ}{q}}\right] + \mathbb{E}\left[\|\mathfrak{s}(Y) - \mathfrak{s}(Y')\|_Q^{\frac{pQ}{q}}\right]
$$
$$
\stackrel{(10.38)}{\leqslant} 2^{\max\left\{\frac{pQ}{q}-1, 3-\frac{pQ}{q}, 1+\frac{p(Q-2)}{q}, 1+\frac{p(2-Q)}{q}\right\}} \mathbb{E}\left[\|\mathfrak{s}(X) - \mathfrak{s}(Y)\|_Q^{\frac{pQ}{q}}\right]
$$
$$
\stackrel{(10.39)}{=} 2^{\max\left\{\frac{pQ}{q}-1, 3-\frac{pQ}{q}, 1+\frac{p(Q-2)}{q}, 1+\frac{p(2-Q)}{q}\right\}} \mathbb{E}\left[\|X - Y\|_q^p\right]. \tag{10.40}
$$

It is in our interest to choose $Q \geqslant q$ so as to minimize the right hand side of (10.40). If $\frac{1}{p} + \frac{1}{q} \leqslant 1$, then $Q = q$ is the optimal choice in (10.40), and therefore we return to (10.38). But, if $\frac{1}{p} + \frac{1}{q} \geqslant 1$, then $Q = 1 + \frac{q}{p} \geqslant q$ is the optimal choice in (10.40) and we arrive at the following estimate which is better than (10.38) in the stated range

$$\frac{1}{p} + \frac{1}{q} \geqslant 1 \implies \mathbb{E}\big[\|X - X'\|_q^p\big] + \mathbb{E}\big[\|Y - Y'\|_q^p\big] \leqslant 2^{\max\left\{\frac{p}{q}, 2 - \frac{p}{q}\right\}} \mathbb{E}\big[\|X - Y\|_q^p\big].$$

(10.41)

The bound (10.41) covers the third and fourth ranges that appear in (10.12), as well as the case $p = q \in [1, 2]$ of the fifth range that appears in (10.12). However, (10.41) is inferior to (10.12) when $1 \leqslant p < q \leqslant 2$. When this occurs, use the fact [12] that L_q is isometric to a subspace of L_p and apply the already established case $p = q$ to the L_p-valued random variables $i(X), i(X'), i(Y), i(Y')$, where $i : L_q \to L_p$ is any isometric embedding.

We will next prove that $\imath_p(L_q) \geqslant 2^{C_{\mathrm{opt}}(p,q)}$, where $C_{\mathrm{opt}}(p, q)$ is given in (10.14). In particular, this will justify the second sharpness assertion of Theorem 10.1.2, namely that (10.38) is sharp when p, q belong to the first, second or fifth ranges that appear in (10.12). Firstly, by considering the special case of (10.7) in which X, Y are i.i.d., we see that $\imath_p(F) \geqslant 1$ for any Banach space F. Next, fix $n \in \mathbb{N}$ and let $r_1, \ldots, r_n, \rho_1, \ldots, \rho_n \in L_q$ be such that $r_1, \ldots, r_n$ and $\rho_1, \ldots, \rho_n$ each form a sequence of i.i.d. symmetric Bernoulli random variables, and the supports of $r_1, \ldots, r_n$ are disjoint from the supports of $\rho_1, \ldots, \rho_n$. For example, one could consider them as the elements of $L_q(\{-1, 1\}^n) \oplus_q L_q(\{-1, 1\}^n)$ that are given by $r_i = (\omega \mapsto \omega_i, 0)$ and $\rho_i = (0, \omega \mapsto \omega_i)$ for each $i \in \{1, \ldots, n\}$. Let X be uniformly distributed over $\{r_1, \ldots, r_n\}$ and Y be uniformly distributed over $\{\rho_1, \ldots, \rho_n\}$. Due to the disjointness of the supports, we have $\|X - Y\|_q^p = (\|X\|_q^q + \|Y\|_q^q)^{p/q} = 2^{p/q}$ point-wise. At the same time, $\mathbb{E}[\|X - X'\|_q^p] + \mathbb{E}[\|Y - Y'\|_q^p] = 2(1 - 1/n)(2^q/2)^{p/q} = (1 - 1/n)2^{1 + p(q-1)/q}$. By letting $n \to \infty$, this shows that necessarily $\imath_p(L_q) \leqslant 2^{1 + p(q-2)/q}$. Finally, if (10.7) holds, then in particular it holds for scalar-valued random variables. By integrating, we see that $\imath_p(F) \geqslant \imath_p(L_p)$ for any Banach space F. But, the case $p = q$ of the above discussion gives $\imath_p(L_p) \geqslant 2^{1 + p(p-2)/p} = 2^{p-1}$, as required.

The bound (10.13) of Theorem 10.1.2 coincides with (10.16). When $p \leqslant q \leqslant 2$, we have $C(p, q) = 1$, $c(p, q) = p - 1$ and thus $m_p(L_q) \leqslant 2^{2-p}$. It therefore remains to check that $\mathfrak{b}_p(L_q) \geqslant 2^{2-p}$ when $p \leqslant q \leqslant 2$. In fact, $\mathfrak{b}_p(F) \geqslant 2^{2-p}$ for every $p \geqslant 1$ and every Banach space $(F, \|\cdot\|_F)$. Indeed, fix distinct $a, b \in F$. Let

X, Y be independent and uniformly distributed over $\{a, b\}$. Then

$$\mathfrak{b}_p(F) \geq \frac{\mathbb{E}\left[\|X - z\|_F^p + \|Y - z\|_F^p\right]}{\mathbb{E}\left[\|X - Y\|_F^p\right]} = \frac{\|a - z\|_F^p + \|b - z\|_F^p}{\frac{1}{2}\|a - b\|_F^p}$$

$$\geq \frac{2\left\|\frac{1}{2}(a - z) - \frac{1}{2}(b - z)\right\|_F^p}{\frac{1}{2}\|a - b\|_F^p} = 2^{2-p},$$

where the penultimate step is an application of the convexity of $\|\cdot\|_F^p$. $\qquad\square$

Remark 10.3.1 Fix $n \in \mathbb{N}$. Following [7], for $a = (a_1, \ldots, a_{2n}) \in \mathbb{C}^{2n}$ denote by $\mathfrak{R}(a) = (\mathfrak{R}(a_1), \ldots, \mathfrak{R}(a_{2n})) \in \mathbb{R}^{2n}$ and $\mathfrak{I}(a) = (\mathfrak{I}(a_1), \ldots, \mathfrak{I}(a_{2n})) \in \mathbb{R}^{2n}$ the vectors of real parts and imaginary parts of the entries of a, respectively. Let $\Lambda(a) \in [0, \infty)$ be the area of the parallelogram that is generated by $\mathfrak{R}(a)$ and $\mathfrak{I}(a)$, i.e.,

$$\Lambda(a) \overset{\text{def}}{=} \sqrt{\|\mathfrak{R}(a)\|_2^2\|\mathfrak{I}(a)\|_2^2 - \langle\mathfrak{R}(a), \mathfrak{I}(a)\rangle}.$$

By [7, Lemma 5.2] there is a linear operator $\mathcal{C} : \mathbb{C}^{2n} \to M_{2^n}(\mathbb{C})$ from $\mathbb{C}^{2n}$ to the space of 2^n by 2^n complex matrices, such that for any $a \in \mathbb{C}^{2n}$ the Schatten-1 norm of the matrix $\mathcal{C}(a)$ satisfies

$$\|\mathcal{C}(a)\|_{\mathsf{S}_1} = \frac{1}{2}\sqrt{\|a\|_2^2 + 2\Lambda(a)} + \frac{1}{2}\sqrt{\|a\|_2^2 - 2\Lambda(a)}. \tag{10.42}$$

Let $e_1, \ldots, e_{2n} \in \mathbb{C}^{2n}$ be the standard basis of $\mathbb{C}^{2n}$ and define $2n$ matrices $x_1, \ldots, x_n, y_1, \ldots, y_n \in M_{2^n}(\mathbb{C})$ by $x_k = \mathcal{C}(e_k)$ and $y_k = \mathcal{C}(ie_{n+k})$ for $k \in \{1, \ldots, n\}$. By (10.42) we have $\|x_j - x_k\|_{\mathsf{S}_1} = \|y_j - y_k\|_{\mathsf{S}_1} = \sqrt{2}$ for distinct $j, k \in \{1, \ldots, n\}$, while $\|x_j - y_k\|_{\mathsf{S}_1} = 1$ for all $j, k \in \{1, \ldots, n\}$. Hence, if we let X and Y be independent and distributed uniformly over $\{x_1, \ldots, x_n\}$ and $\{y_1, \ldots, y_n\}$, respectively, and X', Y' are independent copies of X, Y, respectively, then for every $p \geq 1$ we have

$$\mathbb{E}\left[\|X - Y\|_{\mathsf{S}_1}^p\right] = 1 \quad \text{and} \quad \mathbb{E}\left[\|X - X'\|_{\mathsf{S}_1}^p\right] = \mathbb{E}\left[\|Y - Y'\|_{\mathsf{S}_1}^p\right] = \frac{n-1}{n}2^{\frac{p}{2}}.$$

By letting $n \to \infty$, this implies that $\mathfrak{z}_p(\mathsf{S}_1) \geq 2^{\frac{p}{2}+1}$. In particular, $\mathfrak{z}_1(\mathsf{S}_1) \geq 2\sqrt{2}$.

Remark 10.3.2 Fix $q \geq 1$. Let $(F, \|\cdot\|_F)$ be a Banach space. Assume that F has a linear subspace $G \subseteq F$ that is isometric to L_q (or the Schatten–von Neumann trace class S_q). If $X, Y \in L_p(G)$ are i.i.d. random variables taking values in G, then for $c(p, q)$ as in (10.11), by Theorem 10.1.2 we have

$$\mathbb{E}\left[\|X - Y\|_F^p\right] \geq 2^{c(p,q)} \cdot \inf_{z \in F} \mathbb{E}\left[\|X - z\|_F^p\right]. \tag{10.43}$$

We note that this inequality is optimal despite the fact that the infimum is now taken over z in the larger super-space F. Indeed, in the proof of Theorem 10.1.2 the random variables that established optimality of $c(p, q)$ were symmetric when p, q belong to the first three ranges that appear in (10.11). In these cases, by the convexity of $\| \cdot \|_F^p$, the infimum in the right had side of (10.43) is attained at $z = 0 \in G$. The fact that the term $2^{c(p,q)}$ in the right hand side of (10.43) cannot be replaced by any value greater than 2 needs the following separate treatment. If $\varepsilon \in (0, 1)$ and $\mathbb{P}[X = v] = \varepsilon = 1 - \mathbb{P}[X = 0]$ for some $v \in G$ with $\|v\|_F = 1$, then $\mathbb{E}[\|X - Y\|_F^p] = 2\varepsilon(1 - \varepsilon)$. Next, for any $z \in F$ we have

$$\mathbb{E}[\|X - z\|_F^p] = (1-\varepsilon)\|z\|_F^p + \varepsilon\|v - z\|_F^p \geq (1-\varepsilon)\|z\|_F^p + \varepsilon \left(\max\left\{0, 1 - \|z\|_F\right\}\right)^p$$

$$\geq \min_{r \geq 0}\left((1 - \varepsilon)r^p + \varepsilon(\max\{0, 1 - r\})^p\right) = \frac{\varepsilon(1 - \varepsilon)}{\left(\varepsilon^{\frac{1}{p-1}} + (1 - \varepsilon)^{\frac{1}{p-1}}\right)^{p-1}},$$

where the final step follows by elementary calculus. Therefore,

$$\frac{\mathbb{E}[\|X - Y\|_F^p]}{\inf_{z \in F} \mathbb{E}[\|X - z\|_F^p]} \leq 2\left(\varepsilon^{\frac{1}{p-1}} + (1 - \varepsilon)^{\frac{1}{p-1}}\right)^{p-1} \xrightarrow[\varepsilon \to 0^+]{} 2.$$

Remark 10.3.3 An extrapolation theorem of Pisier [18] asserts that if $(F, \| \cdot \|_F)$ is a Banach lattice that is both p-convex with constant 1 and q-concave with constant 1, where $\frac{1}{p} + \frac{1}{q} = 1$, then there exists a Banach lattice W, a Hilbert space H, and $\theta \in (0, 1]$ such that F is isometric to the complex interpolation space $[W, H]_\theta$. Hence, Theorem 10.1.5 applies in this setting, implying in particular that there is $r \in [1, \infty)$, namely $r = \frac{2}{\theta}$, such that every i.i.d. F-valued random variables $X, Y \in L_r(F)$ satisfy

$$\mathbb{E}[\|X - Y\|_F^r] \geq 2\mathbb{E}[\|X - \mathbb{E}[X]\|_F^r].$$

We will conclude by discussing further bounds in the non-convex range $p < 1$, as well as their limit when $p \to 0^+$. When $p \in (0, 1)$, the topological vector space L_p is not a normed space. Despite this, when we say that a normed space $(F, \| \cdot \|_F)$ admits a linear isometric embedding into L_p we mean (as usual) that there exists a linear mapping $T : F \to L_p$ such that $\|Tx\|_p = \|x\|_F$ for all $x \in F$. This of course forces the L_p quasi-norm to induce a metric on the image of T, so the use of the term "iso**metric**" is not out of place here, though note that it is inconsistent with the standard metric on L_p, which is given by $\|f - g\|_p^p$ for all $f, g \in L_p$. The following proposition treats the case $p \in (0, 2]$, though later we will mainly be interested in the non-convex range $p \in (0, 1)$. Note that the case $p = 1$ implies the stated inequalities for, say, any two-dimensional normed space, since any such space admits [5] an isometric embedding into L_1.

Proposition 10.3.4 *Let $(F, \| \cdot \|_F)$ be a Banach space that admits an isometric linear embedding into L_p for some $p \in (0, 2]$. Let $X, X', Y, Y' \in L_p(F)$ be independent F-valued random vectors such that X' has the same distribution as X and Y' has the same distribution as Y. Then,*

$$\mathbb{E}\left[\|X - X'\|_F^p\right] + \mathbb{E}\left[\|Y - Y'\|_F^p\right] \leqslant 2\mathbb{E}\left[\|X - Y\|_F^p\right], \tag{10.44}$$

and

$$\inf_{z \in F} \mathbb{E}\left[\|X - z\|_F^p + \mathbb{E}\|Y - z\|_F^p\right] \leqslant \min\left\{2, 2^{2-p}\right\} \mathbb{E}\left[\|X - Y\|_F^p\right]. \tag{10.45}$$

The constants 2 and $\min\left\{2, 2^{2-p}\right\}$ in (10.44) and (10.45), respectively, cannot be improved.

Proof By [6, 19] there is a mapping $\mathfrak{s} : F \to L_2$ such that $\|\mathfrak{s}(x) - \mathfrak{s}(y)\|_2 = \|x - y\|_F^{\frac{p}{2}}$ for all $x, y \in F$. By the (trivial) Hilbertian case $p = q = 2$ of Theorem 10.1.2 applied to the L_2-valued random vectors $\mathfrak{s}(X), \mathfrak{s}(Y)$,

$$\mathbb{E}\left[\|X - X'\|_F^p\right] + \mathbb{E}\left[\|Y - Y'\|_F^p\right] = \mathbb{E}\left[\|\mathfrak{s}(X) - \mathfrak{s}(X')\|_2^2\right] + \mathbb{E}\left[\|\mathfrak{s}(Y) - \mathfrak{s}(Y')\|_2^2\right]$$

$$\leqslant 2\mathbb{E}\left[\|\mathfrak{s}(X) - \mathfrak{s}(Y)\|_2^2\right] = 2\mathbb{E}\left[\|X - Y\|_F^p\right].$$

This substantiates (10.44). When $p < 1$ we cannot proceed from here to prove (10.45) by considering the analogue of the mixture constant $m(\cdot)$, namely by bounding the left hand side of (10.5) as we did in the Introduction, since the present L_p integrability assumption on X, Y does not imply that $\mathbb{E}[X]$ and $\mathbb{E}[Y]$ are well-defined elements of F. Instead, let Z' be independent of X, Y and distributed according to the mixture of the laws of X and Y, as in (10.6). The point $z \in F$ will be chosen randomly according to Z', i.e.,

$$\inf_{z \in F} \mathbb{E}\left[\|X - z\|_F^p + \|Y - z\|_F^p\right] \leqslant \mathbb{E}\left[\|X - Z'\|_F^p + \|Y - Z'\|_F^p\right]$$

$$= \frac{1}{2}\mathbb{E}\left[\|X - X'\|_F^p + \|Y - Y'\|_F^p\right] + \mathbb{E}\left[\|X - Y\|_F^p\right]$$

$$\overset{(10.44)}{\leqslant} 2\mathbb{E}\left[\|X - Y\|_F^p\right]. \tag{10.46}$$

For $p \geqslant 1$ we have $\tau_p(F) \leqslant 2$ by (10.44), and $j_p(F) \geqslant j_p(L_p) = p - 1$ by Theorem 10.1.2, so $\mathfrak{b}_p(F) \leqslant 2^{2-p}$, by (10.10).

The sharpness of (10.44) is seen by taking X and Y to be identically distributed. When $p \geqslant 1$, we already saw in the proof of Theorem 10.1.2 that $\mathfrak{b}_p(F) \geqslant 2^{2-p}$ for any Banach space F; thus (10.45) is sharp in this range. The same reasoning as in the proof of Theorem 10.1.2 shows that the factor 2 in (10.45) cannot be improved in the non-convex range $p \in (0, 1)$ as well. Indeed, fix v with $\|v\|_F = 1$ and let

X and Y be uniformly distributed over $\{0, v\}$. Then, $\mathbb{E}\left[\|X - z\|_F^p + \mathbb{E}\|Y - z\|_F^p\right] = \|z\|_F^p + \|v - z\|_F^p \geq (\|z\|_F + \|v - z\|_F)^p \geq \|v\|_F^p = 1$ for every $z \in F$, while $\mathbb{E}\left[\|X - Y\|_F^p\right] = \frac{1}{2}\|v\|_F^p = \frac{1}{2}$.   □

Proposition 10.3.5 below is the limit of Proposition 10.3.4 as $p \to 0^+$. While it is possible to deduce it formally from Proposition 10.3.4 by passing to the limit, a justification of this fact is quite complicated due to the singularity of the logarithm at zero. We will instead proceed via a shorter alternative approach.

Following [13], a real Banach space $(F, \|\cdot\|_F)$ is said to admit a linear isometric embedding into L_0 if there exists a probability space (Ω, μ) and a linear operator $T : F \to \text{Meas}(\Omega, \mu)$, where $\text{Meas}(\Omega, \mu)$ denotes the space of (equivalence classes of) real-valued μ-measurable functions on Ω, such that

$$\forall x \in F, \qquad \|x\|_F = e^{\int_\Omega \log |Tx| d\mu}. \tag{10.47}$$

As shown in [13], every three-dimensional real normed space admits a linear isometric embedding into L_0, so in particular the following proposition applies to any such space.

Proposition 10.3.5 *Let $(F, \|\cdot\|_F)$ be a real Banach space that admits a linear isometric embedding into L_0. Let X, X', Y, Y' be independent F-valued random vectors such that X' has the same distribution as X and Y' has the same distribution as Y. Assume that $\mathbb{E}\left[\log(1 + \|X\|_F)\right] < \infty$ and $\mathbb{E}\left[\log(1 + \|Y\|_F)\right] < \infty$. Then,*

$$e^{\mathbb{E}\left[\log(\|X - X'\|_F \cdot \|Y - Y'\|_F)\right]} \leq e^{2\mathbb{E}\left[\log(\|X - Y\|_F)\right]}, \tag{10.48}$$

and

$$\inf_{z \in F} e^{\mathbb{E}\left[\log(\|X - z\|_F \cdot \|Y - z\|_F)\right]} \leq e^{2\mathbb{E}\left[\log(\|X - Y\|_F)\right]}. \tag{10.49}$$

The multiplicative constant 1 in both of these inequalities is optimal.

Proof (10.49) is a consequence of (10.48) by reasoning analogously to (10.46). Due to the assumed representation (10.47), by Fubini's theorem it suffices to prove (10.48) for real-valued random variables.

So, suppose that X, Y are independent real-valued random variables such that $\mathbb{E}\left[\log(1 + |X|)\right] < \infty$ and $\mathbb{E}\left[\log(1 + |Y|)\right] < \infty$. Note that every nonnegative random variable W with $\mathbb{E}\left[\log(1 + W)\right] < \infty$ satisfies

$$\mathbb{E}\left[\log W\right] = \int_0^\infty \frac{e^{-s} - \mathbb{E}\left[e^{-sW}\right]}{s} ds. \tag{10.50}$$

Indeed, for every $a, b \in [0, \infty)$ with $a \leqslant b$ we have

$$\int_0^\infty \frac{e^{-as} - e^{-bs}}{s} \, ds = \int_0^\infty \left(\int_a^b s e^{-ts} \, dt \right) \frac{ds}{s}$$

$$= \int_a^b \left(\int_0^\infty e^{-ts} \, ds \right) dt = \int_a^b \frac{dt}{t} = \log b - \log a,$$

so that (10.50) follows by applying this identity and the Fubini theorem separately on each of the events $\{W \geqslant 1\}$ and $\{W < 1\}$, taking advantage of the fact that $e^{-s} - e^{-sW}$ is of constant sign on both events.

Let Z, Z' be independent random variables whose law is the mixture of the laws of X, Y as in (10.6). the desired inequality (10.48) is equivalent to the assertion that $\mathbb{E}\left[\log(Z - Z')^2\right] \leqslant \mathbb{E}\left[\log(X - Y)^2\right]$. By two applications of (10.50), once with $W = (X - Y)^2$ and once with $W = (Z - Z')^2$, it suffices to prove that

$$\forall s \geqslant 0, \qquad \mathbb{E}\left[e^{-s(Z-Z')^2}\right] \geqslant \mathbb{E}\left[e^{-s(X-Y)^2}\right].$$

This is so because, using the formula for the Fourier transform of the Gaussian density, we have

$$\mathbb{E}\left[e^{-s(Z-Z')^2}\right] = \mathbb{E}\left[\frac{1}{\sqrt{2\pi}} \int_{-\infty}^\infty e^{it(Z-Z')\sqrt{2s} - \frac{t^2}{2}} \, dt\right]$$

$$= \frac{1}{\sqrt{2\pi}} \int_{-\infty}^\infty \mathbb{E}\left[e^{it\sqrt{2s}Z}\right] \cdot \mathbb{E}\left[e^{-it\sqrt{2s}Z'}\right] e^{-\frac{t^2}{2}} \, dt$$

$$= \frac{1}{\sqrt{2\pi}} \int_{-\infty}^\infty \left|\mathbb{E}\left[e^{it\sqrt{2s}Z}\right]\right|^2 e^{-\frac{t^2}{2}} \, dt \tag{10.51}$$

$$= \frac{1}{\sqrt{2\pi}} \int_{-\infty}^\infty \left|\frac{1}{2}\mathbb{E}\left[e^{it\sqrt{2s}X}\right] + \frac{1}{2}\mathbb{E}\left[e^{it\sqrt{2s}Y}\right]\right|^2 e^{-\frac{t^2}{2}} \, dt$$

$$\geqslant \frac{1}{\sqrt{2\pi}} \int_{-\infty}^\infty \Re\left(\mathbb{E}\left[e^{it\sqrt{2s}X}\right] \cdot \overline{\mathbb{E}\left[e^{it\sqrt{2s}Y}\right]}\right) e^{-\frac{t^2}{2}} \, dt \tag{10.52}$$

$$= \frac{1}{\sqrt{2\pi}} \Re\left(\int_{-\infty}^\infty \mathbb{E}\left[e^{it\sqrt{2s}(X-Y)}\right] e^{-\frac{t^2}{2}} \, dt\right) = \mathbb{E}\left[e^{-s(X-Y)^2}\right], \tag{10.53}$$

where (10.51) uses Fubini and the independence of Z and Z', (10.52) uses the fact that for all $a, b \in \mathbb{C}$ we have $|(a + b)/2|^2 = |(a - b)/2|^2 + \Re(a\bar{b}) \geqslant \Re(a\bar{b})$, the first step of (10.53) uses the independence of X and Y, and the last step of (10.53) uses once more the formula for the Fourier transform of the Gaussian density.

The fact that (10.48) is sharp follows by considering the case when X, Y are i.i.d. and non-atomic. Note that when both X and Y have an atom at the same point, both sides of (10.49) equal 0. The example considered in the proof of Proposition 10.3.4 when $p > 0$ is therefore of no use for establishing the optimality of (10.49), due to the atomic nature of the distributions under consideration. Instead, for an arbitrary $v \in F$ such that $\|v\|_F = 1$, let us consider random vectors $X = (\cos \Theta)v$ and $Y = (\cos \Theta')v$, where Θ and Θ' are independent random variables uniformly distributed on $[0, 2\pi]$.

Observe that for every $\alpha \in \mathbb{R}$ we have

$$
\mathbb{E}\left[\log |\cos \Theta - \cos \alpha|\right] = \mathbb{E}\left[\log \left|2 \sin\left(\frac{\Theta + \alpha}{2}\right) \sin\left(\frac{\Theta - \alpha}{2}\right)\right|\right]
$$
$$
= \log 2 + \mathbb{E}\left[\log \left|\sin\left(\frac{\Theta + \alpha}{2}\right)\right|\right] + \mathbb{E}\left[\log \left|\sin\left(\frac{\Theta - \alpha}{2}\right)\right|\right]
$$
$$
= \log 2 + 2\mathbb{E}\left[\log |\cos \Theta|\right], \tag{10.54}
$$

where the last step of (10.54) holds because, by periodicity, $\left|\sin\left(\frac{\Theta \pm \alpha}{2}\right)\right|$ has the same distribution as $|\cos \Theta|$.

The case $\alpha = \frac{\pi}{2}$ of (10.54) simplifies to give $\mathbb{E}\left[\log |\cos \Theta|\right] = -\log 2$. Hence, (10.54) becomes

$$
\forall \alpha \in \mathbb{R}, \qquad \mathbb{E}\left[\log |\cos \Theta - \cos \alpha|\right] = -\log 2. \tag{10.55}
$$

Consequently,

$$
\forall t \in \mathbb{R}, \qquad \mathbb{E}\left[\log |\cos \Theta - t|\right] \geqslant -\log 2. \tag{10.56}
$$

Indeed, if $t \in [-1, 1]$, then one can write $t = \cos \alpha$ for some $\alpha \in \mathbb{R}$, so that by (10.55) the inequality in (10.56) holds as equality. If $|t| > 1$, then $|\cos \theta - t| \geqslant |\cos \theta - \mathrm{sign}(t)|$ for all $\theta \in [0, 2\pi]$, thus implying (10.56). It also follows from (10.55) that

$$
\mathbb{E}\left[\log \left(\|X - Y\|_F\right)\right] = \mathbb{E}\left[\log |\cos \Theta - \cos \Theta'|\right] \overset{(10.55)}{=} -\log 2.
$$

Next, by the Hahn–Banach theorem, take $\varphi \in F^*$ such that $\|\varphi\|_{F^*} = 1$ and $\varphi(v) = \|v\|_F = 1$. For any $z \in F$,

$$
\mathbb{E}\left[\log \left(\|X - z\|_F\right)\right] = \mathbb{E}\left[\log \left(\|Y - z\|_F\right)\right] \geqslant \mathbb{E}\left[\log |\varphi((\cos \Theta)v - z)|\right]
$$
$$
= \mathbb{E}\left[\log |\cos \Theta - \varphi(z)|\right] \overset{(10.56)}{\geqslant} -\log 2.
$$

This implies the asserted sharpness of (10.49). Note that the above argument that (10.49) cannot hold with a multiplicative constant less than 1 in the right hand side worked for any Banach space F whatsoever.   □

Acknowledgements We are grateful to Oded Regev for pointing us to [7, Lemma 5.2] and for significantly simplifying our initial reasoning for the statement that is proved in Remark 10.3.1.

References

1. A. Andoni, A. Naor, O. Neiman, Snowflake universality of Wasserstein spaces. Ann. Sci. Éc. Norm. Supér. (4) **51**(3), 657–700 (2018)
2. Y. Bartal, N. Linial, M. Mendel, A. Naor, Some low distortion metric Ramsey problems. Discret. Comput. Geom. **33**(1), 27–41 (2005)
3. Y. Benyamini, J. Lindenstrauss, *Geometric Nonlinear Functional Analysis. Vol. 1*. American Mathematical Society Colloquium Publications, vol. 48 (American Mathematical Society, Providence, 2000)
4. J. Bergh, J. Löfström, *Interpolation Spaces. An Introduction* (Springer, Berlin, 1976). Grundlehren der Mathematischen Wissenschaften, No. 223
5. E.D. Bolker, A class of convex bodies. Trans. Amer. Math. Soc. **145**, 323–345 (1969)
6. J. Bretagnolle, D. Dacunha-Castelle, J.-L. Krivine, Lois stables et espaces L^p. Ann. Inst. H. Poincaré Sect. B (N.S.) **2**, 231–259 (1965/1966).
7. J. Briët, O. Regev, R. Saket. Tight hardness of the non-commutative Grothendieck problem. Theory Comput. **13**, Paper No. 15, 24 (2017)
8. A.-P. Calderón, Intermediate spaces and interpolation, the complex method. Studia Math. **24**, 113–190 (1964)
9. A.N. Dranishnikov, G. Gong, V. Lafforgue, G. Yu, Uniform embeddings into Hilbert space and a question of Gromov. Canad. Math. Bull. **45**(1), 60–70 (2002)
10. P. Enflo, On the nonexistence of uniform homeomorphisms between L_p-spaces. Ark. Mat. **8**, 103–105 (1969)
11. A. Eskenazis, M. Mendel, A. Naor, Nonpositive curvature is not coarsely universal. To appear in *Invent. Math.* **217**, 833–886 (2018). Available at https://arxiv.org/abs/1808.02179
12. M.Ĭ. Kadec', Linear dimension of the spaces L_p and l_q. Uspehi Mat. Nauk **13**(6 (84)), 95–98 (1958)
13. N. Kalton, A. Koldobsky, V. Yaskin, M. Yaskina, The geometry of L_0. Can. J. Math. **59**, 1029–1049 (2007)
14. C.J. Lennard, A.M. Tonge, A. Weston, Generalized roundness and negative type. Mich. Math. J. **44**(1), 37–45 (1997)
15. J. Lindenstrauss, L. Tzafriri, *Classical Banach Spaces. I*. Sequence spaces, Ergebnisse der Mathematik und ihrer Grenzgebiete, vol. 92 (Springer, Berlin, 1977)
16. M. Mendel, A. Naor, Euclidean quotients of finite metric spaces. Adv. Math. **189**(2), 451–494 (2004)
17. A. Naor, A phase transition phenomenon between the isometric and isomorphic extension problems for Hölder functions between L_p spaces. Mathematika **48**(1–2), 253–271 (2003), (2001)
18. G. Pisier, La méthode d'interpolation complexe: Applications aux treillis de Banach, in *Séminaire d'Analyse Fonctionnelle (1978–1979)* (École Polytechnique, Palaiseau, 1979), Exp. No. 17, p. 18
19. I.J. Schoenberg, Metric spaces and positive definite functions. Trans. Amer. Math. Soc. **44**(3), 522–536 (1938)
20. B. Simon, *Trace Ideals and Their Applications*. London Mathematical Society Lecture Note Series, vol. 35 (Cambridge University Press, Cambridge, 1979)
21. L. Williams, J.H. Wells, T.L. Hayden, On the extension of Lipschitz-Hölder maps on L^p spaces. Studia Math. **39**, 29–38 (1971)

Chapter 11
The Alon–Milman Theorem for Non-symmetric Bodies

Márton Naszódi

Abstract A classical theorem of Alon and Milman states that any d dimensional centrally symmetric convex body has a projection of dimension $m \geq e^{c\sqrt{\ln d}}$ which is either close to the m-dimensional Euclidean ball or to the m-dimensional cross-polytope. We extended this result to non-symmetric convex bodies.

11.1 Introduction

Some fundamental results from the theory of normed spaces have been shown to hold in the more general setting of non-symmetric convex bodies. Dvoretzky's theorem [3, 7] was extended in [6] and [5]; Milman's Quotient of Subspace theorem [8] and duality of entropy results were extended in [9]. In this note, we extend the Alon–Milman Theorem.

A *convex body* is a compact convex set in $\mathbb{R}^d$ with non-empty interior. We denote the orthogonal projection onto a linear subspace H or $\mathbb{R}^d$ by P_H. For $p = 1, 2, \infty$, the closed unit ball of ℓ_p^d centered at the origin is denoted by $\mathbf{B}_p^d$. Let K and L be convex bodies in $\mathbb{R}^d$ with $L = -L$. We define their *distance* as

$$d(K, L) = \inf\{\lambda > 0 : \ L \subset T(K - a) \subset \lambda L \text{ for some } a \in \mathbb{R}^d \text{ and } T \in GL(\mathbb{R}^d)\}.$$

The author thanks Alexander Litvak and Nicole Tomczak-Jaegermann for the discussions on the topic we had some time ago. The research was partially supported by the National Research, Development and Innovation Office (NKFIH) grant NKFI-K119670 and by the János Bolyai Scholarship of the Hungarian Academy of Sciences.

M. Naszódi (✉)
Alfréd Rényi Institute of Mathematics, Budapest, Hungary

Department of Geometry, Loránd Eötvös University, Budapest, Hungary
e-mail: marton.naszodi@math.elte.hu

© Springer Nature Switzerland AG 2020
B. Klartag, E. Milman (eds.), *Geometric Aspects of Functional Analysis*,
Lecture Notes in Mathematics 2266,
https://doi.org/10.1007/978-3-030-46762-3_11

By compactness, this infimum is attained, and when $K = -K$, it is attained with $a = 0$.

Alon and Milman [1] proved the following theorem in the case when K is centrally symmetric.

Theorem 11.1 *For every $\varepsilon > 0$ there is a constant $C(\varepsilon) > 0$ with the property that in any dimension $d \in \mathbb{Z}^+$, and for any convex body K in $\mathbb{R}^d$, at least one of the following two statements hold:*

(i) *there is an m-dimensional linear subspace H of $\mathbb{R}^d$ such that $d(P_H(K), \mathbf{B}_2^m) <$ $1 + \varepsilon$, for some m satisfying $\ln \ln m \geq \frac{1}{2} \ln \ln d$, or*

(ii) *there is an m-dimensional linear subspace H such that $d(P_H(K), \mathbf{B}_1^m) < 1 + \varepsilon$, for some m satisfying $\ln \ln m \geq \frac{1}{2} \ln \ln d - C(\varepsilon)$.*

The main contribution of the present note is a way to deduce Theorem 11.1 from the original result of Alon and Milman, that is, the centrally symmetric case. By polarity, one immediately obtains

Corollary 11.1 *For every $\varepsilon > 0$ there is a constant $C(\varepsilon) > 0$ with the property that in any dimension $d \in \mathbb{Z}^+$, and for any convex body K in $\mathbb{R}^d$ containing the origin in its interior, at least one of the following two statements hold:*

(i) *there is an m-dimensional linear subspace H of $\mathbb{R}^d$ such that $d(H \cap K, \mathbf{B}_2^m) <$ $1 + \varepsilon$, for some m satisfying $\ln \ln m \geq \frac{1}{2} \ln \ln d$, or*

(ii) *there is an m-dimensional linear subspace H such that $d(H \cap K, \mathbf{B}_\infty^m) < 1 + \varepsilon$, for some m satisfying $\ln \ln m \geq \frac{1}{2} \ln \ln d - C(\varepsilon)$.*

11.2   Proof of Theorem 11.1

For a convex body K in $\mathbb{R}^d$, we denote its polar by $K^* = \{x \in \mathbb{R}^d : \langle x, y \rangle \leq 1 \text{ for all } y \in K\}$. The *support function* of K is $h_K(x) = \sup\{\langle x, y \rangle : y \in K\}$. For basic properties, see [2, 12].

First in Lemma 11.2, by a standard argument, we show that if the *difference body* $L - L$ of a convex body L is close to the Euclidean ball, then so is some linear dimensional section of L. For this, we need Milman's theorem whose proof (cf. [4, 7, 10]) does not use the symmetry of K even if it is stated with that assumption. We use $\mathbb{S}^{d-1}$ to denote the boundary of $\mathbf{B}_2^d$.

Lemma 11.1 (Milman's Theorem) *For every $\varepsilon > 0$ there is a constant $C(\varepsilon) > 0$ with the property that in any dimension $d \in \mathbb{Z}^+$, and for any convex body K in $\mathbb{R}^d$ with $\mathbf{B}_2^d \subseteq K$, there is an m-dimensional linear subspace H of $\mathbb{R}^d$ such that*

$(1 - \varepsilon)r(\mathbf{B}_2^d \cap H) \subseteq K \subseteq (1 + \varepsilon)r(\mathbf{B}_2^d \cap H)$, *for some m satisfying* $m \geq C(\varepsilon)M^2 d$, *where*

$$M = M(K) = \int\limits_{\mathbb{S}^{d-1}} ||x||_K d\sigma(x),$$

and $r = \frac{1}{M}$.

Lemma 11.2 *Let* $\alpha, \varepsilon > 0$ *be given. Then there is a constant* $c = c(\alpha, \varepsilon)$ *with the property that in any dimension* $m \in \mathbb{Z}^+$, *and for any convex body* L *in* $\mathbb{R}^m$ *with* $d(L - L, \mathbf{B}_2^m) < 1 + \alpha$, *there is a* k *dimensional linear subspace* F *of* $\mathbb{R}^m$ *such that* $d(P_F(L), \mathbf{B}_2^k) < 1 + \varepsilon$ *for some* $k \geq cm$.

Proof Let $\delta = d(L - L, \mathbf{B}_2^m)$. We may assume that $\frac{1}{\delta}\mathbf{B}_2^m \subseteq L - L \subseteq \mathbf{B}_2^m$. Thus, for the support function of $L - L$, we have $h_{L-L}(x) \geq \frac{1}{\delta}$ for any $x \in \mathbb{S}^{d-1}$. With the notations of Lemma 11.1, we have

$$M(L^*) = \int\limits_{\mathbb{S}^{d-1}} ||x||_{L^*} d\sigma(x) = \frac{1}{2} \int\limits_{\mathbb{S}^{d-1}} h_L(x) + h_L(-x) d\sigma(x) \qquad (11.1)$$

$$= \frac{1}{2} \int\limits_{\mathbb{S}^{d-1}} h_{L-L}(x) d\sigma(x) \geq \frac{1}{2\delta} \geq \frac{1}{2(1 + \alpha)}.$$

Note that $L^* \supset (L - L)^* \supset \mathbf{B}_2^d$, thus, by Lemma 11.1 and polarity, we obtain that L has a k dimensional projection P_F with $d(P_F L, \mathbf{B}_2^d \cap F) \leq 1 + \varepsilon$ and $k \geq C(\varepsilon)\frac{1}{4(1+\alpha)^2}m$. Here, $C(\varepsilon)$ is the same as in Lemma 11.1. □

The novel geometric idea of our proof is the following. We call a convex body $T = \text{conv}(T_1 \cup \{\pm e\})$ in $\mathbb{R}^m$ a *double cone* if $T_1 = -T_1$ is convex set, span T_1 is an $(m - 1)$-dimensional linear subspace, and $e \in \mathbb{R}^m \setminus \text{span} T_1$. Double cones are *irreducible convex bodies*, that is, for any double cone T, if $T = L - L$ then $L = T/2$, see [11, 13]. We prove a stability version of this fact.

Lemma 11.3 (Stability of Irreducibility of Double Cones) *Let* L *be a convex body in* $\mathbb{R}^m$ *with* $m \geq 2$, *and* T *be a double cone of the form* $T = \text{conv}(T_1 \cup \{\pm e\})$. *Assume that* $T \subseteq L - L \subseteq \delta T$ *for some* $1 \leq \delta < \frac{3}{2}$. *Then*

$$\left(\frac{3}{2} - \delta\right)T \subseteq L - a \subseteq \left(\delta - \frac{1}{2}\right)T.$$

for some $a \in \mathbb{R}^m$.

Proof By the assumptions, $e \in T \subseteq L - L$, thus, by translating L, we may assume that $o, e \in L$. Thus,

$$L \subseteq (L - L) \cap (L - L + e) \subseteq \delta T \cap (\delta T + e). \tag{11.2}$$

We claim that

$$\delta T \cap (\delta T + e) = \frac{e}{2} + \left(\delta - \frac{1}{2}\right) T. \tag{11.3}$$

Indeed, let H_λ denote the hyperplane $H_\lambda = \lambda e + \text{span } T_1$. To prove (11.3), we describe the sections of the right hand side and the left hand side by the hyperplanes H_λ for all relevant values of λ. For any $\lambda \in [-\delta, \delta]$, we have

$$\delta T \cap H_\lambda = \delta(T \cap H_{\lambda/\delta}) = \lambda e + \delta\left(1 - \frac{|\lambda|}{\delta}\right) T_1.$$

For any $\lambda \in [-\delta + 1, \delta + 1]$, we have

$$(\delta T + e) \cap H_\lambda = e + (\delta T \cap H_{\lambda-1}) = \lambda e + \delta\left(1 - \frac{|\lambda - 1|}{\delta}\right) T_1.$$

Thus, for any $\lambda \in [-\delta + 1, \delta]$, we have

$$\delta T \cap (\delta T + e) \cap H_\lambda = \lambda e + \delta\left(1 - \frac{1}{\delta}\max\{|\lambda|, |\lambda - 1|\}\right) T_1.$$

On the other hand, for any $\lambda \in [-\delta + 1, \delta]$, we have

$$(e/2 + (\delta - 1/2)T) \cap H_\lambda = \lambda e + (\delta - 1/2)\left(1 - \frac{|\lambda - 1/2|}{\delta - 1/2}\right) T_1.$$

Combining these two equations yields (11.3).

Thus,

$$T \subseteq L - L = \left(L - \frac{e}{2}\right) - \left(L - \frac{e}{2}\right) \subseteq \left(L - \frac{e}{2}\right) - \left(\delta - \frac{1}{2}\right) T.$$

Using the fact that $T = -T$, and $1 \leq \delta < 3/2$, we obtain

$$\left(\frac{3}{2} - \delta\right) T \subseteq L - \frac{e}{2},$$

finishing the proof of Lemma 11.3. $\qquad\qquad\square$

Now, we are ready to prove Theorem 11.1. With the notations of the theorem, let $D = K - K$, and apply the symmetric version of the theorem for D in place of K. We may assume that $\varepsilon < 1/2$. In case (1), we use Lemma 11.2 and loose a linear factor in the dimension of the almost-Euclidean projection. In case (2), we use Lemma 11.3 with $T = \mathbf{B}_1^m$ and $\delta = 1 + \varepsilon$, and obtain the same dimension for the almost-ℓ_1^m projection.

References

1. N. Alon, V.D. Milman, Embedding of l_∞^k in finite-dimensional Banach spaces. Israel J. Math. **45**(4), 265–280 (1983)
2. S. Brazitikos, A. Giannopoulos, P. Valettas, B.-H. Vritsiou, *Geometry of Isotropic Convex Bodies*. Mathematical Surveys and Monographs, vol. 196 (American Mathematical Society, Providence, 2014)
3. A. Dvoretzky, Some results on convex bodies and Banach spaces, in *Proceedings International Symposium Linear Spaces (Jerusalem, 1960)* (Jerusalem Academic Press, Jerusalem, 1961), pp. 123–160
4. T. Figiel, J. Lindenstrauss, V.D. Milman, The dimension of almost spherical sections of convex bodies. Acta Math. **139**(1–2), 53–94 (1977)
5. Y. Gordon, Gaussian processes and almost spherical sections of convex bodies. Ann. Probab. **16**(1), 180–188 (1988)
6. D.G. Larman, P. Mani, Almost ellipsoidal sections and projections of convex bodies. Math. Proc. Camb. Philos. Soc. **77**, 529–546 (1975)
7. V.D. Milman, A new proof of A. Dvoretzky's theorem on cross-sections of convex bodies. Funkcional. Anal. i Priložen. **5**(4), 28–37 (1971)
8. V.D. Milman, Almost Euclidean quotient spaces of subspaces of a finite-dimensional normed space. Proc. Amer. Math. Soc. **94**(3), 445–449 (1985)
9. V.D. Milman, A. Pajor, Entropy and asymptotic geometry of non-symmetric convex bodies. Adv. Math. **152**(2), 314–335 (2000)
10. V.D. Milman, G. Schechtman, *Asymptotic Theory of Finite-Dimensional Normed Spaces*. Lecture Notes in Mathematics, With an appendix by M. Gromov, vol. 1200 (Springer, Berlin, 1986)
11. M. Naszódi, B. Visy, Sets with a unique extension to a set of constant width, in *Discrete Geometry*. Monographs and Surveys in Pure and Applied Mathematics, vol. 253 (Dekker, New York, 2003), pp. 373–380
12. R. Schneider, *Convex Bodies: The Brunn-Minkowski Theory*. Encyclopedia of Mathematics and its Applications, vol. 151 (Cambridge University Press, Cambridge, Expanded edition, 2014)
13. D. Yost, Irreducible convex sets. Mathematika **38**(1), 134–155 (1991)

Chapter 12
An Interpolation Proof of Ehrhard's Inequality

Joe Neeman and Grigoris Paouris

Abstract We prove Ehrhard's inequality using interpolation along the Ornstein–Uhlenbeck semi-group. We also provide an improved Jensen inequality for Gaussian variables that might be of independent interest.

12.1 Introduction

In [8], A. Ehrhard proved the following Brunn–Minkowski like inequality for convex sets A, B in $\mathbb{R}^n$:

$$\Phi^{-1}\left(\gamma_n(\lambda A + (1-\lambda)B)\right) \geq \lambda \Phi^{-1}(\gamma_n(A)) + (1-\lambda)\Phi^{-1}(\gamma_n(B)), \ \lambda \in [0,1], \tag{12.1}$$

where γ_n is the standard Gaussian measure in $\mathbb{R}^n$ (i.e. the measure with density $(2\pi)^{-n/2}e^{-|x|^2/2}$) and Φ is the Gaussian distribution function (i.e. $\Phi(x) = \gamma_1(-\infty, x)$).

This is a fundamental result of Gaussian space and it is known to have numerous applications (see, e.g., [11]). Ehrhard's result was extended by R. Latała [10] to the case that one of the two sets is Borel and the other is convex. Finally, C. Borell [5] proved that it holds for all pairs of Borel sets. Ehrhard's original proof for convex sets used a Gaussian symmetrization technique. Borell used the heat semi-group and a maximum principle in his proof, which has since been further developed by Barthe and Huet [4]; very recently Ivanisvili and Volberg [9] developed this method into a

J. Neeman
Department of Mathematics, University of Texas at Austin, Austin, TX, USA

G. Paouris (✉)
Department of Mathematics, Texas A&M University, College Station, TX, USA
e-mail: grigoris@math.tamu.edu

© Springer Nature Switzerland AG 2020
B. Klartag, E. Milman (eds.), *Geometric Aspects of Functional Analysis*,
Lecture Notes in Mathematics 2266,
https://doi.org/10.1007/978-3-030-46762-3_12

general technique for proving convolution inequalities. Another proof was recently found by van Handel [14] using a stochastic variational principle.

In this work we will prove Ehrhard's inequality by constructing a quantity that is monotonic along the Ornstein–Uhlenbeck semi-group. In recent years this approach has been developed into a powerful tool to prove Gaussian inequalities such as Gaussian hypercontractivity, the log-Sobolev inequality, and isoperimetry [2]. There is no known proof of Ehrhard inequality using these techniques and the purpose of this note is to fill this gap.

An interpolation proof of the Lebesgue version of Ehrhard's inequality (the Prékopa–Leindler inequality) was presented recently in [7]. This proof uses an "improved reverse Hölder" inequality for correlated Gaussian vectors that was established in [7]. A generalization of the aforementioned inequality also appeared recently [12, 13]. This inequality, while we call an "improved Jensen inequality" for correlated Gaussian vectors, we present and actually also extend in the present note. In Sect. 12.2 we briefly discuss how this inequality implies several known inequalities in probability, convexity and harmonic analysis. Using a "restricted" version of this inequality (Theorem 12.2.2), we will present a proof of Ehrhard's inequality.

The paper is organized as follows: In Sect. 12.2 we introduce the notation and basic facts about the Ornstein–Uhlenbeck semi-group, and we present the proof of the restricted, improved Jensen inequality. In Sect. 12.3 we use Jensen inequality to provide a new proof of Prékopa–Leindler inequality. We will use the main ideas of this proof as a guideline for our proof of Ehrhard's inequality that we present in Sect. 12.4.

12.2 An "Improved Jensen" Inequality

Fix a positive semi-definite $D \times D$ matrix A, and let $X \sim \mathcal{N}(0, A)$. For $t \geq 0$, we define the operator P_t^A on $L_1(\mathbb{R}^D, \gamma_A)$ by

$$(P_t^A f)(x) = \mathbb{E} f(e^{-t} x + \sqrt{1 - e^{-2t}} X).$$

We will use the following well-known (and easily checked) facts:

- the measure γ_A is stationary for P_t^A;
- for any $s, t \geq 0$, $P_s^A P_t^A = P_{s+t}^A$;
- if f is a continuous function having limits at infinity then $P_s^A f$ converges uniformly to $P_t^A f$ as $s \to t$.

We will also use the following "diffusion" formula for P_t^A: let $\Psi : \mathbb{R}^k \to \mathbb{R}$ be a bounded C^2 function. For any bounded, measurable $f = (f_1, \ldots, f_k) : \mathbb{R}^D \to \mathbb{R}^k$, any $x \in \mathbb{R}^D$ and any $0 < s < t$, $P_{t-s}^A \Psi(P_s^A f(x))$ is differentiable in s and satisfies

$$\frac{\partial}{\partial s} P_{t-s}^A \Psi(P_s^A f) = -P_{t-s}^A \sum_{i,j=1}^{k} \partial_i \partial_j \Psi(f) \langle \nabla P_s^A f_i, A \nabla P_s^A f_j \rangle. \tag{12.2}$$

Suppose that $D = \sum_{i=1}^{k} d_i$, where $d_i \geq 1$ are integers. We decompose $\mathbb{R}^D$ as $\prod_{i=1}^{k} \mathbb{R}^{d_i}$ and write Π_i for the projection on the ith component. Given a $k \times k$ matrix M, write $\mathcal{E}_{d_1,\ldots,d_k}(M)$ for the $D \times D$ matrix whose i, j entry is $M_{k,\ell}$ if $\sum_{a<k} d_a < i \leq \sum_{a\leq k} d_a$ and $\sum_{b<\ell} d_b < j \leq \sum_{b\leq \ell} d_b$; that is, each entry $M_{k,\ell}$ of M is expanded into a $d_k \times d_\ell$ block. We write '$\odot$' for the element-wise product of matrices, '$\succcurlyeq$' for the positive semi-definite matrix ordering, and H_J for the Hessian matrix of the function J.

Our starting point in this note is the following inequality, which may be seen as an improved Jensen inequality for correlated Gaussian variables.

Theorem 12.2.1 *Let* $\Omega_1, \ldots, \Omega_k$ *be open intervals in* $\mathbb{R}$*; let* $\Omega = \prod_{i=1}^{k} \Omega_i$ *and let* $X \sim \gamma_A$*. For a bounded,* C^2 *function* $J : \Omega \to \mathbb{R}$*, the following are equivalent:*

(2.1.a) for every $x \in \Omega$*,* $A \odot \mathcal{E}_{d_1,\ldots,d_k}(H_J(x)) \succcurlyeq 0$
(2.1.b) for every k*-tuple of measurable functions* $f_i : \mathbb{R}^{d_i} \to \Omega_i$*,*

$$\mathbb{E} J(f_1(X_1), \ldots, f_k(X_k)) \geq J(\mathbb{E} f_1(X_1), \ldots, \mathbb{E} f_k(X_k)). \tag{12.3}$$

We remark that the restriction that J be bounded can often be lifted. For example, if J is a continuous but unbounded function then one can still apply Theorem 12.2.1 on bounded domains $\Omega_i' \subset \Omega_i$. If J is sufficiently nice (e.g. monotonic, or bounded above) then one can take a limit as Ω_i' exhausts Ω_i (e.g. using the monotone convergence theorem, or Fatou's lemma).

As we have already mentioned, Theorem 12.2.1 is known to have many consequences. However, we do not know how to obtain Ehrhard's inequality using only Theorem 12.2.1; we will first need to extend Theorem 12.2.1 in a few ways. To motivate our first extension, note that the usual Jensen inequality on $\mathbb{R}$ extends easily to the case where some function is convex only on a sub-level set. To be more precise, take a C^2 function $\psi : \mathbb{R}^d \to \mathbb{R}$ and the set $B = \{x \in \mathbb{R}^d : \psi(x) < 0\}$. If B is connected and ψ is convex when restricted to B, one can show that B is convex and hence $\mathbb{E}\psi(X) \geq \psi(\mathbb{E}X)$ for any random vector supported on B. A similar modification may be made to Theorem 12.2.1.

Theorem 12.2.2 *Take the notation and assumptions of Theorem 12.2.1, and assume in addition that* $\{x \in \Omega : J(x) < 0\}$ *is homeomorphic to an open ball. Then the following are equivalent:*

(2.1.a) *for every* $x \in \Omega$ *such that* $J(x) < 0$, $A \odot \mathcal{E}_{d_1,\ldots,d_k}(H_J(x)) \succcurlyeq 0$
(2.1.b) *for every k-tuple of measurable functions* $f_i : \mathbb{R}^{d_i} \to \Omega_i$ *that* γ_A*-a.s. satisfy*
$J(f_1,\ldots,f_k) < 0$,

$$\mathbb{E}J(f_1(X_1),\ldots,f_k(X_k)) \geq J(\mathbb{E}f_1(X_1),\ldots,\mathbb{E}f_k(X_k)).$$

Note that the threshold of zero in the conditions $J(x) < 0$ and $J(f_1,\ldots,f_k) < 0$ is arbitrary, since we may apply the theorem to the function $J(\cdot) - a$ for any $a \in \mathbb{R}$. Of course, taking a sufficiently large recovers Theorem 12.2.1.

Proof Suppose that (2.2.a) holds. By standard approximation arguments, it suffices to prove (2.2.b) for a more restricted class of functions f. Indeed, let F be the set of measurable $f = (f_1,\ldots,f_k)$ satisfying $J(f) < 0$ γ_A-a.s. and let $F_\epsilon \subset F$ be those functions that are continuous, converge to a limit at infinity, and satisfy $J(f) \leq -\epsilon$ γ_A-a.s. Now, every $f \in F$ can be approximated in $L^1(\gamma_A)$ by a sequence $f^{(n)} \in F_{1/n}$: by truncating the values of f outside of a large ball in $\mathbb{R}^D$ and away from the boundary of $\{x : J < 0\}$, we can approximate $f \in F$ in $L^1(\gamma_A)$ by $\tilde{f}$ satisfying the latter two conditions. To ensure continuity, we can use mollifiers: if Tg denotes the convolution of g with a smooth, compactly supported mollifier and $\mathcal{F}$ is a homeomorphism from $\{J < 0\}$ to a ball, then $\mathcal{F}^{-1} \circ (T(\mathcal{F} \circ \tilde{f}))$ is a continuous approximation of $\tilde{f}$ that takes values in $\{J < 0\}$. With these approximations in mind, it suffices to prove (2.2.b) for $f \in F_\epsilon$, where $\epsilon > 0$ is arbitrarily small. From now on, fix $\epsilon > 0$ and fix $f = (f_1,\ldots,f_k) \in F_\epsilon$.

Recalling that $\Pi_i : \mathbb{R}^{d_1} \times \cdots \times \mathbb{R}^{d_k} \to \mathbb{R}^{d_i}$ is the projection onto the ith block of coordinates, define $g_i = f_i \circ \Pi_i$ and $G_{s,t}(x) = P_{t-s}^A J(P_s^A g(x))$. Since $f \in F_\epsilon$, we have $G_{0,0}(x) \leq -\epsilon$ for every $x \in \mathbb{R}^D$. Moreover, since f is continuous and vanishes at infinity, $P_s^A g \to g$ uniformly as $s \to 0$. Since g is bounded, J is uniformly continuous on the range of g and so there exists $\delta > 0$ such that $|G_{s,s}(x) - G_{r,r}(x)| < \epsilon$ for every $x \in \mathbb{R}^D$ and every $|s - r| \leq \delta$.

Now, fix $r \geq 0$ and assume that $G_{r,r} \leq -\epsilon$ pointwise; by the previous paragraph, $G_{s,s} < 0$ pointwise for every $r \leq s \leq r + \delta$. Now we apply the commutation formula (12.2): with $B_s = B_s(x) = A \odot \mathcal{E}_{d_1,\ldots,d_k}(H_J(P_s^A g))$, we have

$$\frac{\partial}{\partial s} G_{s,t} = -P_{t-s}^A \sum_{i,j=1}^{k} \langle \nabla P_s^A g_i, B \nabla P_s^A g_j \rangle$$

(here, we have used the observation that $P_s^A g_i(x)$ depends only on $\Pi_i x$, and so $\nabla P_s^A g_i$ is zero outside the ith block of coordinates). The assumption (2.2.a) implies that B_s is positive semi-definite whenever $G_{s,s} < 0$; since $G_{s,s} < 0$ for every $s \in [r, r+\delta]$, we see that for such s, $\frac{\partial}{\partial s} G_{s,r+\delta} \leq 0$ pointwise. Since $G_{s,r+\delta}$ is continuous in s and $G_{r,r} \leq -\epsilon$, it follows that $G_{s,s} \leq -\epsilon$ pointwise for all $s \in [r, r + \delta]$.

Next, note that $r = 0$ satisfies the assumption $G_{r,r} \leq -\epsilon$ of the previous paragraph. By induction, it follows that $G_{r,r} \leq -\epsilon$ pointwise for all $r \geq 0$. Hence, the matrix B_s is positive semi-definite for all $s \geq 0$ and $x \in \mathbb{R}^D$, which implies that $G_{s,t}(x)$ is non-increasing in s for all $t \geq s$ and $x \in \mathbb{R}^D$. Hence,

$$\mathbb{E}J(f_1(X_1), \ldots, f_k(X_k)) = \lim_{t \to \infty} G_{0,t}(0) \geq \lim_{t \to \infty} G_{t,t}(0) = J(\mathbb{E}f_1, \ldots, \mathbb{E}f_k).$$

This completes the proof of (2.2.b).

Now suppose that (2.2.b) holds. Choose some $v \in \mathbb{R}^D$ and some $y \in \Omega$ with $J(y) < 0$; to prove (2.2.a), it is enough to show that

$$v^T (A \odot \mathcal{E}_{d_1, \ldots, d_k}(H_J(y)))v \geq 0. \tag{12.4}$$

Since Ω is open and J is continuous, there is some $\delta > 0$ such that $y + z \in \Omega$ and $J(y + z) < 0$ whenever $\max_i |z_i| \leq \delta$. For this δ, define $\psi : \mathbb{R} \to \mathbb{R}$ by

$$\psi(t) = \max\{-\delta, \min\{\delta, t\}\}.$$

For $\epsilon > 0$, define $f_{i,\epsilon} : \mathbb{R}^{d_i} \to \Omega_i$ by

$$f_{i,\epsilon}(x) = y_i + \psi(\epsilon \langle x, \Pi_i v \rangle).$$

By (2.2.b),

$$\mathbb{E}J(f_{1,\epsilon}(X_1), \ldots, f_{k,\epsilon}(X_k)) \geq J(\mathbb{E}f_{1,\epsilon}(X_1), \ldots, \mathbb{E}f_{k,\epsilon}(X_k)).$$

Since ψ is odd, $\mathbb{E}f_{i,\epsilon}(X_i) = y_i$ for all $\epsilon > 0$; hence,

$$\mathbb{E}J(f_{1,\epsilon}(X_1), \ldots, f_{k,\epsilon}(X_k)) \geq J(y). \tag{12.5}$$

Taylor's theorem implies that for any z with $y + z \in \Omega$,

$$J(y + z) = J(y) + \sum_{i=1}^{k} \frac{\partial J(y)}{\partial y_i} z_i + \sum_{i,j=1}^{k} \frac{\partial^2 J(y)}{\partial y_i \partial y_j} z_i z_j + \rho(|z|),$$

where ρ is some function satisfying $\epsilon^{-2}\rho(\epsilon) \to 0$ as $\epsilon \to 0$. Now consider what happens when we replace z_i above with $Z_i = \psi(\epsilon \langle X_i, \Pi_i v \rangle)$ and take expectations. One easily checks that $\mathbb{E}Z_i = 0$, $\mathbb{E}\rho(|Z|) = o(\epsilon^2)$, and

$$\mathbb{E}Z_i Z_j = \epsilon^2 (\Pi_i v)^T \mathbb{E}[X_i X_j](\Pi_i v) + o(\epsilon^2);$$

hence,

$$\mathbb{E}J(y + Z) = J(y) + \epsilon^2 \sum_{i,j=1}^{k} \frac{\partial^2 J(y)}{\partial y_i \partial y_j}(\Pi_i v)^T \mathbb{E}[X_i X_j](\Pi_i v) + o(\epsilon^2)$$

$$= J(y) + \epsilon^2 v^T (A \odot \mathcal{E}_{d_1,\ldots,d_k}(H_J(y)))v + o(\epsilon^2).$$

On the other hand, $\mathbb{E}J(y + Z) = \mathbb{E}J(f_{1,\epsilon}(X_1), \ldots, f_{k,\epsilon}(X_k))$, which is at least $J(y)$ according to (12.5). Taking $\epsilon \to 0$ proves (12.4).   □

12.3   A Short Proof of Prékopa–Leindler Inequality

The Prékopa–Leindler inequality states that if $f, g, h : \mathbb{R}^d \to [0, \infty)$ satisfy

$$h(\lambda x + (1 - \lambda)y) \geq f(x)^\lambda g(y)^{1-\lambda}$$

for all $x, y \in \mathbb{R}^d$ and some $\lambda \in (0, 1)$ then

$$\mathbb{E}h \geq (\mathbb{E}f)^\lambda (\mathbb{E}g)^{1-\lambda},$$

where expectations are taken with respect to the standard Gaussian measure on $\mathbb{R}^d$. By applying a linear transformation, the standard Gaussian measure may be replaced by any Gaussian measure; by taking a limit over Gaussian measures with large covariances, the expectations may also be replaced by integrals with respect to the Lebesgue measure.

As M. Ledoux brought to our attention, the Prékopa–Leindler inequality may be seen as a consequence of Theorem 12.2.1; we will present only the case $d = 1$, but the case for general d may be done in a similar way. Alternatively, one may prove the Prékopa–Leindler inequality for $d = 1$ first and then extend to arbitrary d using induction and Fubini's theorem.

Fix $\lambda \in (0, 1)$, let $(X, Y) \sim \mathcal{N}\left(0, \left(\begin{smallmatrix} 1 & \rho \\ \rho & 1 \end{smallmatrix}\right)\right)$ and let $Z = \lambda X + (1 - \lambda)Y$. Let $\sigma^2 = \sigma^2(\rho, \lambda)$ be the variance of Z and let $A = A(\rho, \lambda)$ be the covariance of (X, Y, Z). Note that A is a rank-two matrix, and that it may be decomposed as $A = uu^T + vv^T$ where u and v are both orthogonal to $(\lambda, 1 - \lambda, -1)^T$.

For $\alpha, R \in \mathbb{R}_+$, define $J_{\alpha,R} : \mathbb{R}_+^3 \to \mathbb{R}$ by

$$J_{\alpha,R}(x, y, z) = (x^\lambda y^{1-\lambda} z^{-\alpha})^R.$$

Lemma 12.3.1 *For any λ and ρ, and for any $\alpha < \sigma^2$, there exists $R \in \mathbb{R}_+$ such that $A \odot H_{J_{\alpha,R}} \succeq 0$.*

To see how the Prékopa–Leindler inequality follows from Theorem 12.2.1 and Lemma 12.3.1, suppose that $h(\lambda x + (1 - \lambda)y) \geq f^\lambda(x)g^{1-\lambda}(y)$ for all $x, y \in \mathbb{R}$.

Then $J_{\alpha,R}(f(X), g(Y), h^{1/\alpha}(Z)) \leq 1$ with probability one (because $Z = \lambda X + (1-\lambda)Y$ with probability one). By Theorem 12.2.1, with the R from Lemma 12.3.1 we have

$$1 \geq \mathbb{E}J_{\alpha,R}(f(X), g(Y), h(Z))$$

$$\geq J_{\alpha,R}(\mathbb{E}f(X), \mathbb{E}g(Y), \mathbb{E}h(Z))$$

$$= \left(\frac{(\mathbb{E}f(X))^{\lambda}(\mathbb{E}g(Y))^{1-\lambda}}{(\mathbb{E}h^{1/\alpha}(Z))^{\alpha}} \right)^{R}.$$

In other words, $(\mathbb{E}h^{1/\alpha}(Z))^{\alpha} \geq (\mathbb{E}f)^{\lambda}(\mathbb{E}g)^{1-\lambda}$. This holds for any ρ and any $\alpha < \sigma^2$. By sending $\rho \to 1$, we send $\sigma^2 \to 1$ and so we may take $\alpha \to 1$ also. Finally, note that in this limit Z converges in distribution to $\mathcal{N}(0, 1)$. Hence, we recover the Prékopa–Leindler inequality for the standard Gaussian measure.

Proof of Lemma 12.3.1 By a computation,

$$H_{J_{\alpha,R}} = J_{\alpha,R}(x, y, z) \begin{pmatrix} \frac{\lambda R(\lambda R - 1)}{x^2} & \frac{\lambda R(1-\lambda)R}{xy} & -\frac{\lambda \alpha R^2}{xz} \\ \frac{\lambda R(1-\lambda)R}{xy} & \frac{(1-\lambda)R((1-\lambda)R-1)}{y^2} & -\frac{(1-\lambda)\alpha R^2}{yz} \\ -\frac{\lambda \alpha R^2}{xz} & -\frac{(1-\lambda)\alpha R^2}{yz} & \frac{\alpha R(\alpha R + 1)}{z^2} \end{pmatrix}.$$

We would like to show that $A \odot H_J \succcurlyeq 0$; since elementwise multiplication commutes with multiplication by diagonal matrices, it is enough to show that

$$A \odot \left(\begin{pmatrix} \lambda \\ 1-\lambda \\ -\alpha \end{pmatrix}^{\otimes 2} - \frac{1}{R} \begin{pmatrix} \lambda & 0 & 0 \\ 0 & 1-\lambda & 0 \\ 0 & 0 & -\alpha \end{pmatrix} \right) \geq 0. \qquad (12.6)$$

Let $\theta = (\lambda, 1-\lambda, -\alpha)^T$ and recall that $A = uu^T + vv^T$, where u and v are both orthogonal to $(\lambda, 1-\lambda-1)^T$. Then

$$A \odot (\theta\theta^T) = (u \odot \theta)(u \odot \theta)^T + (v \odot \theta)(v \odot \theta)^T,$$

where $u \odot \theta$ and $v \odot \theta$ are both orthogonal to $(1, 1, \frac{1}{\alpha})^T$ (call this w). In particular, $A \odot (\theta\theta^T)$ is a rank-two, positive semi-definite matrix whose null space is the span of w.

On the other hand, $A \odot \text{diag}(\lambda, 1-\lambda, -\alpha) = \text{diag}(\lambda, 1-\lambda, -\alpha\sigma^2)$ (call this D). Then $w^T D w = 1 - \sigma^2/\alpha < 0$. As a consequence of the following Lemma,

$$A \circ (\theta\theta^T) - \frac{1}{R}D \geq 0$$

for all sufficiently large R. $\square$

Lemma 12.3.2 *Let A be a positive semi-definite matrix and let B be a symmetric matrix. If $u^T Bu \geq \delta |u|^2$ for all $u \in \ker(A)$ and $v^T Av \geq \delta |v|^2$ for all $v \in \ker(A)^\perp$ then $A + \epsilon B \succ 0$ for all $0 \leq \epsilon \leq \frac{\delta^2}{\|B\|^2 + \delta\|B\|}$, where $\|B\|$ is the operator norm of B.*

Proof Any vector w may be decomposed as $w = u + v$ with $u \in \ker(A)$ and $v \in \ker(A)^\perp$. Then

$$w^T(A + \epsilon B)w = u^T Au + \epsilon u^T Bu + 2\epsilon u^T Bv + \epsilon v^T Bv$$

$$\geq \delta |u|^2 - \epsilon\|B\||u|^2 - 2\epsilon\|B\||u||v| + \epsilon\delta|v|^2.$$

Considering the above expression as a quadratic polynomial in $|u|$ and $|v|$, we see that it is non-negative whenever $(\delta - \epsilon\|B\|)\delta \geq \epsilon\|B\|^2$. □

We remark that the preceding proof of the Prékopa–Leindler inequality may be extended in an analogous way to prove Barthe's inequality [3].

12.4   Proof of Ehrhard's Inequality

The parallels between the Prékopa–Leindler and Ehrhard inequalities become obvious when they are both written in the following form. The version of Prékopa–Leindler that we proved above may be restated to say that

$$\left.\begin{array}{c} \exp(R(\lambda \log f(X) + (1 - \lambda)\log g(Y) - \alpha\log h(Z))) \leq 0 \text{ a.s.} \\[2mm] \text{implies} \\[2mm] \exp(R(\lambda \log \mathbb{E}f(X) + (1 - \lambda)\log \mathbb{E}g(Y) - \alpha\log \mathbb{E}h(Z))) \leq 0. \end{array}\right\} \quad (12.7)$$

On the other hand, here we will prove that

$$\left.\begin{array}{c} \Phi\left(R(\lambda\Phi^{-1}(f(X)) + (1 - \lambda)\Phi^{-1}(g(Y)) - \sigma\Phi^{-1}(h(Z)))\right) \leq 0 \text{ a.s.} \\[2mm] \text{implies} \\[2mm] \Phi\left(R(\lambda\Phi^{-1}(\mathbb{E}f(X)) + (1 - \lambda)\Phi^{-1}(\mathbb{E}g(Y)) - \sigma\Phi^{-1}(\mathbb{E}h(Z)))\right) \leq 0. \end{array}\right\} \quad (12.8)$$

(It may not yet be clear why the α in (12.7) has become σ in (12.8); this turns out to be the right choice, as will become clear from the example in Sect. 12.4.1.) This implies Ehrhard's inequality in the same way that (12.7) implies the Prékopa–Leindler inequality. In particular, our proof of (12.7) suggests a strategy for attacking (12.8): define the function

$$J_R(x, y, z) = \Phi\left(R(\lambda\Phi^{-1}(x) + (1 - \lambda)\Phi^{-1}(y) - \sigma\Phi^{-1}(z))\right).$$

(We will drop the parameter R when it can be inferred from the context.) In analogy with our proof of Prékopa–Leindler, we might then try to show that for sufficiently large R, $A \odot H_{J_R} \succcurlyeq 0$. Unfortunately, this is false.

12.4.1   An Example

Recall from the proof of Theorem 12.2.2 that if $A \odot H_J \succcurlyeq 0$ then

$$G_{s,t,R}(x, y) := P_{t-s}^A J_R(P_s^1 f(x), P_s^1 g(y), P_s^{\sigma^2} h(\lambda x + (1 - \lambda)y))$$

is non-increasing in s for every x and y. We will give an example in which $G_{s,t,R}$ may be computed explicitly and it clearly fails to be non-increasing.

From now on, define $f_s = P_s^1 f$, $g_s = P_s^1 g$ and $h_s = P_s^{\sigma^2} h$. Let $f(x) = 1_{\{x \le a\}}$, $g(y) = 1_{\{y \le b\}}$ and $h(z) = 1_{\{z \le c\}}$, where $c \ge \lambda a + (1 - \lambda)b$. A direct computation yields

$$f_s(x) = \Phi \left(\frac{a - e^{-s}x}{\sqrt{1 - e^{-2s}}} \right)$$

$$g_s(y) = \Phi \left(\frac{b - e^{-s}y}{\sqrt{1 - e^{-2s}}} \right)$$

$$h_s(z) = \Phi \left(\frac{c - e^{-s}z}{\sigma \sqrt{1 - e^{-2s}}} \right).$$

Hence,

$$J(f_s(x), g_s(y), h_s(\lambda x + (1 - \lambda y))) = \Phi \left(R \frac{\lambda a + (1 - \lambda)b - c}{\sqrt{1 - e^{-2s}}} \right).$$

If $c > \lambda a + (1 - \lambda)b$ then the above quantity is increasing in s. Since it is also independent of x and y, it remains unchanged when applying P_{t-s}^A. That is,

$$G_{s,t,R} = \Phi \left(R \frac{\lambda a + (1 - \lambda)b - c}{\sqrt{1 - e^{-2s}}} \right)$$

is increasing in s. On the bright side, in this example $G_{s,r,R\sqrt{1-e^{-2s}}}$ is constant. Since Theorem 12.2.1 was not built to consider such behavior, we will adapt it so that the function J is allowed to depend on s.

12.4.2 Allowing J to Depend on t

Recalling the notation of Sect. 12.2, we assume from now on that $\Omega_i \subseteq [0, 1]$ for each i. Then A is a $k \times k$ matrix; let $\sigma_1^2, \ldots, \sigma_k^2$ be its diagonal elements. We will consider functions of the form $J : \Omega \times [0, \infty] \to \mathbb{R}$. We write H_J for the Hessian matrix of J with respect to the variables in Ω, and $\frac{\partial J}{\partial t}$ for the partial derivative of J with respect to the variable in $[0, \infty]$. Let $I : [0, 1] \to \mathbb{R}$ be the function $I(x) = \phi(\Phi^{-1}(x))$.

Lemma 12.4.1 *With the notation above, suppose that* $J : \Omega \times [0, \infty] \to \mathbb{R}$ *is bounded and* C^2, *and take* $(X_1, \ldots, X_k) \sim \gamma_A$. *Let* $\lambda_1, \ldots, \lambda_k$ *be non-negative numbers with* $\sum_i \lambda_i = 1$, *let* $D(x)$ *be the* $k \times k$ *diagonal matrix with* $\lambda_i \sigma_i^2 / I^2(x_i)$ *in position* i, *and take some* $\epsilon \geq 0$. *If* $\frac{\partial J}{\partial t}(x, t) \leq 0$ *and*

$$A \odot H_J(x, t) - (e^{2(t+\epsilon)} - 1) \frac{\partial J(x, t)}{\partial t} D \succcurlyeq 0 \qquad (12.9)$$

for every $x \in \Omega$ *and* $t > 0$ *then for every* k-*tuple of measurable functions* $f_i : \mathbb{R} \to \Omega_i$,

$$\mathbb{E}J(P_\epsilon^{\sigma_1} f_1(X_1), \ldots, P_\epsilon^{\sigma_k} f_k(X_k), 0) \geq J(\mathbb{E}f_1(X_1), \ldots, \mathbb{E}f_k(X_k), \infty). \qquad (12.10)$$

Note that Lemma 12.4.1 has an extra parameter $\epsilon \geq 0$ compared to our previous versions of Jensen's inequality. This is for convenience when applying Lemma 12.4.1: when $\epsilon > 0$ then the function $e^{2(t+\epsilon)} - 1$ is bounded away from zero, which makes (12.9) easier to check.

Proof Write $f_{i,s}$ for $P_{s+\epsilon}^{\sigma_i^2} f_i$ and $f_s = (f_{1,s}, \ldots, f_{k,s})$. Define

$$G_{s,t} = P_{t-s}^A J(f_{1,s}, \ldots, f_{k,s}, s).$$

We differentiate in s, using the commutation formula (12.2). Compared to the proof of Theorem 12.2.2, an extra term appears because the function J itself depends on s:

$$-\frac{\partial}{\partial s} G_{s,t} = P_{t-s} \sum_{i,j=1}^k \partial_i \partial_j J(f_s, s) A_{ij} f'_{i,s} f'_{j,s} - P_{t-s} \frac{\partial J}{\partial s}(f_s, s)$$

$$= P_{t-s} v_s^T (A \odot H_J(f_s, s)) v_s - P_{t-s} \frac{\partial J}{\partial s}(f_s, s),$$

where $v_s = \nabla f_s$. Bakry and Ledoux [1] proved that $|v_{i,s}| \leq \sigma_i^{-1}(e^{2(s+\epsilon)} - 1)^{-1/2} I(f_{i,s})$. Hence,

$$v_s^T D(f_s) v_s = \sum_{i=1}^k \lambda_i \left(\frac{\sigma_i |v_{i,s}|}{I(f_{i,s})} \right)^2 \leq (e^{2(s+\epsilon)} - 1)^{-1},$$

and so

$$-\frac{\partial}{\partial s} G_{s,t} \geq P_{t-s}\left(v_s^T (A \odot H_J(f_s, s))v_s - (e^{2(s+\epsilon)} - 1)\frac{\partial J}{\partial s}(f_s, s)v_s^T D(f_s)v_s\right).$$

Clearly, the argument of P_{t-s} is non-negative pointwise if

$$A \odot H_J(x, s) \succcurlyeq (e^{2(s+\epsilon)} - 1)\frac{\partial J(x, s)}{\partial s}D(x)$$

for all x, s. In this case, $G_{s,t}$ is non-increasing in s and we conclude as in the proof of Theorem 12.2.2. $\qquad\square$

By combining the ideas of Theorem 12.2.2 and Lemma 12.4.1, we obtain the following combined version.

Corollary 12.4.2 *With the notation of Lemma 12.4.1, suppose that $J : \Omega \times [0, \infty] \to \mathbb{R}$ is bounded and C^2, and take $(X_1 \ldots, X_k) \sim \gamma_A$. Let $\lambda_1, \ldots, \lambda_k$ be non-negative numbers with $\sum_i \lambda_i = 1$, let $D(x)$ be the $k \times k$ diagonal matrix with $\lambda_i \sigma_i^2 / I^2(x_i)$ in position i, and take some $\epsilon \geq 0$. Assume that $\{x \in \Omega : J(x, 0) < 0\}$ is homeomorphic to an open ball, that $\frac{\partial J(x,t)}{\partial t} \leq 0$ whenever $J(x, t) < 0$, and that*

$$A \odot H_J(x, t) - (e^{2(t+\epsilon)} - 1)\frac{\partial J(x, t)}{\partial t}D \succcurlyeq 0$$

for every $t \geq 0$ and every x such that $J(x, t) < 0$. Then for every k-tuple of measurable functions $f_i : \mathbb{R} \to \Omega_i$ satisfying $J(P_\epsilon^{\sigma_1^2} f_1, \ldots, P_\epsilon^{\sigma_k^2} f_k, 0) < 0$ γ_A-a.s.,

$$\mathbb{E}J(P_\epsilon^{\sigma_1^2} f_1(X_1), \ldots, P_\epsilon^{\sigma_k^2} f_k(X_k), 0) \geq J(\mathbb{E}f_1(X_1), \ldots, \mathbb{E}f_k(X_k), \infty).$$

Proof As in the proof of Theorem 12.2.2, we can assume that $f = (f_1, \ldots, f_k)$ is bounded, continuous, converges to a constant near infinity, and we can strengthen the assumption

$$J(P_\epsilon^{\sigma_1^2} f_1, \ldots, P_\epsilon^{\sigma_k^2} f_k, 0) < 0$$

to

$$J(P_\epsilon^{\sigma_1^2} f_1, \ldots, P_\epsilon^{\sigma_k^2} f_k, 0) < -\eta$$

for some fixed but arbitrarily small $\eta > 0$. As in the proof of Lemma 12.4.1, we define $f_{i,s} = P_{s+\epsilon}^{\sigma_i^2} f_i$ and

$$G_{s,t} = P_{t-s}^A J(f_{1,s}, \ldots, f_{k,s}, s).$$

The same computation as in Lemma 12.4.1 shows that $\frac{\partial}{\partial s} G_{s,t} \le 0$ whenever $G_{s,s} = J(f_{1,s}, \ldots, f_{k,s}, s) < 0$ (the requirement that $G_{s,s} < 0$ is the only difference so far compared to the proof of Lemma 12.4.1, in which it was shown that $\frac{\partial}{\partial s} G_{s,t} \le 0$ unconditionally).

Now we use the argument from the proof of Theorem 12.2.2: by uniform continuity there exists $\delta > 0$ such that $|G_{s,s}(x) - G_{r,r}(x)| < \eta$ for every $x \in \mathbb{R}^k$ and $|s - r| < \delta$. Hence, if $G_{r,r} \le -\eta$ pointwise then $G_{s,s} < 0$ (and hence $G_{s,r+\delta} < 0$) pointwise for every $s \in [r, r + \delta]$. By the previous paragraph, $G_{s,r+\delta}$ is non-increasing in s for $s \in [r, r+\delta]$, and so $G_{r+\delta,r+\delta} \le G_{r,r} \le -\eta$ pointwise. Since we assumed that $G_{0,0} \le -\eta$, it follows by induction that $\lim_{t \to \infty} G_{t,t} \le G_{0,0}$, which is the required conclusion. $\square$

12.4.3   The Hessian of J

Define $J_R : (0, 1)^3 \to 0$ by

$$J_R(x, y, z) = \Phi\left(R\big(\lambda\Phi^{-1}(x) + (1 - \lambda)\Phi^{-1}(y) - \sigma\Phi^{-1}(z)\big)\right).$$

Let $H_J = H_J(x, y, z)$ denote the 3×3 Hessian matrix of J; let A be the 3×3 covariance matrix of (X, Y, Z). In order to apply Corollary 12.4.2, we will compute the matrix $A \odot H_J$. First, we define some abbreviations: set

$$u = \Phi^{-1}(x) \qquad\qquad \Xi = \lambda u + (1 - \lambda)v - \sigma w$$

$$v = \Phi^{-1}(y) \qquad\qquad \theta = (\lambda, 1 - \lambda, -\sigma)^T$$

$$w = \Phi^{-1}(z) \qquad\qquad \mathcal{I} = \mathrm{diag}(\phi(u), \phi(v), \phi(w))$$

We will use a subscript s to denote that any of the above quantities is evaluated at (f_s, g_s, h_s) instead of (x, y, z). That is $u_s = \Phi^{-1}(f_s)$, $\Xi_s = \lambda u_s + (1-\lambda)v_s - \sigma w_s$, and so on.

Lemma 12.4.3 $H_J = \phi(R\Xi)\mathcal{I}^{-1}\left(R\,\mathrm{diag}(\lambda u, (1 - \lambda)v, -\sigma w) - R^3 \Xi\theta\theta^T\right)\mathcal{I}^{-1}.$

Proof Noting that $\frac{du}{dx} = 1/\phi(u)$, the chain rule gives

$$\frac{d}{dx}\Phi(R\Xi) = R\lambda\frac{\phi(R\Xi)}{\phi(u)} = R\lambda \exp\left(-\frac{R^2\Xi^2 - u^2}{2}\right).$$

Differentiating again,

$$\frac{d^2}{dx^2}\Phi(R\Xi) = R\lambda(u - R^2\Xi\lambda)\frac{\phi(R\Xi)}{\phi^2(u)}.$$

For cross-derivatives,

$$\frac{d^2}{dxdy}\Phi(R\Xi) = -R^3\Xi\lambda(1-\lambda)\frac{\phi(R\Xi)}{\phi(u)\phi(v)}.$$

Putting these together with the analogous terms involving differentiation by z,

$$\frac{H_J}{\phi(R\Xi)} = -R^3\Xi\begin{pmatrix} \frac{\lambda^2}{\phi^2(u)} & \frac{\lambda(1-\lambda)}{\phi(u)\phi(v)} & -\frac{\lambda\sigma}{\phi(u)\phi(w)} \\ \frac{\lambda(1-\lambda)}{\phi(u)\phi(v)} & \frac{(1-\lambda)^2}{\phi^2(v)} & -\frac{(1-\lambda)\sigma}{\phi(v)\phi(w)} \\ -\frac{\lambda\sigma}{\phi(u)\phi(w)} & -\frac{(1-\lambda)\sigma}{\phi(u)\phi(v)} & \frac{\sigma^2}{\phi^2(w)} \end{pmatrix}$$
$$+ R\begin{pmatrix} \frac{\lambda u}{\phi^2(u)} & 0 & 0 \\ 0 & \frac{(1-\lambda)v}{\phi^2(v)} & 0 \\ 0 & 0 & -\frac{\sigma w}{\phi^2(w)} \end{pmatrix}.$$

Recalling the definition of $\mathcal{I}$ and θ, this may be rearranged into the claimed form.

□

Having computed H_J, we need to examine $A \odot H_J$. Recall that A is a rank-two matrix and so it may be decomposed as $A = aa^T + bb^T$. Moreover, the fact that $Z = \lambda X + (1-\lambda)Y$ means that a and b are both orthogonal to $(\lambda, 1-\lambda, -1)^T$. Recalling the definition of θ, this implies that $a \odot \theta$ and $b \odot \theta$ are both orthogonal to $(1, 1, \sigma^{-1})^T$. This observation allows us to deal with the $\theta\theta^T$ term in Lemma 12.4.3:

$$A \odot \theta\theta^T = (aa^T) \odot (\theta\theta^T) + (bb^T) \odot (\theta\theta^T) = (a \odot \theta)^{\otimes 2} + (b \odot \theta)^{\otimes 2}.$$

To summarize:

Lemma 12.4.4 *The matrix $B := A \odot \theta\theta^T$ is positive semidefinite and has rank two. Its kernel is the span of $(1, 1, \frac{1}{\sigma})^T$.*

On the other hand, the diagonal entries of A are 1, 1, and σ^2; hence,

$$A \odot \mathrm{diag}(\lambda u, (1-\lambda)v, -\sigma w) = \mathrm{diag}(\lambda u, (1-\lambda)v, -\sigma^3 w) =: D.$$

Combining this with Lemma 12.4.3, we have

$$A \odot H_J = R\phi(R\Xi)\mathcal{I}^{-1}(D - R^2\Xi B)\mathcal{I}^{-1}. \tag{12.11}$$

Consider the expression above in the light of our earlier proof of Prékopa–Leindler. Again, we have a sum of two matrices (D and $-R^2\Xi B$), one of which is multiplied by a factor (R^2) that we may take to be large. There are two important differences. The first is that the matrix D (whose analogue was constant in the proof of Prékopa–Leindler) cannot be controlled pointwise in terms of B. This difference is closely related to the example in Sect. 12.4.1; we will solve it by making J depend on t in the right way; the $\frac{dJ}{dt}$ term in Corollary 12.4.2 will then cancel out part of D's contribution.

The second difference is that in (12.11), the term that is multiplied by a large factor (namely, $-\Xi B$) is not everywhere positive semi-definite because there exist $(x, y, z) \in \mathbb{R}^3$ such that $\Xi(x, y, z) > 0$. This is the reason that we consider the "restricted" formulation of Jensen's inequality in Theorem 12.2.2 and Corollary 12.4.2.

12.4.4 Adding the Dependence on t

Recall that X and Y have variance 1 and covariance ρ, that $Z = \lambda X + (1 - \lambda)Y$, and that A is the covariance of (X, Y, Z). Recall also the notations u, v, w, Ξ, and their subscripted variants. For $R > 0$, define $r(t) = R\sqrt{1 - e^{-2t-\epsilon}}$ and

$$J_R(x, y, z, t) = \Phi\left(r(t)\left(\lambda\Phi^{-1}(x) + (1 - \lambda)\Phi^{-1}(y) - \sigma\Phi^{-1}(z)\right)\right)$$
$$= \Phi(r(t)\Xi). \tag{12.12}$$

Let $E = \mathrm{diag}(\lambda, 1 - \lambda, \sigma)/(1 + \sigma^{-1})$.

Lemma 12.4.5 *Define* $\Omega_\epsilon = [\Phi(-1/\epsilon), \Phi(1/\epsilon)]^3$. *For every* ρ, λ, *and* ϵ, *there exists* $R > 0$ *such that*

$$A \odot H_J - (e^{2(t+\epsilon)} - 1)\frac{\partial J}{\partial t}\mathcal{I}^{-1}E\mathcal{I}^{-1} \succcurlyeq 0$$

on $\{(x, t) \in \Omega_\epsilon \times [0, \infty) : \Xi(x) \leq -\epsilon\}$.

Proof We computed $\Lambda \odot H_J$ in (12.11) already; applying that formula and noting that $\mathcal{I}^{-1} \succcurlyeq 0$, it suffices to show that

$$r(t)\phi(r(t)\Xi)(D - r^2(t)\Xi B) - (e^{2(t+\epsilon)} - 1)\frac{\partial J}{\partial t}E \succcurlyeq 0$$

whenever $\Xi \leq -\epsilon$. (Recall that $D = \mathrm{diag}(\lambda u, (1 - \lambda)v, -\sigma^3 w)$, and that B is a rank-two positive semidefinite matrix that depends only on ρ and λ, and whose kernel is the span of $(1, 1, \sigma^{-1})^T$). We compute

$$\frac{\partial J}{\partial t} = r'(t)\Xi\phi(r(t)\Xi) = \frac{r(t)}{e^{2t+\epsilon} - 1}\Xi\phi(r(t)\Xi).$$

Now, there is some $\delta = \delta(\epsilon) > 0$ such that

$$\frac{e^{2(t+\epsilon)} - 1}{e^{2t+\epsilon} - 1} \geq 1 + \delta$$

for all $t \geq 0$. For this δ,

$$r(t)\phi(r(t)\Xi)(D - r^2(t)\Xi B) - (e^{2(t+\epsilon)} - 1)\frac{\partial J}{\partial t}E$$

$$\succcurlyeq r(t)\phi(r(t)\Xi)(D - (1+\delta)\Xi E - r^2(t)\Xi B);$$

Hence, it suffices to show that $D - (1+\delta)\Xi E - r^2(t)\Xi B \succcurlyeq 0$. Since $\Xi \leq -\epsilon$, it suffices to show that $r^2(t)\epsilon B + D - (1+\delta)\Xi E \succcurlyeq 0$. Now, B is a rank-two positive semi-definite matrix depending only on λ and ρ. Its kernel is spanned by $\theta = (1, 1, \sigma^{-1})^T$. Note that $\theta^T D\theta = \Xi$ and $\theta^T E\theta = 1$. Hence,

$$\theta^T(D - (1+\delta)\Xi E)\theta = -\delta\Xi \geq \delta\epsilon > 0.$$

Next, note that we can bound the norm of $D - (1+\delta)\Xi E$ uniformly: on Ω_ϵ, $\|D\| \leq 1/\epsilon$ and $|\Xi| \leq 2/\epsilon$. All together, if we assume (as we may) that $\delta \leq 1$ then $\|D + (1+\delta)\Xi E\| \leq 5/\epsilon$. By Lemma 12.3.2, if $\eta > 0$ is sufficiently small then

$$\epsilon B + \eta(D - (1+\delta)\Xi E) \succcurlyeq 0.$$

To complete the proof, choose R large enough so that $R^2(1 - e^\epsilon) \geq 1/\eta$; then $r^2(t) \geq 1/\eta$ for all t.   □

Finally, we complete the proof of (12.8) by a series of simple approximations. First, let C_a denote the set of continuous functions $\mathbb{R} \to [0, 1]$ that converge to a at $\pm\infty$, and note that it suffices to prove (12.8) in the case that $f, g \in C_0$ and $h \in C_1$. Indeed, any measurable $f, g : \mathbb{R} \to [0, 1]$ may be approximated (pointwise at γ_1-almost every point) from below by functions in C_0, and any measurable $h : \mathbb{R} \to [0, 1]$ may be approximated from above by functions in C_1. If we can prove (12.8) for these approximations, then it follows (by the dominated convergence theorem) for the original f, g, and h.

Now consider $f, g \in C_0$ and $h \in C_1$ satisfying $\Xi(f, g, h) \leq 0$ pointwise. For $\delta > 0$, define

$$f_\delta = \Phi(-1/\delta) \vee f \wedge \Phi(1/(3\delta))$$

$$g_\delta = \Phi(-1/\delta) \vee g \wedge \Phi(1/(3\delta))$$

$$h_\delta = \Phi\left(-\frac{1}{3\delta} \vee (\Phi^{-1}(h) + \delta) \wedge \frac{1}{\delta}\right).$$

If $\delta > 0$ is sufficiently small then $\Xi(f_\delta, g_\delta, h_\delta) \leq -\delta$ pointwise; moreover, f_δ, g_δ, and h_δ all take values in $[\Phi(-1/\delta), \Phi(1/\delta)]$, are continuous, and have limits at $\pm\infty$. Since $f_\delta \to f$ as $\delta \to 0$ (and similarly for g and h), it suffices to show that

$$\lambda\Phi^{-1}(\mathbb{E}f_\delta) + (1-\lambda)\Phi^{-1}(\mathbb{E}g_\delta) \leq \sigma\Phi^{-1}(\mathbb{E}h_\delta) \qquad (12.13)$$

for all sufficiently small $\delta > 0$.

Since f_δ has limits at $\pm\infty$, it follows that $P_\epsilon f_\delta \to f_\delta$ uniformly as $\epsilon \to 0$ (similarly for g_δ and h_δ). By taking ϵ small enough (at least as small as $\delta/2$), we can ensure that $\Xi(P_\epsilon^1 f_\delta, P_\epsilon^1 g_\delta, P_\epsilon^{\sigma^2} h_\delta) < -\epsilon$ pointwise. Now we apply Corollary 12.4.2 with $\Omega_i = [\Phi(-1/\epsilon), \Phi(1/\epsilon)]$, the function J defined in (12.12), $a = \frac{1}{2}$, and with $(\lambda_1, \lambda_2, \lambda_3) = (\lambda, 1 - \lambda, \sigma^{-1})/(1 + \sigma^{-1})$. Lemma 12.4.5 implies that the condition of Corollary 12.4.2 is satisfied. We conclude that

$$\frac{1}{2} \geq J_R(\mathbb{E}f_\delta, \mathbb{E}g_\delta, \mathbb{E}h_\delta, \infty)$$

$$= \Phi\left(R\left(\lambda\Phi^{-1}(\mathbb{E}f_\delta) + (1 - \lambda)\Phi^{-1}(\mathbb{E}g_\delta) - \sigma\Phi^{-1}(\mathbb{E}h_\delta)\right)\right),$$

which implies (12.13) and completes the proof of (12.8).

Acknowledgements We thank F. Barthe and M. Ledoux for helpful comments and for directing us to related literature.

We would also like to thank R. van Handel for pointing out to us that (12.8) corresponds more directly to a generalized form of Ehrhard's inequality contained in Theorem 1.2 of [6].

The first named author is supported by a fellowship from the Alfred P. Sloan Foundation. The second named author is supported by CAREER NSF-115171 grant and NSF grant DMS-1812240.

References

1. D. Bakry, M. Ledoux, Levy-Gromov's isoperimetric inequality for an infinite dimensional diffusion generator. Invent. Math. **123**(1), 259–281 (1996)
2. D. Bakry, I. Gentil, M. Ledoux, *Analysis and Geometry of Markov Diffusion Operators* (Springer, Berlin, 2013)
3. F. Barthe, On a reverse form of the Brascamp-Lieb inequality. Invent. Math. **134**, 335–361 (1998)
4. F. Barthe, N. Huet, On Gaussian Brunn-Minkowski inequalities. Studia Math. **191**(3), 283–304 (2009)
5. C. Borell, The Ehrhard inequality. C.R. Math. Acad. Sci. Paris **337**(10), 663–666 (2003)
6. C. Borell, Minkowski sums and Brownian exit times. Ann. Fac. Sci. Toulouse: Math. **16**(1), 37–47 (2007)
7. W.-K. Chen, N. Dafnis, G. Paouris, Improved Holder and reverse Holder inequalities for Gaussian random vectors. Adv. Math. **280**, 643–689 (2015)
8. A. Ehrhard, Symétrisation dans l'espace de Gauss. Math. Scand. **53**, 281–301 (1983)
9. P. Ivanisvili, A. Volberg, Bellman partial differential equation and the hill property for classical isoperimetric problems. Preprint. `arXiv:1506.03409` (2015)
10. R. Latala, A note on the Ehrhard inequality. Studia Math. **118**, 169–174 (1996)
11. R. Latala, On some inequalities for Gaussian measures. Proc. ICM **II**, 813–822 (2002)
12. M. Ledoux, Remarks on Gaussian noise stability, Brascamp-Lieb and Slepian inequalities, in *Geometric Aspects of Functional Analysis* (Springer, Berlin, 2014), pp. 309–333
13. J. Neeman, A multi-dimensional version of noise stability, Electron. Commun. Probab. **19**(72), 1–10 (2014)
14. R. van Handel, The Borell-Ehrhard Game. Probab. Theory Relat. Fields **170**(3–4), 555–585 (2018)

Chapter 13
Bounds on Dimension Reduction in the Nuclear Norm

Oded Regev and Thomas Vidick

Abstract For all $n \geq 1$, we give an explicit construction of $m \times m$ matrices $A_1, \ldots, A_n$ with $m = 2^{\lfloor n/2 \rfloor}$ such that for any d and $d \times d$ matrices $A'_1, \ldots, A'_n$ that satisfy

$$\|A'_i - A'_j\|_{S_1} \leq \|A_i - A_j\|_{S_1} \leq (1 + \delta)\|A'_i - A'_j\|_{S_1}$$

for all $i, j \in \{1, \ldots, n\}$ and small enough $\delta = O(n^{-c})$, where $c > 0$ is a universal constant, it must be the case that $d \geq 2^{\lfloor n/2 \rfloor - 1}$. This stands in contrast to the metric theory of commutative ℓ_p spaces, as it is known that for any $p \geq 1$, any n points in ℓ_p embed exactly in ℓ_p^d for $d = n(n-1)/2$.

Our proof is based on matrices derived from a representation of the Clifford algebra generated by n anti-commuting Hermitian matrices that square to identity, and borrows ideas from the analysis of nonlocal games in quantum information theory.

The author "Oded Regev" was supported by the Simons Collaboration on Algorithms and Geometry and by the National Science Foundation (NSF) under Grant No. CCF-1814524. Any opinions, findings, and conclusions or recommendations expressed in this material are those of the authors and do not necessarily reflect the views of the NSF.

The author "Thomas Vidick" was supported by NSF CAREER Grant CCF-1553477, a CIFAR Azrieli Global Scholar award, and the Institute for Quantum Information and Matter, an NSF Physics Frontiers Center (NSF Grant PHY-1733907).

O. Regev (✉)
Courant Institute of Mathematical Sciences, New York University, New York, NY, USA

T. Vidick
Department of Computing and Mathematical Sciences, California Institute of Technology, Pasadena, CA, USA
e-mail: vidick@cms.caltech.edu

© Springer Nature Switzerland AG 2020
B. Klartag, E. Milman (eds.), *Geometric Aspects of Functional Analysis*,
Lecture Notes in Mathematics 2266,
https://doi.org/10.1007/978-3-030-46762-3_13

279

13.1 Introduction

For $p \geq 1$ let ℓ_p denote the space of real-valued sequences $x \in \mathbb{R}^N$ with finite p-th norm $\|x\|_p = (\sum_i |x_i|^p)^{1/p}$. For any $n \geq 1$ and any $x_1, \ldots, x_n \in \ell_2$ there exist $y_1, \ldots, y_n \in \ell_2^n$ such that $\|x_i - x_j\|_2 = \|y_i - y_j\|_2$ for all $i, j \in \{1, \ldots, n\}$. This is immediate from the fact that any n-dimensional subspace of Hilbert space is isometric to ℓ_2^n. In fact, there even exist such $y_1, \ldots, y_n$ in ℓ_2^{n-1} by considering the $n - 1$ vectors $x_2 - x_1, \ldots, x_n - x_1$. We can equivalently describe this as saying that any n points in ℓ_2 can be isometrically embedded into ℓ_2^{n-1}. By considering the n-point set $\{0, e_1, \ldots, e_{n-1}\} \subseteq \mathbb{R}^{n-1}$, where e_i is the i-th canonical basis vector, the dimension $n - 1$ is easily seen to be the best possible for isometric embeddings.

The Johnson–Lindenstrauss lemma [12] establishes the striking fact that if we allow a small amount of error $\delta > 0$, a much better "dimension reduction" is possible. Namely, for any $n \geq 1$, any points $x_1, \ldots, x_n \in \ell_2$, and any $0 < \delta < 1$, there exist n points $y_1, \ldots, y_n \in \ell_2^d$ with $d = O(\delta^{-2} \log n)$ and such that for all $i, j \in \{1, \ldots, n\}$,

$$\|y_i - y_j\|_2 \leq \|x_i - x_j\|_2 \leq (1 + \delta)\|y_i - y_j\|_2 . \tag{13.1}$$

This can be described as saying that any n points in ℓ_2 can be embedded into ℓ_2^d with (bi-Lipschitz) distortion at most $1 + \delta$. We remark that this bound on d was recently shown to be tight [13] for essentially all values of δ for which the bound is nontrivial.

The situation for other norms is not as well understood. Ball [4] showed that for any $p \geq 1$ and any integer $n \geq 1$, any n points in ℓ_p embed isometrically into ℓ_p^d for $d = n(n - 1)/2$. He also showed that for $1 \leq p < 2$ this is essentially the best possible result. However, if we allow some $1 + \delta$ distortion as in (13.1), the situation again changes considerably. Specifically, for $p = 1$, Talagrand [26] (improving slightly the dependence on δ in an earlier result by Schechtman [22]) showed that for any $0 < \delta < 1$, one can embed any n points in ℓ_1 into ℓ_1^d with $d \leq C\delta^{-2} n \log n$ where here and in what follows C is a universal constant that might vary at each occurrence.[1] See also [6, 22, 27] for extensions to other p and more details. The bound was improved by Newman and Rabinovich [18] to $d \leq Cn/\delta^2$ (see [16]), and if we allow large enough distortion $D > 1$, the bound can be further reduced to $d \leq Cn/D$ [3]. In terms of lower bounds, Brinkman and Charikar [7] showed that there exist n points in ℓ_1 (in fact, in ℓ_1^n) such that any embedding with distortion $D > 1$ into ℓ_1^d requires $d \geq n^{C/D^2}$. For embeddings with distortion $1 + \delta$, Andoni et al. [2] showed a bound of $d \geq n^{1-C/\log(1/\delta)}$. See also [14, 21] for alternative proofs.

---

[1] In fact, he showed that one can even embed any n-dimensional subspace of ℓ_1 into ℓ_1^d with distortion $1 + \delta$.

Let S_1 be the space of bounded linear operators on a separable Hilbert space with finite Schatten-1 (or nuclear) norm $\|A\|_{S_1} = \sum_i \sigma_i(A)$, where $\{\sigma_i(A)\}$ are the singular values of A. We also write S_1^m for the space of linear operators acting on an m-dimensional Hilbert space, equipped with the Schatten-1 norm. Our main theorem shows that dimension reduction in this noncommutative analogue of ℓ_1 is strikingly different from that in ℓ_p spaces. Namely, there are n points that require *exponential dimension* in any embedding with sufficiently low distortion. In contrast, Ball's result mentioned above [4] shows that in ℓ_p, any n points embed isometrically into dimension $n(n-1)/2$.

Theorem 13.1 *For any $n \geq 1$, there exist $(2n+2)$ points in S_1^m, where $m = 2^{\lfloor n/2 \rfloor}$, such that any embedding into S_1^d with distortion $1 + \delta$ for $0 \leq \delta \leq Cn^{-c}$ requires $d \geq 2^{\lfloor n/2 \rfloor - 1}$, where $c, C > 0$ are universal constants.*

The space S_1 is a major object of study in many areas of mathematics and physics; see [17] for further details and references. One area where it plays an especially important role is quantum mechanics, and specifically quantum information. This area, and specifically the theory of Bell inequalities and nonlocal games, served as an inspiration for our proof and the source of our techniques.

The best previously known bound on dimension reduction in S_1 is due to Naor et al. [17], who proved a result analogous to that of Brinkman and Charikar [7]. Namely, they showed that there exist n points in S_1^n for which any embedding into S_1^d with distortion $D > 1$ requires $d \geq n^{C/D^2}$.[2] The set of points they use is Brinkman-Charikar use the diamond graph. Naor et al. use Laakso graphs, which are very similar, and they comment that the same holds also for the diamond graphs. I therefore think it's OK not to say "essentially".the one used by Brinkman and Charikar [7] through the natural identification of ℓ_1^n with the subspace of diagonal matrices in S_1^n. The effort then goes into showing that the bound in [7], which only applies to embeddings into diagonal matrices, also applies to arbitrary matrices.

In Lemma 13.19 we show that for any $0 < \delta < 1$ the metric space induced by the $(2n+2)$ points from Theorem 13.1 can be embedded with distortion $(1+\delta)$ in S_1^d for $d = n^{O(1/\delta^2)}$. Therefore, in order to We can still hope to prove that for $\delta = 0.001$, our set of points requires dimension n^{100}. Funny how in our case there really is a threshold, where either the dimension is exponential, or we don't get anything. Having something more smooth seems to require "different techniques" to bound the dimension of ε-representations of $C(n)$ for large-ish ε. obtain exponential lower bounds with constant δ one would have to use a different set of points.

Proof Overview Due to Ball's upper bound [4], our set of points cannot be in ℓ_1, and in particular, cannot be the set used in previous work [7, 17]. Instead, we

---

[2] Their result is actually much stronger, and incomparable to Theorem 13.1: they show that there is no embedding into any n^{C/D^2}-dimensional subspace of S_1 (and in fact, they even allow quotients of S_1).

introduce a new set of $2n + 2$ points in S_1^m, for $m = 2^{\lfloor n/2 \rfloor}$, and show that any embedding with $(1 + \delta)$ distortion for small enough δ requires almost as large a dimension. To achieve this we use metric conditions on the set of points to derive algebraic relations on any operators that (approximately) satisfy the conditions. We then conclude by applying results on the dimension of (approximate) representations of a suitable algebra.

We now describe our construction. Let n be an even integer. For a matrix A and an integer i, let $A^{\otimes i}$ denote the tensor product of i copies of A. Let

$$X = \begin{pmatrix} 0 & 1 \\ 1 & 0 \end{pmatrix}, \quad Y = \begin{pmatrix} 0 & i \\ -i & 0 \end{pmatrix}, \text{ and } \quad Z = \begin{pmatrix} 1 & 0 \\ 0 & -1 \end{pmatrix}.$$

For $i \in \{1, \ldots, n/2\}$ let $C_{2i-1} = X^{\otimes(i-1)} \otimes Z \otimes \mathrm{Id}^{\otimes(n/2-i)}$ and $C_{2i} = X^{\otimes(i-1)} \otimes Y \otimes \mathrm{Id}^{\otimes(n/2-i)}$. Then the matrices $C_1, \ldots, C_n$ are Hermitian operators in S_1^d, where $d = 2^{n/2}$.[3] Moreover, $C_i^2 = \mathrm{Id}$ for each $i \in \{1, \ldots, n\}$ and $\{C_i, C_j\} = C_i C_j + C_j C_i = 0$ for $i \neq j \in \{1, \ldots, n\}$. For $i \in \{1, \ldots, n\}$ let $P_{i,+}$ (resp., $P_{i,-}$) be the projection on the $+1$ (resp., -1) eigenspace of C_i. Using that $P_{i,+}$ and $P_{i,-}$ are orthogonal trace 0 projections that sum to identity, it is immediate that

$$\forall i \in \{1, \ldots, n\},$$

$$\frac{1}{d}\|P_{i,+}\|_{S_1} = \frac{1}{d}\|\mathrm{Id} - P_{i,+}\|_{S_1} = \frac{1}{d}\|P_{i,-}\|_{S_1} = \frac{1}{d}\|\mathrm{Id} - P_{i,-}\|_{S_1} = \frac{1}{2}, \tag{13.2}$$

and

$$\forall i \in \{1, \ldots, n\}, \quad \frac{1}{d}\|P_{i,+} - P_{i,-}\|_{S_1} = 1. \tag{13.3}$$

Finally, using the anti-commutation property, it follows by an easy calculation that

$$\forall i \neq j \in \{1, \ldots, n\}, \quad \forall q, r \in \{+, -\}, \quad \frac{1}{d}\|P_{i,q} - P_{j,r}\|_{S_1} = \frac{\sqrt{2}}{2}. \tag{13.4}$$

Our main result is that (13.2)–(13.4) characterize the algebraic structure of any operators that satisfy those metric relations, even up to distortion $(1 + \delta)$ for small enough $\delta = O(n^{-c})$. Using labels O and σ to represent 0 and Id/d, and X_i and Y_i to represent $P_{i,+}/d$ and $P_{i,-}/d$ respectively, we show the following.

---

[3]For a construction over the reals, consider $C_{2i-1}' = C_{2i-1} \otimes \mathrm{Id}$ and $C_{2i}' = C_{2i} \otimes Y$. For even values of n congruent to 4 or 6 mod 8 the doubling of the dimension is necessary [19].

Theorem 13.2 *Let* $n, d \geq 1$ *be integers,* $0 \leq \delta \leq 1$, *and* O, σ *and* $X_1, Y_1, \ldots, X_n, Y_n$ *operators on* $\mathbb{C}^d$ *satisfying that for all* $i \in \{1, \ldots, n\}$,

$$1 - \delta \leq \|\sigma - O\|_{S_1} \leq 1 + \delta \, ,$$

$$\|X_i - O\|_{S_1} + \|\sigma - X_i\|_{S_1} \leq 1 + \delta \, ,$$

$$\|Y_i - O\|_{S_1} + \|\sigma - Y_i\|_{S_1} \leq 1 + \delta \, ,$$

$$\|X_i - Y_i\|_{S_1} \geq 1 - \delta \, ,$$

and for all $1 \leq i < j \leq n$,

$$\min \left\{ \|X_i - X_j\|_{S_1}, \|X_i - Y_j\|_{S_1}, \|Y_i - X_j\|_{S_1}, \|Y_i - Y_j\|_{S_1} \right\} \geq (1 - \delta) \frac{\sqrt{2}}{2} \, . \tag{13.5}$$

Then there is a universal constant $C > 0$ *and for* $i \in \{1, \ldots, n\}$ *orthogonal projections* $P_{i,+}$ *and* $P_{i,-}$ *on* $\mathbb{C}^{d'}$ *for some* $d' \leq d$ *such that* $P_{i,+} + P_{i,-} = \mathrm{Id}$ *such that if* $A_i = P_{i,+} - P_{i,-}$ *then*

$$\forall i \neq j \in \{1, \ldots, n\} \, , \qquad \frac{1}{d'} \left\| A_i A_j + A_j A_i \right\|_{S_2}^2 \leq C n^2 \delta^{1/16} \, . \tag{13.6}$$

Note that the theorem does not assume that the X_i and Y_i are positive semidefinite, nor even that they are Hermitian; our proof shows that the metric constraints are sufficient to impose these conditions, up to a small approximation error. Similarly, while we think of O as the zero matrix and of σ as the scaled identity matrix, these conditions are not imposed a priori and have to be derived (which is very easy in the case of O but less so in the case of σ). The proof of the theorem explicitly shows how to construct the projections $P_{i,+}$, $P_{i,-}$ from X_i, Y_i, O, and σ.

Theorem 13.1 follows from Theorem 13.2 by applying known lower bounds on the dimension of (approximate) representations of the Clifford algebra that is generated by n Hermitian anti-commuting operators[4]; we give an essentially self-contained treatment in Sect. 13.6.

The proof of Theorem 13.2 is inspired by the theory of self-testing in quantum information theory. We interpret conditions such as (13.5) as requirements on the trace distance (which, up to a factor 2 scaling, is the name used for the nuclear norm in quantum information) between post-measurement states that result from the measurement of one half of a bipartite quantum entangled state. This allows us to draw an analogy between metric conditions such as those in Theorem 13.2 and constraints expressed by nonlocal games such as the CHSH game. Although this interpretation can serve as useful intuition for the proof, we give a self-contained proof that makes no reference to quantum information. We note that the relevance

---

[4]Note that the norm in (13.6) is the Schatten-2 norm.

of dimension reduction for Schatten-1 spaces for quantum information has been recognized before; e.g., Harrow et al. [10] show limitations on dimension reduction maps that are restricted to be quantum channels (a result mostly superseded by [17]).

Open Questions We are currently not aware of *any* upper bound on the dimension d required to embed any n points in S_1 into S_1^d with, say, constant distortion. Proving such a bound would be interesting.

Regarding possible improvements to our main theorem, our result requires the distortion of the embedding to be sufficiently small; specifically, δ needs to be at most inverse polynomial in n. It is open whether our result can be extended to larger distortions.

The connection with quantum information and nonlocal games suggests that additional strong lower bounds may be achievable. For example, is it possible to adapt the results from [11, 23] to construct a constant number of points in S_1 such that any embedding with distortion $(1 + \delta)$ in S_1^d requires $d \geq 2^{1/\delta^c}$ for some constant $c > 0$?

We are not aware of results specifically addressing other Schatten spaces. Nevertheless, here are some statements that follow easily from known results. First, any set of n points in S_2 trivially embeds into $S_2^{\lceil \sqrt{n-1} \rceil}$ by first embedding the points isometrically into ℓ_2^{n-1}, as discussed earlier. Second, for S_∞, it is well known that any n point metric isometrically embeds in ℓ_∞^{n-1} and hence also in S_∞^{n-1}; it is possible that this could be improved. If we allow some distortion, a result by Matoušek [15] shows that for $D \geq 1$, an arbitrary n-point metric space embeds with distortion D in ℓ_∞^k, for some $k = O(Dn^{1/\lfloor (D+1)/2 \rfloor} \ln n)$ (see also [1] for more general trade-offs between distortion and dimension for embedding arbitrary metric space into ℓ_p, $1 \leq p \leq \infty$).

13.2 Preliminaries

For a matrix $A \in \mathbb{C}^{d \times d}$ we write $\|A\|_{S_1}$ for the Schatten-1 norm (the sum of the singular values). For the Schatten 2-norm (also known as the Frobenius norm) we use $\|A\|_F$ instead of $\|A\|_{S_2}$, and introduce the dimension-normalized norm $\|A\|_f = d^{-1/2}\|A\|_F$. We write $\|A\|_{S_\infty}$ for the operator norm (the largest singular value). We often consider terms of the form $\|T\sigma^{1/2}\|_F$ for a Hermitian matrix T and a positive semidefinite matrix σ; notice that the square of this norm equals $\mathrm{Tr}(T^2\sigma)$. For A, B square matrices we write $[A, B] = AB - BA$ and $\{A, B\} = AB + BA$ for the commutator and anti-commutator respectively. We write $U(d)$ for the set of unitary matrices in $\mathbb{C}^{d \times d}$. We use the term "observable" to refer to any Hermitian operator that squares to identity, and the term "projection" to refer to an orthogonal projection.

We will often use that for any A and B,

$$\|AB\|_{S_1} \leq \|A\|_{S_\infty}\|B\|_{S_1},$$

and similarly with Schatten-1 replaced by the Frobenius norm (see, e.g., [5, (IV.40)]).

Lemma 13.3 (Cauchy–Schwarz) *For all matrices A, B,*

$$\|AB\|_{S_1} \le \|A\|_F \|B\|_F .$$

Proof By definition,

$$\|AB\|_{S_1} = \sup_U \text{Tr}(UAB) \le \sup_U \|UA\|_F \|B\|_F = \|A\|_F \|B\|_F ,$$

where the supremum is over all unitary matrices, and the inequality follows from the Cauchy–Schwarz inequality.   □

13.3   Certifying Projections

In this section we prove Proposition 13.4, showing that metric constraints on a triple of operators (X, Y, σ), where σ is assumed to be positive semidefinite of trace 1, can be used to enforce that the pair (X, Y) is close to a "resolution of the identity," in the sense that there exists a pair (P, Q) of orthogonal projections such that $P + Q = \text{Id}$ and $X \approx \sigma^{1/2} P \sigma^{1/2}$, $Y \approx \sigma^{1/2} Q \sigma^{1/2}$. The proposition also shows that P, Q approximately commute with σ.

Proposition 13.4 *Let σ be positive semidefinite with trace 1. Suppose that X, Y satisfy the following constraints, for some $0 \le \delta \le 1$:*

$$\|X\|_{S_1} + \|\sigma - X\|_{S_1} \le 1 + \delta , \tag{13.7}$$

$$\|Y\|_{S_1} + \|\sigma - Y\|_{S_1} \le 1 + \delta , \tag{13.8}$$

$$\|X - Y\|_{S_1} \ge 1 - \delta . \tag{13.9}$$

Then there exist orthogonal projections P, Q such that $P + Q = \text{Id}$ and

$$\max \left\{ \|X - \sigma^{1/2} P \sigma^{1/2}\|_{S_1}, \|Y - \sigma^{1/2} Q \sigma^{1/2}\|_{S_1} \right\} = O(\delta^{1/8}) , \tag{13.10}$$

$$\max \left\{ \|[P, \sigma^{1/2}]\|_F , \|[Q, \sigma^{1/2}]\|_F \right\} = O(\delta^{1/8}) . \tag{13.11}$$

For intuition regarding Proposition 13.4, consider the case where $\delta = 0$, and where X, Y, σ are one-dimensional, i.e., scalar complex numbers, $X = x, Y = y$, and $\sigma = 1$. Then the first two conditions (13.7) and (13.8) imply that x, y are real and $x, y \in [0, 1]$. The third condition (13.9) then implies that $x, y \in \{0, 1\}$ and $x + y = 1$. The proof of Proposition 13.4 follows the same outline, adapted to higher-dimensional operators. The main idea is to argue that the projections P, Q

on the positive and negative eigenspace of $X - Y$ respectively approximately block-diagonalize X, Y, and σ.

The proof is broken down into a sequence of lemmas. The first lemma shows that X is close to its Hermitian part.

Lemma 13.5 (Hermitianity) *Let σ be positive semidefinite such that $\mathrm{Tr}(\sigma) = 1$, and X such that (13.7) holds, for some $0 \leq \delta \leq 1$. Then $\|X - X_h\|_{S_1} \leq 3\sqrt{\delta}$, where $X_h = \frac{1}{2}(X + X^*)$ is the Hermitian part of X.*

Proof By (13.7),

$$\Re(\mathrm{Tr}(X)) = 1 - \Re(\mathrm{Tr}(\sigma - X)) \geq 1 - \|\sigma - X\|_{S_1} \geq \|X\|_{S_1} - \delta . \quad (13.12)$$

Let $X = X_h + X_a$ be the decomposition of X into Hermitian and anti-Hermitian parts. Then $\Re(\mathrm{Tr}(X_a)) = 0$, so $\mathrm{Tr}(X_h) \geq \|X\|_{S_1} - \delta$. Let W be a unitary such that $\mathrm{Tr}(WX_a) = \|X_a\|_{S_1}$. Note that replacing $W \mapsto (W - W^*)/2$ we may assume that W is anti-Hermitian (of operator norm at most 1), so (iW) is Hermitian. Let $0 \leq \alpha \leq 1$ be a parameter to be determined. Then all eigenvalues of $\mathrm{Id} + \alpha W$ are in the complex interval $[1 - \alpha i, 1 + \alpha i]$ and therefore $U = (\mathrm{Id} + \alpha W)/(1 + \alpha^2)^{1/2}$ has operator norm at most 1. Then

$$\|X\|_{S_1} \geq |\mathrm{Tr}(UX)| \geq \Re\big(\mathrm{Tr}(UX_h) + \mathrm{Tr}(UX_a)\big)$$

$$= \frac{1}{(1 + \alpha^2)^{1/2}}\big(\mathrm{Tr}(X_h) + \alpha\|X_a\|_{S_1}\big)$$

$$\geq \frac{1}{(1 + \alpha^2)^{1/2}}\big(\|X\|_{S_1} - \delta + \alpha\|X_a\|_{S_1}\big) ,$$

which shows that $\|X_a\|_{S_1} \leq \alpha\|X\|_{S_1} + \delta/\alpha$. Choosing $\alpha = \sqrt{\delta}$ and using $\|X\|_{S_1} \leq (1 + \delta)$ gives $\|X_a\|_{S_1} \leq 3\sqrt{\delta}$. $\qquad\square$

Lemma 13.6 *Let X and Y be Hermitian matrices satisfying*

$$\|X\|_{S_1} + \|Y\|_{S_1} \leq 1 + \delta ,$$

$$\|X - Y\|_{S_1} \geq 1 - \delta ,$$

$$\mathrm{Tr}(X^-) \leq \delta, \text{ and } \mathrm{Tr}(Y^-) \leq \delta$$

for some $0 \leq \delta \leq 1$ where X^- denotes the negative part of X in the decomposition $X = X^+ - X^-$ and similarly for Y. Then, if P denotes the projection on the positive eigenspace of $X - Y$ and $Q = \mathrm{Id} - P$, we have

$$\mathrm{Tr}(PX) \geq \|X\|_{S_1} - 4\delta , \quad \mathrm{Tr}(QY) \geq \|Y\|_{S_1} - 4\delta .$$

Proof We have

$$1 - \delta \leq \|X - Y\|_{S_1} = \mathrm{Tr}(P(X - Y)) - \mathrm{Tr}(Q(X - Y))$$

$$\leq \mathrm{Tr}(PX) + \mathrm{Tr}(QY) + 2\delta$$

$$\leq \mathrm{Tr}(PX) + \|Y\|_{S_1} + 2\delta$$

$$\leq \mathrm{Tr}(PX) + 1 + 3\delta - \|X\|_{S_1},$$

where in the second inequality we used that $\mathrm{Tr}(PY) \geq -\mathrm{Tr}(Y^-) \geq -\delta$, which holds for any projection P, and similarly for $\mathrm{Tr}(QX)$. As a result, we get that

$$\mathrm{Tr}(PX) \geq \|X\|_{S_1} - 4\delta,$$

and similarly for $\mathrm{Tr}(QY)$.   □

Lemma 13.7 *Let X be a Hermitian matrix and P a projection satisfying*

$$\|X\|_{S_1} \leq 1,$$

$$\mathrm{Tr}(X^-) \leq \delta, \tag{13.13}$$

$$\mathrm{Tr}(PX) \geq \|X\|_{S_1} - \delta, \tag{13.14}$$

for some $0 \leq \delta \leq 1$. Then,

$$\|PXP - X\|_{S_1} \leq O(\sqrt{\delta}).$$

Proof The assumption (13.13) is equivalent to $\|X - X^+\|_{S_1} \leq \delta$, which implies that $\|PXP - PX^+P\|_{S_1} \leq \delta$. Therefore, by the triangle inequality, it suffices to prove that

$$\|PX^+P - X^+\|_{S_1} \leq O(\sqrt{\delta}). \tag{13.15}$$

Using the Cauchy–Schwarz inequality,

$$\|(\mathrm{Id} - P)X^+\|_{S_1}^2 \leq \|(\mathrm{Id} - P)(X^+)^{1/2}\|_F^2 \|(X^+)^{1/2}\|_F^2$$

$$= \mathrm{Tr}((\mathrm{Id} - P)X^+)\|X^+\|_{S_1}$$

$$= (\mathrm{Tr}(X^+) - \mathrm{Tr}(PX) - \mathrm{Tr}(PX^-))\|X^+\|_{S_1}$$

$$\leq \delta\|X\|_{S_1} \leq \delta,$$

where the second line uses that $(\mathrm{Id} - P)$ is a projection and the fourth uses $\mathrm{Tr}(X^+) \leq \|X\|_{S_1}$ for the first term and (13.14) for the second. To conclude, use the triangle inequality to write

$$\|PX^+P - X^+\|_{S_1} \leq \|(P - \mathrm{Id})X^+P\|_{S_1} + \|X^+(\mathrm{Id} - P)\|_{S_1} \leq 2\|(\mathrm{Id} - P)X^+\|_{S_1}.$$

$\square$

Lemma 13.8 *Let σ, X, and Y satisfy the assumptions of Proposition 13.4 for some $0 \leq \delta \leq 1$. Then there exist orthogonal projections P, Q such that $P + Q = \mathrm{Id}$ and*

$$\|X - P\sigma P\|_{S_1} \leq O(\delta^{1/4}) \text{ and } \|Y - Q\sigma Q\|_{S_1} \leq O(\delta^{1/4}). \tag{13.16}$$

Moreover, there exists a positive semidefinite ρ that commutes with P and Q and that satisfies $\|\rho - \sigma\|_{S_1} \leq O(\delta^{1/4})$.

Proof Using Lemma 13.5, we can replace X and Y with their Hermitian parts, and have Eqs. (13.7)–(13.9) still hold with $O(\sqrt{\delta})$ in place of δ. By summing Eqs. (13.7) and (13.8), and noting by the triangle inequality that $\|\sigma - X\|_{S_1} + \|\sigma - Y\|_{S_1} \geq \|X - Y\|_{S_1} \geq 1 - O(\sqrt{\delta})$, we get that $\|X\|_{S_1} + \|Y\|_{S_1} \leq 1 + O(\sqrt{\delta})$. Moreover,

$$\mathrm{Tr}(X^-) = \|X^+\|_{S_1} - \mathrm{Tr}(X)$$

$$\leq 1 + O(\sqrt{\delta}) - \|\sigma - X\|_{S_1} - \mathrm{Tr}(X)$$

$$\leq 1 + O(\sqrt{\delta}) - \mathrm{Tr}(\sigma - X) - \mathrm{Tr}(X) = O(\sqrt{\delta})$$

and similarly for Y. We can therefore apply Lemma 13.6 and obtain that if P is the projection on the positive eigenspace of $X - Y$ and $Q = \mathrm{Id} - P$,

$$\mathrm{Tr}(PX) \geq \|X\|_{S_1} - O(\sqrt{\delta}) \text{ and } \mathrm{Tr}(QY) \geq \|Y\|_{S_1} - O(\sqrt{\delta}).$$

Applying Lemma 13.7 to X (scaled by a factor at most $(1 + \delta)$ so that the condition $\|X\|_{S_1} \leq 1$ is satisfied) and P, we get that

$$\|PXP - X\|_{S_1} = O(\delta^{1/4}) \text{ and } \|QYQ - Y\|_{S_1} = O(\delta^{1/4}). \tag{13.17}$$

Notice that the set of constraints in Eqs. (13.7)–(13.9) is invariant under replacing the pair (X, Y) with $(\sigma - Y, \sigma - X)$. Moreover, our assumption that X and Y are Hermitian implies that $\sigma - X$ and $\sigma - Y$ are also Hermitian. Therefore, the exact same argument as above applies also to $\sigma - X$ and $\sigma - Y$ and we conclude that

$$\|P(\sigma - Y)P - (\sigma - Y)\|_{S_1} = O(\delta^{1/4}) \text{ and}$$

$$\|Q(\sigma - X)Q - (\sigma - X)\|_{S_1} = O(\delta^{1/4}). \tag{13.18}$$

Notice that we used here the fact that $(\sigma - Y) - (\sigma - X) = X - Y$ and therefore the projections P and Q obtained when we apply Lemma 13.6 to X and Y are identical to those obtained when we apply it to $\sigma - Y$ and $\sigma - X$.

From (13.18), and since $PQ = QP = 0$, we obtain that

$$\|P\sigma P - PXP\|_{S_1} = \|PQ(\sigma - X)QP - P(\sigma - X)P\|_{S_1}$$
$$\leq \|Q(\sigma - X)Q - (\sigma - X)\|_{S_1} = O(\delta^{1/4}) \,.$$

Together with (13.17) and the triangle inequality, this proves (13.16).

To prove the last part of the lemma, let $\tilde{\rho} = PXP + Q(\sigma - X)Q$. Using again that $PQ = QP = 0$ and that P, Q are projections, we see that $\tilde{\rho}$ commutes with P and Q. By Eqs. (13.17) and (13.18) and the triangle inequality, $\|\tilde{\rho} - \sigma\|_{S_1} = O(\delta^{1/4})$. Finally, we define ρ to be the positive part of $\tilde{\rho}$, which due to the block diagonal form of $\tilde{\rho}$ still commutes with P and Q. We have $\|\rho - \sigma\|_{S_1} = O(\delta^{1/4})$ since

$$\|\rho - \tilde{\rho}\|_{S_1} = \frac{1}{2}(\|\tilde{\rho}\|_{S_1} - \mathrm{Tr}(\tilde{\rho})) \leq \frac{1}{2}(\|\sigma\|_{S_1} - \mathrm{Tr}(\sigma)) + O(\delta^{1/4}) = O(\delta^{1/4}) \,,$$

where the last equality uses that σ is positive semidefinite. $\qquad\square$

We conclude by giving the proof of Proposition 13.4.

Proof of Proposition 13.4 Let P, Q, and ρ be as guaranteed by Lemma 13.8. Using the Powers-Stormer inequality $\|\sqrt{R} - \sqrt{S}\|_F \leq \|R - S\|_{S_1}^{1/2}$ for positive semidefinite R, S (see, e.g., [5, (X.7)]), it follows that

$$\|\rho^{1/2} - \sigma^{1/2}\|_F \leq \|\rho - \sigma\|_{S_1}^{1/2} = O(\delta^{1/8}) \,. \tag{13.19}$$

As a result, using the triangle inequality and Cauchy–Schwarz,

$$\|\sigma^{1/2}P\sigma^{1/2} - \rho^{1/2}P\rho^{1/2}\|_{S_1} \leq \|(\sigma^{1/2} - \rho^{1/2})P\sigma^{1/2}\|_{S_1}$$
$$+ \|\rho^{1/2}P(\sigma^{1/2} - \rho^{1/2})\|_{S_1} \leq O(\delta^{1/8}) \,,$$

where we used that $\|P\sigma^{1/2}\|_F \leq \|\sigma^{1/2}\|_F = 1$ and $\|P\rho^{1/2}\|_F \leq \|\rho^{1/2}\|_F = 1 + O(\delta^{1/4})$. But ρ commutes with P and therefore $\rho^{1/2}P\rho^{1/2} = P\rho P$, and we complete the proof of (13.10) by noting that

$$\|P\rho P - P\sigma P\|_{S_1} \leq \|\rho - \sigma\|_{S_1} = O(\delta^{1/4}) \,.$$

To prove (13.11), notice that by (13.19) and the triangle inequality,

$$\|P\sigma^{1/2} - \sigma^{1/2}P\|_F \leq \|P\rho^{1/2} - \rho^{1/2}P\|_F + O(\delta^{1/8}) \,,$$

but the latter norm is zero since P commutes with ρ. $\qquad\square$

Fig. 13.1 Vectors satisfying
the metric constraints

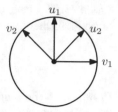

13.4 Certifying Anti-commutation

In this section we prove Proposition 13.11. The proposition shows that assuming two pairs of operators (X_1, Y_1) and (X_2, Y_2) satisfying the assumptions of Proposition 13.4 satisfy additional metric constraints, the corresponding projections (P_1, Q_1) and (P_2, Q_2) are such that the operators $P_1 - Q_1$ and $P_2 - Q_2$ have small anti-commutator, in the appropriate norm. For intuition, consider the case of operators in two dimensions, and $\sigma = \mathrm{Id}$. Then, Proposition 13.4 shows that we can think of (X_1, Y_1) and (X_2, Y_2) as two pairs of orthogonal projections. Assuming that these projections are of rank 1 (as would follow from the constraint (13.22) below), we can think of them as two pairs of orthonormal bases (u_1, v_1) and (u_2, v_2) of $\mathbb{C}^2$. Suppose we were to impose that these vectors satisfy the four Euclidean conditions

$$\|u_1 - u_2\|_2^2 = \|u_1 - v_2\|_2^2 = \|v_1 - u_2\|_2^2 = \|v_1 + v_2\|_2^2 = 2 - \sqrt{2}.$$
(13.20)

By expanding the squares, it is not hard to see that these conditions imply that the bases must form an angle of $\frac{\pi}{4}$ as shown in Fig. 13.1.[5] In particular, the reflection operators $A_i = u_i u_i^* - v_i v_i^*$, $i \in \{1, 2\}$, anti-commute. Proposition 13.11 adapts this observation to the trace norm between matrices in any dimension, and small error. We start with two technical claims.

Claim 13.9 Let $A, B \neq 0$ be such that $\Re(\mathrm{Tr}(A^* B)) \geq (1 - \delta)\|A\|_F \|B\|_F$ for some $0 \leq \delta \leq 1$. Let $\alpha = \|A\|_F / \|B\|_F$. Then $\|A - \alpha B\|_F \leq \sqrt{2\delta}\|A\|_F$.

Proof Expand

$$\begin{aligned}
\|A - \alpha B\|_F^2 &= \|A\|_F^2 + \alpha^2 \|B\|_F^2 - 2\alpha\Re(\mathrm{Tr}(A^* B)) \\
&\leq \|A\|_F^2 + \alpha^2 \|B\|_F^2 - 2\alpha(1 - \delta)\|A\|_F \|B\|_F \\
&= 2\delta\|A\|_F^2 .
\end{aligned}$$

□

---

[5]These conditions underlie the rigid properties of the famous CHSH inequality from quantum information [25, 28].

Claim 13.10 Let R be Hermitian and σ positive semidefinite such that $\mathrm{Tr}(\sigma) = 1$. Suppose further that $\|\sigma^{1/2}R\sigma^{1/2}\|_{S_1} \geq (1-\delta)\sqrt{\mu}$, where $\mu = \mathrm{Tr}(R^2\sigma)$. Then

$$\left\|(R^2 - \mu\mathrm{Id})\sigma^{1/2}\right\|_F^2 = O\left(\sqrt{\delta}\|R\|_{S_\infty}^2\right)\mu \, .$$

Proof Let U be a unitary such that $U\sigma^{1/2}R\sigma^{1/2} = |\sigma^{1/2}R\sigma^{1/2}|$ as given by the polar decomposition. Let $A = R\sigma^{1/2}$ and $B = \sigma^{1/2}U$, and notice that $\|A\|_F = \sqrt{\mu}$ and $\|B\|_F = 1$. Then

$$\mathrm{Tr}(A^*B) = \mathrm{Tr}(\sigma^{1/2}R\sigma^{1/2}U) = \mathrm{Tr}|\sigma^{1/2}R\sigma^{1/2}| \geq (1-\delta)\sqrt{\mu} \, ,$$

by assumption. Applying Claim 13.9 it follows that

$$\|R\sigma^{1/2} - \sqrt{\mu}\sigma^{1/2}U\|_F^2 \leq 2\delta\mu \, . \tag{13.21}$$

By the triangle inequality,

$$\begin{aligned}
\|R\sigma R - \mu\sigma\|_{S_1} &\leq \|(R\sigma^{1/2} - \sqrt{\mu}\sigma^{1/2}U)\sigma^{1/2}R\|_{S_1} \\
&\quad + \|\sqrt{\mu}\sigma^{1/2}U(\sqrt{\mu}U^*\sigma^{1/2} - \sigma^{1/2}R)\|_{S_1} \\
&\leq 2\sqrt{2\delta}\mu \, ,
\end{aligned}$$

where the second line uses the Cauchy–Schwarz inequality and (13.21). Thus

$$\begin{aligned}
\mathrm{Tr}\left((R^2 - \mu\mathrm{Id})^2\sigma\right) &= \mathrm{Tr}(R^4\sigma) - 2\mu\mathrm{Tr}(R^2\sigma) + \mu^2 \\
&= \mathrm{Tr}(R^2(R\sigma R - \mu\sigma)) \\
&\leq 2\sqrt{2\delta}\|R\|_{S_\infty}^2\mu \, .
\end{aligned}$$

$\square$

Proposition 13.11 *Let σ be positive semidefinite such that $\mathrm{Tr}(\sigma) = 1$. Let X_1, Y_1 and X_2, Y_2 be operators satisfying the assumptions of Proposition 13.4 for some $0 \leq \delta \leq 1$, and P_1, Q_1 and P_2, Q_2 be as in the conclusion of the proposition. Suppose further that*[6]

$$\min\left\{\|X_1 - X_2\|_{S_1}, \|X_1 - Y_2\|_{S_1}, \|Y_1 - X_2\|_{S_1}, \|Y_1 - Y_2\|_{S_1}\right\} \geq (1-\delta)\frac{\sqrt{2}}{2} \, . \tag{13.22}$$

---

[6]The reason that the "+" sign in the last term in (13.20) is replaced by a "−" in (13.22) is that one should think of X_i, Y_j as the projections on u_i, v_j.

For $i \in \{1, 2\}$ let $A_i = P_i - Q_i$. Then A_1, A_2 are observables[7] such that

$$\left\| \{A_1, A_2\} \sigma^{1/2} \right\|_F = O\left(\delta^{1/32}\right) . \tag{13.23}$$

Proof Using first (13.10) and then the Cauchy–Schwarz inequality and $\mathrm{Tr}(\sigma) = 1$,

$$\frac{1 - 2\delta}{2} \leq \|X_1 - X_2\|_{S_1}^2 \leq \left\| \sigma^{1/2} (P_1 - P_2) \sigma^{1/2} \right\|_{S_1}^2 + O\left(\delta^{1/8}\right)$$

$$\leq \mathrm{Tr}\left((P_1 - P_2)^2 \sigma\right) + O\left(\delta^{1/8}\right) \tag{13.24}$$

and similarly for the three other pairs ($X_1 - Y_2$, $Y_1 - X_2$, and $Y_1 - Y_2$). Summing those four inequalities, we get

$$2(1 - 2\delta) \leq \mathrm{Tr}\left((P_1 - P_2)^2 \sigma\right) + \mathrm{Tr}\left((P_1 - Q_2)^2 \sigma\right) + \mathrm{Tr}\left((Q_1 - P_2)^2 \sigma\right)$$

$$+ \mathrm{Tr}\left((Q_1 - Q_2)^2 \sigma\right) + O\left(\delta^{1/8}\right)$$

$$= 2\left(\mathrm{Tr}\left((P_1 - P_2)^2 \sigma\right) + \mathrm{Tr}\left((P_1 + P_2 - \mathrm{Id})^2 \sigma\right)\right) + O\left(\delta^{1/8}\right)$$

$$= 2 + O\left(\delta^{1/8}\right) ,$$

where the first equality uses $Q_1 - Q_2 = P_2 - P_1$ and $Q_1 - P_2 = Q_2 - P_1$, and the second uses $\mathrm{Tr}(\sigma) = 1$. Therefore all inequalities in (13.24) must be equalities, up to $O(\delta^{1/8})$. In particular, both

$$\left\| \sigma^{1/2} (P_1 - P_2) \sigma^{1/2} \right\|_{S_1}^2 \qquad \text{and} \qquad \mathrm{Tr}\left((P_1 - P_2)^2 \sigma\right)$$

are within $O(\delta^{1/8})$ of $\frac{1}{2}$. Applying Claim 13.10 with $R = P_1 - P_2$ it follows that

$$\left\| \left((P_1 - P_2)^2 - \frac{1}{2}\mathrm{Id}\right) \sigma^{1/2} \right\|_F^2 = O\left(\delta^{1/16}\right) , \tag{13.25}$$

and similar bounds for the three other pairs. To conclude the proof, use the triangle inequality, Eq. (13.25), and the observation that by writing $A_1 = 2P_1 - \mathrm{Id}$ and $A_2 = 2P_2 - \mathrm{Id}$,

$$\{A_1, A_2\} = 4P_1 P_2 + 4P_2 P_1 - 4P_1 - 4P_2 + 2\mathrm{Id}$$

$$= 2\left((P_1 - Q_2)^2 - (P_1 - P_2)^2\right) .$$

$\square$

---

[7]Recall that an observable is a Hermitian operator that squares to identity.

13.5 Replacing σ with Identity

The anti-commutation relations obtained in Proposition 13.11 involve the arbitrary positive semidefinite operator σ. In this section we show that up to a small loss of parameters we may without loss of generality assume that $\sigma = \mathrm{Id}$. Intuitively, this follows from the approximate commutation relation

$$\left\|[A, \sigma^{1/2}]\right\|_F = O\left(\delta^{1/8}\right), \tag{13.26}$$

which follows immediately from the definition of the observable $A = P - Q$ and (13.11). If σ has two eigenvalues with a big gap between them, then it is not hard to see that A satisfying (13.26) must have a corresponding approximate block structure, in which case we can restrict to one of the blocks and obtain $\sigma = \mathrm{Id}$ as desired. The difficulty is in carefully handling the general case, where some eigenvalues of σ might be closely spaced. The following lemma does this, using an elegant argument borrowed from [24].

Lemma 13.12 *Let σ be a positive semidefinite matrix with trace 1, and $T_1, \ldots, T_k$ and $X_1, \ldots, X_\ell$ Hermitian operators such that $X_j^2 = \mathrm{Id}$ for all $j \in \{1, \ldots, \ell\}$. Let*

$$\varepsilon = \frac{1}{k} \sum_{i=1}^{k} \left\|T_i \sigma^{1/2}\right\|_F^2 \qquad and \qquad \delta = \frac{1}{\ell} \sum_{j=1}^{\ell} \left\|[X_j, \sigma^{1/2}]\right\|_F^2.$$

Then there exists a nonzero projection R such that

$$\frac{1}{k} \sum_{i=1}^{k} \left\|T_i R\right\|_F^2 = O(\varepsilon)\mathrm{Tr}(R) \qquad and \qquad \frac{1}{\ell} \sum_{j=1}^{\ell} \left\|[X_j, R]\right\|_F^2 = O(\delta^{1/2})\mathrm{Tr}(R).$$

Proof The proof relies on two simple claims. For a Hermitian matrix ρ and $a \geq 0$, let $\chi_{\geq a}(\rho)$ denote the projection on the direct sum of eigenspaces of ρ with eigenvalues at least a. The first claim appears as [24, Lemma 5.6].

Claim 13.13 Let ρ be positive semidefinite. Then

$$\int_0^{+\infty} \chi_{\geq \sqrt{a}}(\rho)\, da = \rho^2.$$

The second is due to Connes [8, Lemma 1.2.6]. We state the claim as it appears in [24, Lemma 5.5].

Claim 13.14 ([8], Lemma 1.2.6) Let ρ, ρ' be positive semidefinite. Then

$$\int_0^{+\infty} \left\|\chi_{\geq \sqrt{a}}(\rho) - \chi_{\geq \sqrt{a}}(\rho')\right\|_F^2\, da \leq \|\rho - \rho'\|_F \|\rho + \rho'\|_F.$$

Both claims can be proven by direct calculation, writing out the spectral decomposition of ρ, ρ' and using Fubini's theorem (exchanging summation indices). The proof is given in [24].

Applying Claim 13.13 with $\rho = \sigma^{1/2}$,

$$\frac{1}{k} \int_0^{+\infty} \sum_{i=1}^k \left\| T_i \, \chi_{\geq \sqrt{a}}(\sigma^{1/2}) \right\|_F^2 \, da = \frac{1}{k} \sum_{i=1}^k \left\| T_i \sigma^{1/2} \right\|_F^2$$

$$= \varepsilon \int_0^{+\infty} \mathrm{Tr}\big(\chi_{\geq \sqrt{a}}(\sigma^{1/2})\big) \, da \,, \qquad (13.27)$$

where the first equality uses $\| T_i \, \chi_{\geq \sqrt{a}}(\sigma^{1/2}) \|_F^2 = \mathrm{Tr}(T_i^2 \, \chi_{\geq \sqrt{a}}(\sigma^{1/2}))$ and the second equality follows from Claim 13.13 and $\mathrm{Tr}(\sigma) = 1$. Applying Claim 13.14 with $\rho = \sigma^{1/2}$ and $\rho' = X_j \sigma^{1/2} X_j$, and using that X_j is Hermitian and unitary,

$$\frac{1}{\ell} \int_0^{+\infty} \sum_{j=1}^\ell \left\| [X_j, \chi_{\geq \sqrt{a}}(\sigma^{1/2})] \right\|_F^2 \, da \leq \frac{1}{\ell} \sum_{j=1}^\ell \left\| [X_j, \sigma^{1/2}] \right\|_F \left\| \{X_j, \sigma^{1/2}\} \right\|_F$$

$$\leq O(\delta^{1/2})$$

$$= O(\delta^{1/2}) \int_0^{+\infty} \mathrm{Tr}\big(\chi_{\geq \sqrt{a}}(\sigma^{1/2})\big) \, da \,,$$

$$(13.28)$$

where the second inequality that by the triangle inequality, $\| \{X_j, \sigma^{1/2}\} \|_F \leq \| X_j \sigma^{1/2} \|_F + \| \sigma^{1/2} X_j \|_F = 2$, and Jensen's inequality. Adding $(1/\varepsilon)$ times (13.27) and $(1/\delta^{1/2})$ times (13.28), there exists an $a \geq 0$ such that both inequalities are satisfied simultaneously (up to a multiplicative constant factor) with a nonzero right-hand side, for that a. Then $R = \chi_{\geq \sqrt{a}}(\sigma^{1/2})$ is a projection that satisfies the conclusions of the lemma. $\qquad \square$

Combining Proposition 13.11 and Lemma 13.12, we obtain the following.

Proposition 13.15 *Let* $n, d \geq 1$ *be integers,* $0 \leq \delta \leq 1$, $X_1, Y_1, \ldots, X_n, Y_n$ *operators on* $\mathbb{C}^d$, *and* σ *positive semidefinite of trace* 1, *such that for each* $i \in \{1, \ldots, n\}$, σ, X_i, Y_i *satisfy* (13.7)–(13.9), *and such that for each* $i \neq j \in \{1, \ldots, n\}$, (X_i, Y_i, X_j, Y_j) *satisfy* (13.22). *Then there exist a* $d' \leq d$ *and observables* $A_1', \ldots, A_n'$ *on* $\mathbb{C}^{d'}$ *such that*

$$\frac{2}{n(n-1)} \sum_{1 \leq i < j \leq n} \left\| \{A_i', A_j'\} \right\|_f^2 = O(\delta^{1/16}) \,.$$

Proof Applying Proposition 13.11 and (13.26) we deduce the existence of observables $A_1, \ldots, A_n$ on $\mathbb{C}^d$ such that

$$\forall i \neq j \in \{1, \ldots, n\}, \quad \left\| \{A_i, A_j\} \sigma^{1/2} \right\|_F^2 = O\left(\delta^{1/16}\right), \tag{13.29}$$

$$\forall i \in \{1, \ldots, n\}, \quad \left\| [A_i, \sigma^{1/2}] \right\|_F^2 = O\left(\delta^{1/4}\right). \tag{13.30}$$

(Note that this uses that for each $i \in \{1, \ldots, n\}$, the projections P_i, Q_i used to define $A_i = P_i - Q_i$ depend on X_i and Y_i only.) Next apply Lemma 13.12 with $T_{ij} = \{A_i, A_j\}$ and $X_i = A_i$. The lemma gives a projection R on $\mathbb{C}^d$ such that

$$\frac{2}{n(n-1)} \sum_{1 \leq i < j \leq n} \left\| \{A_i, A_j\} R \right\|_F^2 = O\left(\delta^{1/16}\right) \mathrm{Tr}(R), \tag{13.31}$$

$$\frac{1}{n} \sum_{i \in \{1, \ldots, n\}} \left\| [A_i, R] \right\|_F^2 = O\left(\delta^{1/8}\right) \mathrm{Tr}(R). \tag{13.32}$$

For $i \in \{1, \ldots, n\}$ let $\tilde{A}_i = R A_i R$, and define the observable $A_i' = R \, \mathrm{sgn}(\tilde{A}_i) R$. Using the inequality $(\mathrm{sgn}(x) - x)^2 \leq (x^2 - 1)^2$ valid for all $x \in [-1, 1]$, we see that

$$\begin{aligned}
\left\| A_i' - \tilde{A}_i \right\|_F &\leq \left\| \tilde{A}_i^2 - R \right\|_F \\
&= \left\| R[A_i, R] A_i R \right\|_F \\
&\leq \left\| [A_i, R] \right\|_F,
\end{aligned} \tag{13.33}$$

where we used $A_i^2 = \mathrm{Id}$ and $R^2 = R$. For any i, j, expanding

$$\{A_i', A_j'\} = \{\tilde{A}_i, \tilde{A}_j\} + \{A_i' - \tilde{A}_i, \tilde{A}_j\} + \{A_i', A_j' - \tilde{A}_j\}$$

and using $\|\{A, B\}\|_F \leq 2\|A\|_{S_\infty} \|B\|_F$ for any A, B, and $\|A_i'\|_{S_\infty} \leq 1$, $\|\tilde{A}_j\|_{S_\infty} \leq 1$, we obtain by the triangle inequality

$$\begin{aligned}
\left\| \{A_i', A_j'\} \right\|_F &\leq \left\| \{\tilde{A}_i, \tilde{A}_j\} \right\|_F + 2\left(\left\| A_i' - \tilde{A}_i \right\|_F + \left\| A_j' - \tilde{A}_j \right\|_F \right) \\
&\leq \left\| \{\tilde{A}_i, \tilde{A}_j\} \right\|_F + 2\left(\left\| [A_i, R] \right\|_F + \left\| [A_j, R] \right\|_F \right) \\
&\leq \left\| \{A_i, A_j\} R \right\|_F + 4\left(\left\| [A_i, R] \right\|_F + \left\| [A_j, R] \right\|_F \right),
\end{aligned}$$

where the second inequality uses (13.33), and the third uses the definition of $\tilde{A}_i$ and $\tilde{A}_j$. Squaring this inequality and using Cauchy–Schwarz gives

$$\left\| \{A_i', A_j'\} \right\|_F^2 \leq O\left(\left\| \{A_i, A_j\} R \right\|_F^2 + \left\| [A_i, R] \right\|_F^2 + \left\| [A_j, R] \right\|_F^2 \right).$$

Averaging over all pairs $i \neq j$ and using (13.31) and (13.32) proves the proposition.

$\square$

Proof of Theorem 13.2 By subtracting O from all the operators, we can assume without loss of generality that O is zero. Let U be a unitary such that $\sigma = U|\sigma|$, as given by the polar decomposition. Multiplying all operators on the left by U^{-1}, we may further assume that σ is positive semidefinite. Dividing by $\|\sigma\|_{S_1}$, we may assume that $\text{Tr}(\sigma) = 1$, and δ is replaced by $\delta' = O(\delta)$. Equation (13.6) now follows from Proposition 13.15.

$\square$

13.6   Dimension Bounds

The following lemma shows that pairwise approximately anti-commuting observables only exist in large dimension. The observation is not new; see, e.g., [20, 23]. We give a proof that closely follows [23]. Theorem 13.1 follows immediately by combining the lemma with Theorem 13.2, provided $\delta^{1/32} \leq C/n^3$ for some sufficiently small constant C.

Lemma 13.16 *Let $n \geq 2$ and $d \geq 1$ be integers, $0 \leq \varepsilon \leq 1$, and $A_1, \ldots, A_n$ observables on $\mathbb{C}^d$ such that*

$$\forall i \neq j \in \{1, \ldots, n\}, \quad \left\|\{A_i, A_j\}\right\|_f \leq \varepsilon. \tag{13.34}$$

Then there are universal constants $c, C > 0$ such that if $n^2 \varepsilon \leq c$ then $d \geq (1 - Cn^4\varepsilon^2)2^{\lfloor n/2 \rfloor}$.

Proof The idea for the proof is that if $\varepsilon = 0$, then the A_i would induce a representation of the (finite) finitely presented group

$$C(n) = \langle J, x_1, \ldots, x_n : Jx_i = x_i J, J^2 = x_i^2 = 1,$$

$$x_i x_j = J x_j x_i \text{ for all } i \neq j \in \{1, \ldots, n\}\rangle$$

such that moreover, the representation maps J to $-\text{Id}$. Depending on the parity of n, the group $C(n)$ has either one or two irreducible representations such that $J \mapsto -\text{Id}$, each of dimension $2^{\lfloor n/2 \rfloor}$, implying a corresponding lower bound on the dimension d of the A_i. The goal for the proof is to extend this lower bound to $\varepsilon > 0$. This is done in [23] (see Lemma 3.1 and Lemma 3.4). There are two steps: first, we use A_i satisfying (13.34) to define an approximate homomorphism on $C(n)$ such that $J \mapsto -\text{Id}$. Second, we use a stability theorem due to Gowers and Hatami [9] to argue that any such approximate homomorphism is close to an exact one, and hence must have large dimension.

The first step is given by the following claim, a slightly simplified version of [23, Lemma 3.4].

Claim 13.17 (Lemma 3.4 in [23]) Let $A_1, \ldots, A_n$ satisfy the conditions of Lemma 13.16. For any $x = J^a x_{i_1} \cdots x_{i_k} \in C(n)$, where $1 \leq i_1 < \cdots < i_k \leq n$, define $\phi(x) = (-1)^a A_{i_1} \cdots A_{i_k}$. Then ϕ is an $\eta = n^2\varepsilon$-homomorphism from $C(n)$ to $U(d)$, i.e., for every $x, y \in C(n)$ it holds that $\|\phi(xy) - \phi(x)\phi(y)\|_f \leq \eta$.

Proof Any element of $C(n)$ has a unique representation of the form described in the claim. Let $x, y \in C(n)$ such that $x = J^a x_{i_1} \cdots x_{i_k}$ and $y = J^b x_{j_1} \cdots x_{j_\ell}$. To write xy in canonical form involves at most n^2 application of the anti-commutation relations to sort the $\{x_i, x_j\}$ (together with a number of commutations of J with the x_i, that we need not count since in our representation $\phi(J) = -\text{Id}$ commutes with all A_i), and finally at most n application of the relations $x_i^2 = 1$. When considering $\phi(x)$ and $\phi(y)$, the only operation that is not exact is the anti-commutation between different A_i, A_j. Using the triangle inequality, $\|\phi(xy) - \phi(x)\phi(y)\|_f \leq n^2\varepsilon$, as desired.   □

The second step of the proof is given by the following lemma from [23], which builds on [9].

Lemma 13.18 (Lemma 3.1 in [23]) *Let ϕ be a map from $C(n)$ to the set of unitaries in d dimensions such that ϕ is an η-homomorphism for some $0 \leq \eta \leq 1$. Suppose furthermore that $\|\phi(J) - \text{Id}\|_f > 42\eta$. Then $d \geq (1 - 4\eta^2)2^{\lfloor n/2 \rfloor}$.*

The proof of the lemma first applies the results from [9] to argue that ϕ must be close to an exact representation of $C(n)$, and then concludes using that all irreducible representations of $C(n)$ that send J to $(-\text{Id})$ have dimension $2^{\lfloor n/2 \rfloor}$.

Combining Claim 13.17 and Lemma 13.18 proves Lemma 13.16.   □

We conclude this section by showing that the metric induced by the $(2n + 2)$ points from Theorem 13.1 can be embedded with constant distortion in a Schatten-1 space of polynomial dimension. The construction is inspired by a result of Tsirelson [28] in quantum information.

Lemma 13.19 *For $n \geq 1$ define the metric $d_n(\cdot, \cdot)$ on the $(2n + 2)$ points O, σ, $X_1, \ldots, X_n, Y_1, \ldots, Y_n$ by*

$$d_n(O, \sigma) = 1$$

$$d_n(O, X_i) = d_n(\sigma, X_i) = d_n(O, Y_i) = d_n(\sigma, Y_i) = 1/2$$

$$d_n(X_i, Y_i) = 1$$

$$d_n(X_i, X_j) = d_n(X_i, Y_j) = d_n(Y_i, Y_j) = 1/\sqrt{2} \,,$$

for all $i \neq j$. Then for all $n \geq 1$ and $0 < \delta < 1$ there exists a $(1 + \delta)$ distortion embedding of $d_n(\cdot, \cdot)$ into S_1^d with $d = n^{O(1/\delta^2)}$.

Proof For simplicity, assume that n is even. By the Johnson–Lindenstrauss lemma [12] there are n unit vectors $x_1, \ldots, x_n \in \mathbb{R}^d$ for $d \leq C \log n / \delta^2$ such that the inner products $|x_i \cdot x_j| \leq \delta/4$ for all $i \neq j$. Let $C_1, \ldots, C_d$ be a real representation of the Clifford algebra, i.e., real symmetric matrices such that $\{C_i, C_j\} = C_i C_j + C_j C_i = 2\delta_{ij} \mathrm{Id}$ for all i, j, where δ_{ij} is the Kronecker coefficient. As already mentioned in the introduction, there always exists such a representation of dimension $2^{d'}$ for $d' \leq \lceil d/2 \rceil + 1$. For $i \in \{1, \ldots, n\}$ let $A_i' = \sum_{k=1}^{d} (x_i)_k C_k$. It is easily verified that A_i' is symmetric such that $(A_i')^2 = \mathrm{Id}$, and moreover

$$\forall i \neq j \in \{1, \ldots, n\}, \quad \left(A_i' - A_j'\right)^2 = \left(2 - 2 x_i \cdot x_j\right) \mathrm{Id} . \tag{13.35}$$

Let $A_i' = P_{i,+}' - P_{i,-}'$ be the spectral decomposition, and $X_i = 2^{-d'} P_{i,+}'$, $Y_i = 2^{-d'} P_{i,-}'$. Let $\sigma = 2^{-d'} \mathrm{Id}$ and $O = 0$. Then $\|\sigma - O\|_{\mathsf{S}_1} = 1$. Using that A_i' has trace 0, we also have

$$\|X_i - O\|_{\mathsf{S}_1} = \|Y_i - O\|_{\mathsf{S}_1} = \|\sigma - X_i\|_{\mathsf{S}_1} = \|\sigma - Y_i\|_{\mathsf{S}_1} = \frac{1}{2} ,$$

and $\|X_i - Y_i\|_{\mathsf{S}_1} = 1$, for all $i \in \{1, \ldots, n\}$. It only remains to consider the distance between different i and j. Using that $X_i - X_j = 2^{-d'-1}(A_i' - A_j')$, the condition $|x_i \cdot x_j| \leq \delta/4$ for $i \neq j$, and (13.35), it follows that

$$\left(1 - \frac{\delta}{4}\right) \frac{\sqrt{2}}{2} \leq \|X_i - X_j\|_{\mathsf{S}_1} \leq \left(1 + \frac{\delta}{4}\right) \frac{\sqrt{2}}{2} .$$

Similar bounds hold for pairs of the form $(X_i - Y_j)$ and $(Y_i - Y_j)$. We therefore obtain an embedding in S_1^d with distortion at most $(1 + \delta/4)(1 - \delta/4)^{-1} \leq (1 + \delta)$.

□

Acknowledgements We are grateful to IPAM and the organizers of the workshop "Approximation Properties in Operator Algebras and Ergodic Theory" where this work started. We also thank Assaf Naor for useful comments and encouragement.

References

1. I. Abraham, Y. Bartal, O. Neiman, Advances in metric embedding theory. Adv. Math. **228**(6), 3026–3126 (2011)
2. A. Andoni, M.S. Charikar, O. Neiman, H.L. Nguyen, Near linear lower bound for dimension reduction in ℓ_1, in *2011 IEEE 52nd Annual Symposium on Foundations of Computer Science—FOCS 2011* (IEEE Computer Society, Los Alamitos, 2011), pp. 315–323
3. A. Andoni, A. Naor, O. Neiman, On isomorphic dimension reduction in ℓ_1. Technical Report (2018)
4. K. Ball. Isometric embedding in l_p-spaces. Eur. J. Combin. **11**(4), 305–311 (1990)
5. R. Bhatia, *Matrix Analysis*. Number 169 in Graduate Texts in Mathematics. (Springer, New York, 1997)
6. J. Bourgain, J. Lindenstrauss, V. Milman, Approximation of zonoids by zonotopes. Acta Math. **162**(1–2), 73–141 (1989)

7. B. Brinkman, M. Charikar, On the impossibility of dimension reduction in ℓ_1. J. ACM **52**(5), 766–788 (2005)
8. A. Connes, Classification of injective factors. Cases II_1, II_∞, III_λ, $\lambda \neq 1$. Ann. Math. (2) **104**(1), 73–115 (1976)
9. T. Gowers, O. Hatami, Inverse and stability theorems for approximate representations of finite groups. Preprint. arXiv:1510.04085 (2015)
10. A.W. Harrow, A. Montanaro, A.J. Short, Limitations on quantum dimensionality reduction. Int. J. Quantum Inf. **13**(4), 1440001, 19 (2015)
11. Z. Ji, D. Leung, T. Vidick, A three-player coherent state embezzlement game. Preprint. arXiv:1802.04926 (2018)
12. W.B. Johnson, J. Lindenstrauss, Extensions of Lipschitz mappings into a Hilbert space. Contemp. Math. **26**(189–206), 1 (1984)
13. K.G. Larsen, J. Nelson, Optimality of the Johnson-Lindenstrauss lemma. In *58th Annual IEEE Symposium on Foundations of Computer Science—FOCS 2017* (IEEE Computer Society, Los Alamitos, 2017), pp. 633–638
14. J.R. Lee, A. Naor, Embedding the diamond graph in L_p and dimension reduction in L_1. Geom. Funct. Anal. **14**(4), 745–747 (2004)
15. J. Matoušek, On the distortion required for embedding finite metric spaces into normed spaces. Isr. J. Math. **93**(1), 333–344 (1996)
16. A. Naor, Sparse quadratic forms and their geometric applications [following Batson, Spielman and Srivastava]. *Astérisque*, (348):Exp. No. 1033, viii, 189–217. Séminaire Bourbaki, vol. 2010/2011. Exposés 1027–1042 (2012)
17. A. Naor, G. Pisier, G. Schechtman, Impossibility of dimension reduction in the nuclear norm, in *Proceedings 29th ACM-SIAM Symp. Discrete Algorithms (SODA 18)* (SIAM, Philadelphia, 2018). arXiv:1710.08896. https://arxiv.org/abs/1710.08896
18. I. Newman, Y. Rabinovich, Finite volume spaces and sparsification. Technical Report (2010). arXiv:1002.3541. https://arxiv.org/abs/1002.3541
19. S. Okubo, Real representations of finite Clifford algebras. I. Classification. J. Math. Phys. **32**(7), 1657–1668 (1991)
20. D. Ostrev, T. Vidick, Entanglement of approximate quantum strategies in XOR games. Quantum Inf. Comput. **18**(7–8), 0617–0631 (2018)
21. O. Regev, Entropy-based bounds on dimension reduction in L_1. Isr. J. Math. **195**(2), 825–832 (2013). arXiv:1108.1283. https://arxiv.org/abs/1108.1283
22. G. Schechtman, More on embedding subspaces of L_p in l_r^n. Compos. Math. **61**(2), 159–169 (1987)
23. W. Slofstra, A group with at least subexponential hyperlinear profile. Preprint arXiv:1806.05267 (2018)
24. W. Slofstra, T. Vidick. Entanglement in non-local games and the hyperlinear profile of groups, in *Annales Henri Poincaré* (Springer, Berlin, 2018)
25. S.J. Summers, R. Werner, Maximal violation of Bell's inequalities is generic in quantum field theory. Commun. Math. Phys. **110**(2), 247–259 (1987)
26. M. Talagrand, Embedding subspaces of L_1 into l_1^N. Proc. Amer. Math. Soc. **108**(2), 363–369 (1990)
27. M. Talagrand, Embedding subspaces of L_p in l_p^N, in *Geometric aspects of functional analysis (Israel, 1992–1994)*. Operator Theory: Advances and Applications, vol. 77 (Birkhäuser, Basel, 1995), pp. 311–325
28. B.S. Tsirelson, Quantum analogues of the Bell inequalities. The case of two spatially separated domains. J. Soviet Math. **36**, 557–570 (1987)

Chapter 14
High-Dimensional Convex Sets Arising in Algebraic Geometry

Yanir A. Rubinstein

Dedicated to Bo Berndtsson on the occasion of his 68th birthday

Abstract We introduce an asymptotic notion of positivity in algebraic geometry that turns out to be related to some high-dimensional convex sets. The dimension of the convex sets grows with the number of birational operations. In the case of complex surfaces we explain how to associate a linear program to certain sequences of blow-ups and how to reduce verifying the asymptotic log positivity to checking feasibility of the program.

14.1 Introduction

Convex sets have long been known to appear in algebraic geometry. A well-known example whose origins can be traced to Newton and Minding are the convex polytopes associated to toric varieties [6, 8, 18], also known as Delzant polytopes in the symplectic geometry literature [3]. In recent years, this notion has been further extended to any projective variety, the so-called Newton–Okounkov bodies (or 'nobodies'). In the most basic level, avoiding a formal definition, such a body is a compact convex body (not necessarily a polytope) in $\mathbb{R}^n$ associated to two pieces of data: a nested sequence of subvarieties inside a projective variety of complex

Research supported by NSF grant DMS-1515703. I am grateful to I. Cheltsov and J. Martinez-Garcia for collaboration over the years on these topics, to G. Livshyts for the invitation to speak in the High-dimensional Seminar in Georgia Tech, to the editors for the invitation to contribute to this volume, and to a referee for a careful reading and catching many typos.

Y. A. Rubinstein (✉)
University of Maryland, College Park, MD, USA
e-mail: yanir@alum.mit.edu

© Springer Nature Switzerland AG 2020
B. Klartag, E. Milman (eds.), *Geometric Aspects of Functional Analysis*,
Lecture Notes in Mathematics 2266,
https://doi.org/10.1007/978-3-030-46762-3_14

dimension n, and a line bundle over the variety. Among other things, beautiful relations between the notion of *volume* in algebraic geometry and the volume of these bodies have been proved [12, 14].

The purpose of this article, motivated by a talk in the High-dimensional Seminar at Georgia Tech in December 2018, is to associate another type of convex bodies to projective varieties. The main novelty is that this time the convex bodies can have *unbounded dimension* while the projective variety has *fixed dimension* (which, for most of the discussion, will be in fact 2 (i.e., real dimension 4)). In fact, the asymptotic behavior of the bodies as the dimension grows (on the convex side) corresponds to increasingly complicated birational operations such as blow-ups (on the algebraic side). Rather than volume, we will be interested in intersection properties of these bodies. This gives the first relation between algebraic geometry and asymptotic convex geometry that we are aware of.

This article will be aimed at geometers on both sides of the story—convex and algebraic. Therefore, it will aim to recall at least some elementary notions on both sides. Clearly, a rather unsatisfactory compromise had to be made on how much background to provide, but it is our hope that at least the gist of the ideas are conveyed to experts on both sides of the story.

14.1.1 Organization

We start by introducing asymptotic log positivity in Sect. 14.2. It is a generalization of the notion of positivity of divisors in algebraic geometry, and the new idea is that it concerns pairs of divisors in a particular way. In Sect. 14.3 we associate with this new notion of positivity a convex body, the body of ample angles. In Sect. 14.4 we explain how two previously defined classes of varieties (asymptotically log Fano varieties and asymptotically log canonically polarized varieties) fit in with this picture. The problem of classifying two-dimensional asymptotically log Fano varieties has been posed in 2013 by Cheltsov and the author and is recalled (Problem 14.4.2) as well as the progress on it so far. In Sect. 14.5 we make further progress on this problem by making a seemingly new connection between birational geometry and linear programming, in the process explaining how birational blow-up operations yield convex bodies of increasingly high dimension. Our main results, Theorems 14.5.5 and 14.5.6, first reduce the characterization of "tail blow-ups" (Definition 14.5.3) that preserve the asymptotic log Fano property to checking the feasibility of a certain linear program and, second, show that the linear program can be simplified. The proof, which is the heart of this note, involves associating a linear program to the sequence of blow-ups and characterizing when it is feasible. The canonically polarized case will be discussed elsewhere. A much more extensive classification of asymptotically log del Pezzo surfaces is the topic of a forthcoming work and we refer the reader to Remark 14.5.10 for the relation between Theorems 14.5.5 and 14.5.6 and that work.

This note is dedicated to Bo Berndtsson, whose contributions to the modern understanding and applications of convexity and positivity on the one hand, and whose generosity, passion, curiosity, and wisdom on the other hand, have had a lasting and profound influence on the author over the years.

14.2 Asymptotic Log Positivity

The key new algebraic notion that gives birth to the convex bodies alluded to above is *asymptotic log positivity*. Before introducing this notion let us first pause to explain the classical notion of positivity, absolutely central to algebraic geometry, on which entire books have been written [13].

14.2.1 Positivity

Consider a projective manifold X, i.e., a smooth complex manifold that can be embedded in some complex projective space $\mathbb{P}^N$. In algebraic geometry, one is often interested in notions of positivity. Incidentally, these notions are complex generalizations/analogues of notions of convexity. In discussing these notions one interchangeably switches between line bundles, divisors, and cohomology classes.[1] Complex codimension 1 submanifolds of X are locally defined by a single equation. Formal sums (with coefficients in $\mathbb{Z}$) of such submanifolds is a *divisor* (when the formal sums are taken with coefficients in $\mathbb{Q}$ or $\mathbb{R}$ this is called a $\mathbb{Q}$-divisor or a $\mathbb{R}$-divisor). By the Poincaré duality between homology and cohomology, a (homology class of a) divisor D gives rise to a cohomology class $[D]$ in $H^2(M, \mathbb{F})$ with $\mathbb{F} \in \{\mathbb{Z}, \mathbb{Q}, \mathbb{R}\}$. On the other hand a line bundle is, roughly, a way to patch up local holomorphic functions on X to a global object (a 'holomorphic section' of the bundle). The zero locus of such a section is then a formal sum of complex hypersurfaces, a divisor. E.g., the holomorphic sections of the hyperplane bundle in $\mathbb{P}^N$ are linear equations in the projective coordinates $[z_0 : \ldots : z_N]$, whose associated divisors are the hyperplanes $\mathbb{P}^{N-1} \subset \mathbb{P}^N$. The associated cohomology class, denoted $[H]$, is the generator of $H^2(\mathbb{P}^N, \mathbb{Z}) \cong \mathbb{Z}$. The anticanonical bundle of $\mathbb{P}^N$, on the other hand, is represented by $[(N+1)H]$ and its holomorphic sections are homogeneous polynomials of degree $N + 1$ in $z_0, \ldots, z_N$. Either way, both of these bundles are prototypes of positive ones, a notion we turn to describe.

Now perhaps the simplest way to define positivity, at least for a differential geometer, is to consider the cohomology class part of the story. A class Ω in $H^2(X, \mathbb{Z})$ admits a representative ω (written $\Omega = [\omega]$), a real 2-form, that can

---

[1]A great place to read about this trinity is the cult classic text of Griffiths–Harris [7, §1.1] that was written when the latter was a graduate student of the former.

be written locally as $\sqrt{-1}\sum_{i,j=1}^{n} g_{i\bar{j}}dz^i \wedge \overline{dz^j}$ with $[g_{i\bar{j}}]$ a positive Hermitian matrix, and $z_1, \ldots, z_n$ are local holomorphic coordinates on X. Since a cohomology class can be associated to both line bundles and divisors, this gives a definition of positivity for all three. As a matter of terminology one usually speaks of a divisor being 'ample', while a cohomology class is referred to as 'positive'. For line bundles one may use either word. A line bundle is called negative (the divisor 'anti-ample') if its dual is positive.

The beauty of positivity is that it can be defined in many equivalent ways. Starting instead with the line bundle L, we say L is positive if it admits a smooth Hermitian metric h with positive curvature 2-form $-\sqrt{-1}\partial\bar{\partial}\log h =: c_1(L, h)$. By Chern–Weil theory the cohomology class $c_1(L) = [c_1(L, h)]$ is independent of h.

14.2.2 Asymptotic Log Positivity

We define asymptotic log positivity/negativity similarly, but now we will consider pairs (L, D) and allow for asymptotic corrections along a divisor D (in algebraic geometry the word log usually refers to considering the extra data of a divisor). Let $D = D_1 + \ldots + D_r$ be a divisor on X. We say that $(L, D = D_1 + \ldots + D_r)$ is *asymptotically log positive/negative* if $L - \sum_{i=1}^{r}(1 - \beta_i)D_i$ is positive/negative for all $\beta = (\beta_1, \ldots, \beta_r) \in U \subset (0, 1)^r$ with $0 \in \overline{U}$. For the record, let us give a precise definition as well as two slight variants.

Definition 14.2.1 Let L be a line bundle over a normal projective variety X, and let $D = D_1 + \ldots + D_r$ be a divisor, where $D_i, i = 1, \ldots, r$ are distinct $\mathbb{Q}$-Cartier prime Weil divisors on X.

- We call (L, D) asymptotically log positive/negative if $c_1(L) - \sum_{i=1}^{r}(1 - \beta_i)[D_i]$ is positive/negative for all $\beta = (\beta_1, \ldots, \beta_r) \in U \subset (0, 1)^r$ with $0 \in \overline{U}$.
- We say (L, D) is strongly asymptotically log positive/negative if $c_1(L) - \sum_{i=1}^{r}(1 - \beta_i)[D_i]$ is positive/negative for all $\beta = (\beta_1, \ldots, \beta_r) \in (0, \epsilon)^r$ for some $\epsilon > 0$.
- We say (L, D) is log positive/negative if $c_1(L) - [D]$ is positive/negative.

Note that log positivity implies strong asymptotic log positivity which implies asymptotic log positivity (ALP). None of the reverse implications hold, in general.

The usual notion of positivity can be recovered (by openness of the positivity property) if one required the β_i to be close to 1. By requiring the β_i to hover instead near 0 we obtain a notion that is rather different, but more flexible and still recovers positivity. Indeed, asymptotic log positivity generalizes positivity, as L is positive if and only if (L, D_1) is asymptotically log positive where D_1 is a divisor associated to L. However, the ALP property allows us to 'break' L into pieces and then put different weights along them, so that (L, D) could be ALP even if L itself is not positive. Let us give a simple example.

Example 14.2.2 Let X be the blow-up of $\mathbb{P}^2$ at a point $p \in \mathbb{P}^2$. Let f be a hyperplane containing p and let $\pi^{-1}(f)$ denote the total transform (i.e., the pullback), the union of two curves: the exceptional curve $Z_1 \subset X$ and another curve $F \subset X$ (such that $\pi(Z_1) = p, \pi(F) = f$). Downstairs f is ample, but $\pi^{-1}(f)$ fails to be positive along the exceptional curve Z_1. However, $(\pi^{-1}(f), Z_1)$ is ALP.

This example is not quite illustrative, though, since it is really encoded in a classical object in algebraic geometric called the Seshadri constant. In fact in the example above one does not need to take β small, rather it is really $1 - \beta$ that is the 'small parameter' (and, actually, any $\beta \in (0, 1)$ works, reflecting that the Seshadri constant is 1 here).

A better example is as follows.

Example 14.2.3 Let $X = \mathbb{F}_n$ be the n-th Hirzebruch surface, $n \in \mathbb{N}$. Let $-K_X$ be the anticanonical bundle. It is positive if and only if $n = 0, 1$. In general, $-K_X$ is linearly equivalent to the divisor $2Z_n + (n + 2)F$ where Z_n is the unique $-n$-curve on X (i.e., $Z_n^2 = -n$) and F is a fiber (i.e., $F^2 = 0$). A divisor of the form $aZ_n + bF$ is ample if and only if $b > na$. Thus $(-K_X, Z_n)$ is ALP precisely for $\beta \in (0, \frac{2}{n})$.

14.3   The Body of Ample Angles

The one-dimensional convex body $(0, \frac{2}{n})$ of Example 14.2.3 is the simplest that occurs in our theory. Let us define the bodies that are the topic of the present note.

Let $D = \sum_{i=1}^{r} D_i$, and denote

$$L_{\beta,D} := L - \sum_{i=1}^{r} (1 - \beta_i) D_i. \tag{14.3.1}$$

The problem of determining whether a given pair $(L, D = \sum_{i=1}^{r} D_i)$ is ALP amounts to determining whether the set

$$AA_{\pm}(X, L, D) := \{\beta = (\beta_1, \ldots, \beta_r) \in (0, 1)^r : \pm L_{\beta,D} \text{ is ample}\} \tag{14.3.2}$$

satisfies

$$0 \in \overline{AA_{\pm}(X, L, D)}.$$

Thus, this set is a fundamental object in the study of asymptotic log positivity.

Definition 14.3.1 We call $AA_+(X, L, D)$ the body of ample angles of (X, L, D), and $AA_-(X, L, D)$ the body of anti-ample angles of (X, L, D).

Remark 14.3.2 The body of ample angles encodes both asymptotic log positivity and the classical notion of nefness. Indeed, if $(1, \ldots, 1) \in \overline{AA_\pm(X, L, D)}$ then $\pm L$ is numerically effective (nef). Moreover, one can define a variant of Definitions 14.2.1 and Definitions 14.3.1 where a given $\alpha(1, \ldots, 1) \in \mathbb{R}^n$ is the asymptotic limit instead of the origin (and this could be useful in some situations, e.g., 'wall-crossing' for pairs $(\mathbb{P}^n, dH)$, but observe that just amounts to studying the asymptotic log positivity of $(L + \alpha D, D)$.

Lemma 14.3.3 *When nonempty,* $AA_\pm(X, L, D)$ *is an open convex body in* $\mathbb{R}^r$.

Proof Suppose $AA_+(X, L, D)$ is nonempty. Openness is clear since positivity (and, hence, ampleness) is an open condition on $H^2(X, \mathbb{R})$. For convexity, suppose that $\beta, \gamma \in AA_+(X, L, D) \subset \mathbb{R}^r$. Then, for any $t \in (0, 1)$,

$$L_{t\beta+(1-t)\gamma, D} = L - \sum_{i=1}^{r}(1 - t\beta_i - (1-t)\gamma_i)D_i$$

$$= (t + 1 - t)L - \sum_{i=1}^{r}(t + 1 - t - t\beta_i - (1-t)\gamma_i)D_i$$

$$= t\Big[L - \sum_{i=1}^{r}(1 - \beta_i)D_i\Big] + (1-t)\Big[L - \sum_{i=1}^{r}(1 - \gamma_i)D_i\Big]$$

is positive since the positive cone within $H^2(X, \mathbb{R})$ is convex. If $\beta, \gamma \in AA_-(X, L, D) \subset \mathbb{R}^r$ we get

$$-L_{t\beta+(1-t)\gamma, D} = t(-L_{\beta, D}) + (1-t)(-L_{\gamma, D}),$$

so by the same reasoning $t\beta + (1 - t)\gamma \in AA_-(X, L, D)$. $\square$

Remark 14.3.4 One may wonder why we require AA(X,L,D) to be contained in the unit cube. Indeed, that is not an absolute must. However, we are most interested in the "small angle limit" as $\beta \to 0 \in \mathbb{R}^r$. Still, we require the coordinates to be positive (and not, say, limit to 0 from any orthant) since, geometrically, the β_i can sometimes be interpreted as the cone angle associated to a certain class of Kähler edge metrics. One could in principle allow the whole positive orthant, still. But in this article we restrict to the cube for practical reasons.

There are many interesting questions one can ask about these convex bodies. For instance, how do they transform under birational operations? We now turn to describe a special, but important, situation where we will be able to use tools of convex optimization to say something about this question.

14.4 Asymptotically Log Fano/Canonically Polarized Varieties

Perhaps the most important line bundles in algebraic geometry are the canonical bundle of X, denoted K_X, and its dual, the anticanonical bundle, denoted $-K_X$. These two bundles give rise to two extremely important classes of varieties:

- Fano varieties are those for which $-K_X$ is positive [5, 10],
- Canonically polarized (general type; minimal) varieties are those for which K_X is positive [16] (big; nef). Traditionally, algebraic geometers have been trying to *classify* varieties with positivity properties of $-K_X$ and to *characterize* varieties with positivity properties of K_X. The subtle difference in terminology here stems from the fact that positivity properties of $-K_X$ (think 'positive Ricci curvature') are rare and can sometimes be classified into a list in any given dimension, while positivity or bigness of K_X is much more common, and hence a complete list is impossible, although one can characterize such X sometimes in terms of certain traits. Be it as it may, the importance of these two classes of varieties stems from the fact that, in some very rough sense, the Minimal Model Program stipulates that all projective varieties can be built from minimal/general type and Fano pieces. Put differently, given a projective variety K_X might not have a sign, but one should be able to perform algebraic surgeries (referred to as *birational operations* or *birational maps*) on it to eliminate the 'bad regions' of X where K_X is not well-behaved. Typically, these birational maps will make K_X more positive (in some sense the common case, hence the terminology 'general type'), except in some rare cases when K_X is essentially negative to begin with.

14.4.1 Asymptotic Logarithmic Positivity Associated to (Anti)Canonical Divisors

Thus, given the classical importance of positivity of $\pm K_X$, one may try to extend this to the logarithmic setting.

One may pose the following question:

Question 14.4.1 What are all triples (X, D, β) such that $\beta \in \mathrm{AA}_{\pm}(X, -K_X, D)$?

It turns out that the negative case of this question is too vast to classify, and even the positive case is out of reach unless we make some further assumptions. We now try to at least give some feeling for why this may be so, referring to [19, Question 8.1] for some further discussion. At the end of the day, we will distill from Question 14.4.1, Problem 14.4.6 which we will then take up in the rest of this article.

First, without some restrictions on the parameter β Question 14.4.1 becomes too vast of a generalization which does not seem to be extremely useful. For this reason,[2] we concentrate on the *asymptotic* logarithmic regime, where β is required to be arbitrarily close to the origin.

Definition 14.4.2 ([1, Definition 1.1],[19, Definition 8.13]) (X, D) is (strongly) asymptotically log Fano/canonically polarized if $(-K_X, D)$ is (strongly) asymptotically log positive/negative.

Remark 14.4.3 Definition 14.4.2 is a special case, but, in fact, the main motivation for Definition 14.2.1. The first, when $L = -K_X$, was introduced by Cheltsov and the author [1]. The second, when $L = K_X$, was introduced by the author [19].

Remark 14.4.4 When $(-K_X, D)$ is log positive one says (X, D) is log Fano, a definition due to Maeda [15]. By openness, log Fano is the most restrictive class, a subset of strongly asymptotically log Fano (ALF), itself a subset of ALF.

Remark 14.4.5 There is a beautiful differential geometric interpretation of Definition 14.4.2 in terms of Ricci curvature: (X, D) is asymptotically log Fano/general type if and only if X admits a Kähler metric with edge singularities of arbitrarily small angle β_i along each component D_i of the complex 'hypersurface' D, and moreover the Ricci curvature of this Kähler metric is positive/negative elsewhere. The only if part is an easy consequence of the definition [4, Proposition 2.2], the if part is a generalization of the Calabi–Yau theorem conjectured by Tian [20] and proved in [11, Theorem 2] when $D = D_1$, see also [9] for a different approach in the general case (cf. [17]). When (X, D) is asymptotically log canonically polarized the statement can even be improved to the existence of a Kähler–Einstein edge metric. We refer to [19] for exposition and a survey of these and other results.

Thus, the most basic first step to understand Question 14.4.1 becomes the following, posed in [1].

Problem 14.4.6 Classify all ALF pairs (X, D) with dim $X = 2$ and D having simple normal crossings.

Asymptotically log Fano varieties in dimension 2 are often referred to as *asymptotically log del Pezzo surfaces*. The simple normal crossings (snc) assumption is a standard one in birational geometry and is also the case that is of interest for the study of Kähler edge metrics.

---

[2]Another important reason is that the asymptotic logarithmic regime is closely related to understanding differential-geometric limits, as $\beta \to 0$, towards Calabi–Yau fibrations as conjectured in [1, 19].

14.4.2 Relation to the Body of Ample Angles

The problem of determining whether a given pair $(X, D = \sum_{i=1}^{r} D_i)$ is ALF amounts to determining whether the set $AA_+(X, -K_X, D)$ satisfies

$$0 \in \overline{AA_+(X, -K_X, D)}.$$

Thus, the body of ample angles is a fundamental object in the theory of asymptotically log Fano varieties. This can also be rephrased in terms of intersection properties: there exists $\epsilon_0 > 0$ such that $AA_+(X, -K_X, D) \cap B(0, \epsilon) \neq \emptyset$ for all $\epsilon \in (0, \epsilon_0)$, where $B(0, \epsilon)$ is the ball of radius ϵ centered at the origin in $\mathbb{R}^r$.

If one replaces "ALF" by "strongly ALF" in Problem 14.4.6 the problem has been solved [1, Theorems 2.1,3.1]. However, it turns out that in the strong regime $AA_+(X, -K_X, D) \subset \mathbb{R}^4$ [1, Corollary 1.3]. In sum, the general case is out of reach using only the methods of [1]: in fact, in this note we will exhibit ALF pairs (which are necessarily not strongly ALF) for which $AA_+(X, -K_X, D)$ has arbitrary large dimension and outline a strategy for classifying all ALF pairs.

Before describing our approach to Problem 14.4.6, let us pause to state an open problem concerning these bodies (for X of any dimension).

Problem 14.4.7 How does $AA_\pm(X, -K_X, D)$ behave under birational maps of X?

14.5 Convex Optimization and Classification in Algebraic Geometry

We finally get to the heart of this note where we show how birational operations on X lead to high-dimensional convex bodies.

To emphasize that we are in dimension 2, from now on we use the notation (S, C) instead of (X, D). Also, since we are in the 'Fano regime' we will drop the subscript '+' and simply denote the body of ample angles

$$AA(S, C).$$

We denote the twisted canonical class by (recall (14.3.1))

$$K_{\beta,S,C} := K_S + \sum_{i=1}^{r} (1 - \beta_i) C_i.$$

The Nakai–Moishezon criterion stipulates that $\beta \in AA(S, C)$ if and only if

$$K_{\beta,S,C}^2 > 0 \text{ and } K_{\beta,S,C}.Z < 0 \text{ for every irreducible algebraic curve } Z \text{ in } X.$$
$$(14.5.1)$$

The first is a single quadratic equation in β while the second is a possibly infinite system of linear equations in β. We aim to reduce both of these to a finite system of linear equations.

To that end let us fix some ALF surface (S, C), i.e., suppose $0 \in \overline{AA(S, C)}$. We now ask:

Question 14.5.1 What are all ALF pairs that can be obtained as blow-ups of (S, C)?

It turns out that there are infinitely-many such pairs; the complete analysis is quite involved. In this article we will exhibit a particular type of (infinitely-many) such blow-ups that yields bodies of ample angles of arbitrary dimension.

14.5.1   Tail Blow-Ups

A snc divisor c in a surface is called a chain if $c = c_1 + \ldots + c_r$ with $c_1.c_2 = \ldots = c_{r-1}.c_r = 1$ and otherwise $c_i.c_j = 0$ for all $i \neq j$. In our examples each c_i will be a smooth $\mathbb{P}^1$. The singular points of c are the $r - 1$ intersection points; all other points on c are called its smooth points.

Definition 14.5.2 We say that (S, C) is a single tail blow-up of (s, c) if S is the blow-up of s at a smooth point of $c_1 \cup c_r$, and $C = \pi^{-1}(c)$.

Note that C has $r + 1$ components, the 'new' component being the exceptional curve $E = \pi^{-1}(p)$ where $p \in c_1 \cup c_r$. If, without loss of generality, $p \in c_r$ then $E.\tilde{c}_i = \delta_{ir}$, so

$$C = \tilde{c}_1 + \ldots + \tilde{c}_r + E$$

is still a chain.

As a very concrete example, we could take $S = \mathbb{F}_n$ and $C = Z_n + F$ (recall Example 14.2.3; when $n = 0$ this is simply $S = \mathbb{P}^1 \times \mathbb{P}^1$ and $C = \{p\} \times \mathbb{P}^1 + \mathbb{P}^1 \times \{q\}$, the snc divisor (with intersection point (p, q))). There are two possible single tail blow-ups: blowing-up a smooth point either on Z_n or on F.

14.5.2   Towards a Classification of Nested "Tail" Blow-Ups

In the notation of the previous paragraph, if (S, C) is still ALF we could perform another tail blow-up, blowing up a point on $c_1 \cup E$, and potentially repeat the process any number of times. We formalize this in a definition.

Definition 14.5.3 We say that (S, C) is an ALF tail blow-up of an ALF pair (s, c) if (S, C) is ALF and is obtained from (s, c) as an iterated sequence of single tail blow-ups that result in ALF pairs in all intermediate steps.

In other words, an ALF tail blow-up is a sequence of single tail blow-ups that preserve asymptotic log positivity.

Problem 14.5.4 Classify all ALF tail blow-ups of ALF surfaces $(\mathbb{F}_n, c)$.

The following result reduces the characterization of ALF tail blow-ups to the feasibility of a certain linear program.

Define

$$
\text{LP}(S, C) := \{\beta_x \in (0, 1)^{r+x} : K_{\beta_x, S, C}.Z < 0
$$

$$
\text{for every } Z \subset S \text{ such that } \pi(Z) \subset s \text{ is a}
$$

$$
\text{curve intersecting } c \text{ at finitely-many points}
$$

$$
\text{and passing through the blow-up locus, and}
$$

$$
K_{\beta_x, S, C}.C_i < 0, \quad i = 1, \ldots, r + x.\}
$$

$$
(14.5.2)
$$

Theorem 14.5.5 *Let (s, c) be an ALF pair. An iterated sequence of x single tail blow-ups $\pi : S \to s$ of (s, c) is an ALF tail blow-up if only if (i) $x \le (K_s + c)^2$, and (ii) $0 \in \overline{\text{LP}(S, C)}$.*

In fact, we will also show the following complementary result that shows that (essentially) the only obstacle to completely characterizing tail blow-ups are the (possibly) singular curves Z passing through the blow-up locus in the definition of $\text{LP}(S, C)$.

Define

$$
\widetilde{\text{LP}}(S, C) := \{\beta_x \in (0, 1)^{r+x} : K_{\beta_x, S, C}.C_i < 0, \quad i = 1, \ldots, r + x\}.
$$

$$
(14.5.3)
$$

Theorem 14.5.6 *One always has $0 \in \overline{\widetilde{\text{LP}}(S, C)}$.*

Before we embark on the proofs, a few remarks are in place.

Remark 14.5.7 Observe that $(K_s + c)^2 \ge 0$. Indeed, since (s, c) is ALF $-K_s - c$ is nef (as a limit of ample divisors), so $(K_s + c)^2 \ge 0$.

Remark 14.5.8 The proof will demonstrate that one can drop "that result in ALF pairs in all intermediate steps" from Definition 14.5.3, since it follows from the fact that both (s, c) and (S, C) are ALF (a sort of 'interpolation' property).

Remark 14.5.9 We may assume that c is a connected chain of $\mathbb{P}^1$'s. Indeed, when (s, c) is ALF, c is either a cycle or a union of disjoint chains [1, Lemma 3.5] and each component is a $\mathbb{P}^1$ [1, Lemmas 3.2]. The former is irrelevant for us since there are no tails. For the latter, we may assume that c is connected (i.e., one chain) since the only disconnected case, according to the classification results [1, Theorems 2.1,3.1], is $(\mathbb{F}_n, c_1 + c_2)$ with $c_1 = Z_n$ and $c_2 \in |Z_n + nF|$ and then $(K_{\mathbb{F}_n} + c_1 + c_2)^2 = 0$ so

no tail blow-ups are allowed by Remark 14.5.16. To see that, let $c_1 \in |aZ_n + bF|$
and $c_2 \in |AZ_n + BF|$. Since c_1, c_2 are effective, $b \geq na$, $B \geq nA$. By assumption
$c_1 \cap c_2 = \emptyset$ so $0 = c_1.c_2 = -naA + aB + bA$, i.e., $bA = a(nA - B)$. Since the
right hand side is nonpositive and the left hand side is nonnegative they must both
be zero, leading to $b = 0$, $B = nA$ ($A = 0$ is impossible since it would force $B = 0$,
and $a = 0$ is excluded by $b = 0$). Thus we see $c_1 \in |aZ_n|$, $c_2 \in |A(Z_n + nF)|$.
There are no smooth irreducible representatives of $|aZ_n|$ unless $a = 1$ and similarly
for $|A(Z_n + nF)|$ unless $A = 1$.

Remark 14.5.10 Theorems 14.5.5 and 14.5.6 are mainly given for illustrative
reasons, i.e., to explicitly show how tools of convex programming can be used
in this context. As we show in a forthcoming extensive, but unfortunately long
and tedious, classification work the case of tail blow-ups is in fact the "worst" in
terms of preserving asymptotic log positivity. We will give there a classification of
asymptotically log del Pezzo surfaces that completely avoids tail blow-ups since
condition (ii) in Theorem 14.5.5 is difficult to control, in general. Thus, the present
note and are somewhat complementary. It is still an interesting open problem to
classify all ALF tail blow-ups.

14.5.3   The Set-Up

We start with an ALF pair $(s, c = c_1 + \ldots + c_r)$ and perform $v + h$ single tail
blow-ups of which

$$h \ (\text{'högra'}) \text{ tail blow-ups on the "right tail" } c_r \tag{14.5.4}$$

with associated blow-down map

$$\pi_H = \pi_1 \circ \cdots \circ \pi_h \tag{14.5.5}$$

and exceptional curves

$$\text{exc}(\pi_i) = H_i, \quad i = 1, \ldots, h, \tag{14.5.6}$$

and of which

$$v \ (\text{'vänster'}) \text{ tail blow-ups on the "left tail" } c_1 \tag{14.5.7}$$

with blow-down map

$$\pi_V = \pi_{h+1} \circ \cdots \circ \pi_{h+v} \tag{14.5.8}$$

and exceptional curves

$$\text{exc}(\pi_{h+j}) = V_j, \quad i = 1, \ldots, v, \tag{14.5.9}$$

with new angles $\eta \in (0, 1)^h$ and $v \in (0, 1)^v$, respectively. Finally, we set

$$\eta_0 := \beta_r, \quad v_0 := \beta_1. \tag{14.5.10}$$

An induction argument shows:

Lemma 14.5.11 *With the notation* (14.5.4)–(14.5.10), *if* $v, h > 0$,

$$-K_{(\beta,v,\eta),S,(\pi_H \circ \pi_V)^{-1}(c)}$$

$$= -\pi_V^* \pi_H^* K_{\beta,s,c} - \sum_{i=1}^{h} (1 - \eta_i + \eta_{i-1}) \pi_V^* \pi_h^* \cdots \pi_{i+1}^* H_i$$

$$- \sum_{j=1}^{v} (1 - v_j + v_{j-1}) \pi_{h+v}^* \cdots \pi_{h+1+j}^* V_j. \tag{14.5.11}$$

If $v = 0$,

$$- K_{(\beta,\eta,v),S,\pi_H^{-1}(c)} = -\pi_H^* K_{\beta,s,c} - \sum_{i=1}^{h} (1 - \eta_i + \eta_{i-1}) \pi_h^* \cdots \pi_{i+1}^* H_i. \tag{14.5.12}$$

If $h = 0$,

$$- K_{(\beta,\eta,v),S,\pi_V^{-1}(c)} = -\pi_V^* \pi_H^* K_{\beta,s,c} - \sum_{j=1}^{v} (1 - v_j + v_{j-1}) \pi_{v+h}^* \cdots \pi_{h+1+j}^* V_j. \tag{14.5.13}$$

Before giving the proof, let us recall two elementary facts about blow-ups. Let $\pi : S_2 \to S_1$ be the blow-up at a smooth point p on a surface S_1. Then,

$$K_{S_2} = \pi^* K_{S_1} + E, \tag{14.5.14}$$

where $E = \pi^{-1}(p)$ [7, p. 187], and for every divisor $F \subset S_1$,

$$\widetilde{F} = \begin{cases} \pi^* F, & \text{if } p \notin F, \\ \pi^* F - E, & \text{otherwise.} \end{cases} \tag{14.5.15}$$

Proof Using (14.5.14), if $v = 0$,

$$K_S = \pi_h^*\Big(\pi_{h-1}^*\big(\cdots \big(\pi_1^*(K_s+H_1)+H_2\big)+\ldots+H_{h-2}\big)+H_{h-1}\Big)+H_h. \qquad (14.5.16)$$

Similarly, if $h = 0$,

$$K_S = \pi_v^*\Big(\pi_{v-1}^*\big(\cdots \big(\pi_1^*(K_s+V_1)+V_2\big)+\ldots+V_{v-2}\big)+V_{v-1}\Big)+V_v. \qquad (14.5.17)$$

If $v, h > 0$,

$$K_S = \pi_{v+h}^*\bigg(\pi_{v+h-1}^*\bigg(\cdots \Big(\pi_{h+1}^*\big(\pi_h^*\big(\cdots \big(\pi_1^*(K_s+H_1)+H_2\big)$$

$$+\ldots+\big)+H_h\big)+V_1\Big)+\ldots+V_{v-2}\bigg)+V_{v-1}\bigg)+V_v.$$

$$(14.5.18)$$

Using (14.5.15) and (14.5.10), if $v = 0$,

$$\sum_{i=1}^{r+h}(1-\beta_i)C_i = \sum_{i=1}^{r-1}(1-\beta_i)\pi_H^* c_i + (1-\beta_r)\pi_h^* \cdots \pi_2^*(\pi_1^* c_r - H_1)$$

$$+ (1-\eta_1)\pi_h^* \cdots \pi_3^*(\pi_2^* H_1 - H_2) + \ldots$$

$$+ (1-\eta_{h-1})(\pi_H^* H_{h-1} - H_h) + (1-\eta_h)H_h$$

$$= \sum_{i=1}^{r}(1-\beta_i)\pi_H^* c_i + \sum_{i=1}^{h}(\eta_{i-1} - \eta_i)\pi_h^* \cdots \pi_{i+1}^* H_i, \qquad (14.5.19)$$

if $h = 0$,

$$\sum_{i=1}^{r+v}(1-\beta_i)C_i = (1-\beta_1)\pi_{v+h}^* \cdots \pi_{h+2}^*(\pi_{h+1}^* c_1 - V_1) + \sum_{i=2}^{r}(1-\beta_i)\pi_V^* c_i$$

$$+ (1-\nu_1)\pi_{v+h}^* \cdots \pi_{h+2}^*(\pi_{h+1}^* V_1 - V_2) + \ldots$$

$$+ (1-\nu_{v-1})(\pi_{v+h}^* V_{v-1} - V_v) + (1-\nu_v)V_v$$

$$= \sum_{i=1}^{r}(1-\beta_i)\pi_V^* c_i + \sum_{i=1}^{v}(\nu_{i-1} - \nu_i)\pi_v^* \cdots \pi_{i+1}^* V_i,$$

$$(14.5.20)$$

and if $v, h > 0$,

$$\sum_{i=1}^{r+v+h} (1 - \beta_i)C_i = (1 - \beta_1)\pi_{v+h}^* \cdots \pi_{h+2}^*(\pi_{h+1}^* \pi_H^* c_1 - V_1)$$

$$+ \sum_{i=2}^{r-1}(1 - \beta_i)\pi_V^* \pi_H^* c_i + (1 - \beta_r)\pi_V^* \pi_h^* \cdots \pi_2^*(\pi_1^* c_r - H_1)$$

$$+ (1 - \eta_1)\pi_V^* \pi_h^* \cdots \pi_3^*(\pi_2^* H_1 - H_2) + \ldots$$

$$+ (1 - \eta_{h-1})\pi_V^*(\pi_h^* H_{h-1} - H_h) + (1 - \eta_h)\pi_V^* H_h$$

$$+ (1 - v_1)\pi_{v+h}^* \cdots \pi_{h+2}^*(\pi_{h+1}^* V_1 - V_2) + \ldots$$

$$+ (1 - v_{v-1})(\pi_{v+h}^* V_{v-1} - V_v) + (1 - v_v)V_v$$

$$= \sum_{i=1}^{r}(1 - \beta_i)\pi_V^* \pi_H^* c_i + \sum_{i=1}^{h}(\eta_{i-1} - \eta_i)\pi_V^* \pi_h^* \cdots \pi_{i+1}^* H_i$$

$$+ \sum_{i=1}^{v}(v_{i-1} - v_i)\pi_v^* \cdots \pi_{i+1}^* V_i.$$

$$(14.5.21)$$

Thus, (14.5.18) and (14.5.21) imply (14.5.11), (14.5.16) and (14.5.19) imply (14.5.12), and (14.5.17) and (14.5.20) imply (14.5.13).   □

Remark 14.5.12 In principle, as we will see below, the blow-ups on the left and on the right do not interact.

14.5.4   The Easy Direction and the Sub-critical Case

We start with a simple observation. The easy direction of Theorem 14.5.5 is contained in the next lemma:

Lemma 14.5.13 *Let (s, c) be an ALF pair. Let (S, C) be obtained from (s, c) via an iterated sequence of x single tail blow-ups of (s, c). Then (S, C) is not ALF if $x > (K_s + c)^2$.*

Proof If c does not contain a tail, there is nothing to prove. By Remark 14.5.16, we may assume that c is a single chain. Let $\pi : S \to s$ denote the blow-up of a point on a tail c_r with exceptional curve $E =: C_{r+1}$. Then,

$$-K_{(\beta,\beta_{r+1}),S,C+E} = -\pi^*K_s - E - \sum_{i=1}^{r}(1 - \beta_i)\widetilde{c_i} - (1 - \beta_{r+1})E$$

$$= -\pi^*K_s - E - \sum_{i=1}^{r-1}(1 - \beta_i)\pi^*c_i$$

$$- (1 - \beta_r)(\pi^*c_r - E) - (1 - \beta_{r+1})E$$

$$= -\pi^*K_{\beta,s,c} - (1 + \beta_r - \beta_{r+1})E.$$

In particular, since $E^2 = -1$, $K^2_{(0,0),S,C+E} = K^2_{0,s,c} - 1$. An induction (or directly using Lemma 14.5.11) thus shows that $(K_S + C)^2 = (K_s + c)^2 - x$, which shows that $-K_S - C$ cannot be nef if $x > (K_s + c)^2$, so (S, C) cannot be ALF, by Remark 14.5.7.   □

14.5.5 Dealing with the Quadratic Constraint and the Critical Case

Let

$$\beta_x = (\beta, \beta_{r+1}, \ldots, \beta_{r+x}) \in \mathbb{R}^{r+x}.$$

The proof of Lemma 14.5.13 also shows that

$$K^2_{\beta_x,S,C} = K^2_{\beta,s,c} - x + f(\beta_x),$$

where $f : \mathbb{R}^{r+x} \to \mathbb{R}$ is a quadratic polynomial with no constant term and whose coefficients are integers bounded by a constant depending only on $r + x$. Thus, we also obtain some information regarding the converse to Lemma 14.5.13:

Corollary 14.5.14 *Let (s, c) be an ALF pair. Let (S, C) be obtained from (s, c) via an iterated sequence of x single tail blow-ups of (s, c). Then $K^2_{\beta,S,C} > 0$ for all sufficiently small (depending only on r, x, hence only on r, s, c) $\beta_x \in \mathbb{R}^{r+x}$ if $x < (K_s + c)^2$.*

This corollary is useful since it implies the quadratic inequality in (14.5.1) can be completely ignored except, perhaps, in the borderline case $x = (K_s + c)^2$.

The next result treats precisely that borderline case:

Proposition 14.5.15 *Let (s, c) be an ALF pair. Let (S, C) be obtained from (s, c) via an iterated sequence of $x := (K_s + c)^2$ single tail blow-ups of (s, c). Then*

$$K^2_{\beta,S,C} = f(\beta_x), \tag{14.5.22}$$

where $f : \mathbb{R}^{r+x} \to \mathbb{R}$ *is a quadratic polynomial with no constant term and whose coefficients are integers bounded by a constant depending only on* $r + x$, *and moreover it contains linear terms with positive coefficients and no linear terms with negative coefficients. In particular,* $K^2_{\beta,S,C} > 0$ *for all sufficiently small (depending only on* r, x, *hence only on* r, s, c) $\beta_x \in (0, 1)^{r+x}$.

Remark 14.5.16 The key for later will be (14.5.22) rather than the conclusion about $K^2_{\beta,S,C} > 0$ for all sufficiently small angles. In fact, the latter conclusion (at the end of Proposition 14.5.15) is not precise enough to conclude that the quadratic inequality in (14.5.1) can be ignored as one needs that it holds *simultaneously* with the intersection inequalities of (14.5.1). The exact form of (14.5.22) implies that (14.5.22) can be satisfied together with any *linear* constraints on β_x, which will be the key, and the reason that, ultimately, the quadratic inequality in (14.5.1) *can* be ignored.

Proof We use the notation of Sect. 14.5.3. We wish to show that

$$K^2_{(\beta,\delta,\gamma),S,(\pi_H \circ \pi_V)^{-1}(c)} > 0, \quad \text{for some small } (\beta, \delta, \gamma) \in (0, 1)^{r+h+v}$$

(14.5.23)

(recall $x = h + v = (K_s + c)^2$). We compute,

$$K^2_{(\beta,\delta,\gamma),S,(\pi_H \circ \pi_V)^{-1}(c)} = K_{\beta,s,c} - \sum_{i=1}^{h}(1 - \delta_i + \delta_{i-1})^2 - \sum_{j=1}^{v}(1 - \gamma_j + \gamma_{j-1})^2$$

$$= (K_s + c)^2 - 2\sum_{i=1}^{r}\beta_i c_i.(K_s + c) + \sum_{i=1}^{r}\beta_i^2 c_i^2$$

$$- h + 2\sum_{i=1}^{h}\delta_i - 2\sum_{i=1}^{h}\delta_{i-1} - v + 2\sum_{j=1}^{v}\gamma_j - 2\sum_{j=1}^{v}\gamma_{j-1}$$

$$- \sum_{i=1}^{h}(\delta_i - \delta_{i-1})^2 - \sum_{j=1}^{v}(\gamma_j - \gamma_{j-1})^2$$

$$= -2\sum_{i=1}^{r}\beta_i c_i.(K_s + c) + 2\delta_h - 2\beta_r$$

$$+ 2\gamma_v - 2\beta_1 - O(\beta^2, \delta^2, \gamma^2)$$

$$= 2\beta_1 + 2\beta_r + 2\delta_h - 2\beta_r + 2\gamma_v - 2\beta_1 - O(\beta^2, \delta^2, \gamma^2)$$

$$= 2\delta_h + 2\gamma_v - O(\beta^2, \delta^2, \gamma^2),$$

(14.5.24)

since, by Remark 14.5.9, all c_i are smooth rational curves and c is a single chain, so by adjunction

$$c_i.(K_s + c)$$

$$= \begin{cases} c_i.(K_s + c_i) + c_i.c_{i-1} + c_i.c_{i+1} = -2 + 1 + 1 = 0, & \text{if } i = 2, \ldots, r-1, \\ c_r.(K_s + c_r) + c_r.c_{r-1} = -2 + 1 = -1, & \text{if } i = r, \\ c_1.(K_s + c_1) + c_1.c_2 = -2 + 1 = -1, & \text{if } i = 1. \end{cases}$$

$$(14.5.25)$$

This is clearly positive for $(\beta, \delta, \gamma) = \epsilon(1, \ldots, 1)$ for ϵ small enough. This proves the Proposition.   □

Remark 14.5.17 As alluded to in the remark preceding the proof, one indeed can make $2\delta_h + 2\gamma_v - O(\beta^2, \delta^2, \gamma^2)$ positive under any linear constraints on β, δ, γ without imposing any new linear constraints as the coefficients of the only non-zero linear terms are positive.

14.5.6   Proof of Theorem 14.5.5

First, suppose either (i) or (ii) does not hold. If (i) fails then Lemma 14.5.13 shows that (S, C) is not ALF. If (ii) fails then (S, C) is not ALF by Definition 14.4.2.

Second, if both (i) and (ii) hold then Corollary 14.5.14, Proposition 14.5.15, and the Nakai–Moishezon criterion show that (S, C) is ALF if and only if $K_{\beta_x, S, C}.Z < 0$ for every irreducible curve $Z \subset S$. Naturally, we distinguish between three types of curves Z:

(a) $\pi(Z)$ does not pass through the blow-up locus,
(b) $\pi(Z)$ is contained in the blow-up locus,
(c) $\pi(Z)$ is a curve passing through the blow-up locus.

Curves of type (a) can be ignored: Indeed, then $\pi(Z)$ is a curve in s and $Z = \pi^*\pi(Z)$ (hence, does not intersect any of the exceptional curves) so by Lemma 14.5.11,

$$K_{\beta, S, C}.Z = \pi^* K_{(\beta_1, \ldots, \beta_r), s, c}.\pi^*\pi(Z) = K_{(\beta_1, \ldots, \beta_r), s, c}.\pi(Z).$$

As (s, c) is ALF, this intersection number is negative.

Next, curves of type (c) are covered by condition (ii) by the definition of $LP(S, C)$. Finally, since curves of type (b) are, by definition of the tail blow-up, components of the new boundary C, hence there are at most $x + 2$ (i.e., finitely-many) of them, certainly contained in the finitely-many inequalities:

$$K_{\beta_x,S,C}.C_i < 0, \quad i = 1, \ldots, r + x, \tag{14.5.26}$$

which are once again covered by the definition of LP(S, C). This concludes the proof of Theorem 14.5.5.

14.5.7   Reduction of the Linear Intersection Constraints

In this subsection we explain how to essentially further reduce the linear intersection constraints, i.e., we prove Theorem 14.5.6. To that purpose, we show that curves of type (b) can be handled directly. This shows that the only potential loss of asymptotic logarithmic positivity occurs from curves of type (c) (observe that as in the previous subsection, curves of type (a) can be ignored).

Proof of Theorem 14.5.6 It suffices to check that the system of $2r + 2x$ inequalities

$$K_{\beta_x,S,C}.C_i < 0, \quad i = 1, \ldots, r + x,$$
$$\beta_i > 0, \quad i = 1, \ldots, r + x, \tag{14.5.27}$$

admit a solution along some ray emanating from the origin in $\mathbb{R}^{r+x}$.

Let us first write these inequalities carefully and by doing so eliminate some unnecessary ones.

Using Lemma 14.5.11 we compute, starting with the tails, which turn out to pose no constraints, to wit,

$$-K_{(\beta,\delta,\gamma),S,(\pi_H \circ \pi_V)^{-1}(c)}.V_v = 1 - \gamma_v + \gamma_{v-1} > 0,$$
$$-K_{(\beta,\delta,\gamma),S,(\pi_H \circ \pi_V)^{-1}(c)}.\pi_V^* H_h = 1 - \delta_h + \delta_{h-1} > 0.$$

Next, we intersect with the other new boundary curves (if $h, v > 0$ there are $h+v-2$ such, if $h = 0$ there are $v - 1$ such, if $v = 0$ there are $h - 1$ such),

$$-K_{(\beta,\delta,\gamma),S,(\pi_H \circ \pi_V)^{-1}(c)}.\pi_{h+v}^* \cdots \pi_{h+j+1}^* (\pi_{h+j}^* V_{j-1} - V_j)$$
$$= (1 - \gamma_{j-1} + \gamma_{j-2}) - (1 - \gamma_j + \gamma_{j-1})$$
$$= \gamma_j - 2\gamma_{j-1} + \gamma_{j-2}, \quad j = 2, \ldots, v.$$
$$-K_{(\beta,\delta,\gamma),S,(\pi_H \circ \pi_V)^{-1}(c)}.\pi_V^* \pi_h^* \cdots \pi_{i+1}^* (\pi_i^* H_{i-1} - H_i)$$
$$= (1 - \delta_{i-1} + \delta_{i-2}) - (1 - \delta_i + \delta_{i-1})$$
$$= \delta_i - 2\delta_{i-1} + \delta_{i-2}, \quad i = 2, \ldots, h. \tag{14.5.28}$$

Finally, we intersect with the two 'old tails' (or only one if $\min\{h, v\} = 0$), and use (14.5.25),

$$-K_{(\beta,\delta,\gamma),S,(\pi_H \circ \pi_V)^{-1}(c)} \cdot \pi_{h+v}^* \cdots \pi_{h+2}^* (\pi_{h+1}^* \pi_H^* c_1 - V_1)$$
$$= -K_{\beta,s,c} \cdot c_1 - (1 - \gamma_1 + \beta_1)$$
$$= 1 + \beta_1 c_1^2 - (1 - \gamma_1 + \beta_1)$$
$$= \gamma_1 + (c_1^2 - 1)\beta_1,$$
$$-K_{(\beta,\delta,\gamma),S,(\pi_H \circ \pi_V)^{-1}(c)} \cdot \pi_V^* \pi_h^* \cdots \pi_2^* (\pi_1^* c_r - H_1)$$
$$= -K_{\beta,s,c} \cdot c_r - (1 - \delta_1 + \beta_r)$$
$$= 1 + \beta_r c_r^2 - (1 - \delta_1 + \beta_r)$$
$$= \delta_1 + (c_r^2 - 1)\beta_r. \tag{14.5.29}$$

Equations (14.5.28)–(14.5.29) are $h + v$ linear equations that together with the $r + h + v$ constraints

$$\beta_x = (\beta, \delta, \gamma) \in \mathbb{R}_+^{r+h+v}$$

can be encoded by a $(r + h + v)$-by-$(r + 2h + 2v)$ matrix inequality:

$$(\beta, \delta, \gamma) \mathrm{LP}(S, (\pi_H \circ \pi_V)^{-1}(c)) > 0, \tag{14.5.30}$$

where the inequality symbol means that each component of the vector is positive (typical notation in linear optimization, see, e.g., [2]) with

$$\mathrm{LP}(S, (\pi_H \circ \pi_V)^{-1}(c)) := \begin{cases} \begin{pmatrix} v_r & v_1 & T & I_{r+h+v} \end{pmatrix} & \text{if } h, v > 0, \\[2ex] \begin{pmatrix} v_r & T & I_{r+h} \end{pmatrix} & \text{if } h > 0, v = 0, \\[2ex] \begin{pmatrix} v_1 & T & I_{r+v} \end{pmatrix} & \text{if } h = 0, v > 0, \end{cases}$$

where

$$v_r = (0, \overset{r-1}{\ldots}, 0, c_r^2 - 1, 1, \overset{h-1}{0, \ldots, 0}, \overset{v}{0, \ldots, 0})^T \in \mathbb{Z}^{h+v+r},$$

$$v_1 = (c_1^2 - 1, \overset{r-1}{0, \ldots, 0}, \overset{h}{0, \ldots, 0}, 1, \overset{v-1}{0, \ldots, 0})^T \in \mathbb{Z}^{h+v+r},$$

$$T = \begin{cases} \begin{pmatrix} T_r & T_1 \\ T_h & 0_{h,v-1} \\ 0_{v,h-1} & T_v \end{pmatrix} \in \mathrm{Mat}_{r+h+v,h+v-2}, & \text{if } h, v > 0 \\[2em] \begin{pmatrix} T_r \\ T_h \end{pmatrix} \in \mathrm{Mat}_{r+h,h-1}, & \text{if } h > 0, v = 0 \\[1.5em] \begin{pmatrix} T_1 \\ T_v \end{pmatrix} \in \mathrm{Mat}_{r+v,v-1}, & \text{if } h = 0, v > 0 \end{cases}$$

with

$$T_r = \begin{pmatrix} 0 & 0 & \dots & 0 \\ \vdots & \vdots & \dots & \vdots \\ 0 & 0 & \dots & 0 \\ 1 & 0 & \dots & 0 \end{pmatrix} \in \mathrm{Mat}_{r,h-1}, \qquad T_1 = \begin{pmatrix} 1 & 0 & \dots & 0 \\ 0 & 0 & \dots & 0 \\ \vdots & \vdots & \dots & \vdots \\ 0 & 0 & \dots & 0 \end{pmatrix} \in \mathrm{Mat}_{r,v-1},$$

$$T_h = \begin{pmatrix} -2 & 1 & \dots & 0 \\ 1 & -2 & \dots & 0 \\ 0 & 1 & \dots & 0 \\ \vdots & \vdots & \dots & \vdots \\ 0 & \dots & 0 & 1 \\ 0 & 0 & \dots & -2 \\ 0 & \dots & 0 & 1 \end{pmatrix} \in \mathrm{Mat}_{h,h-1}, \qquad T_v = \begin{pmatrix} -2 & 1 & \dots & 0 \\ 1 & -2 & \dots & 0 \\ 0 & 1 & \dots & 0 \\ \vdots & \vdots & \dots & \vdots \\ 0 & \dots & 0 & 1 \\ 0 & 0 & \dots & -2 \\ 0 & \dots & 0 & 1 \end{pmatrix} \in \mathrm{Mat}_{v,v-1},$$

(here, we use the convention that T_r and T_h are the empty matrix if $h < 2$ and similarly for T_1 and T_v if $v < 2$).

By Gordan's Theorem [2, p. 136], the inequalities (14.5.30) hold if and only if the only solution $y \in \mathbb{R}_+^{r+2h+2v}$ to

$$\mathrm{LP}(S, (\pi_H \circ \pi_V)^{-1}(c))y = 0$$

is $y = 0 \in \mathbb{R}_+^{r+2h+2v}$. We treat first the (easy) cases

$$(h, v) \in \{(1, 0), (0, 1), (2, 0), (0, 2), (1, 1), (2, 1), (1, 2)\}$$

separately.

The case $(1, 0)$ imposes only the inequality $\delta_1 + (c_r^2 - 1)\beta_r > 0$ which is feasible. Similarly, the case $(0, 1)$ imposes only $\gamma_1 + (c_1^2 - 1)\beta_1 > 0$. The case $(1, 1)$ imposes both of these inequalities, but they are independent, hence feasible.

The case $(2, 0)$ imposes the inequalities

$$\delta_1 + (c_r^2 - 1)\beta_r > 0, \quad \delta_2 - 2\delta_1 + \beta_r > 0, \tag{14.5.31}$$

which are equivalent via a Fourier–Motzkin elimination [2, §4.4] to $\delta_2 + \beta_r > 2(1 - c_r^2)\beta_r$, i.e., $\delta_2 > (1 - 2c_r^2)\beta_r$, which is feasible. The case $(0, 2)$ is handled similarly. The case $(2, 2)$ is feasible for the same reasons: both sets of inequalities are feasible and independent. The case $(2, 1)$ (and similarly $(1, 2)$) also follows since it imposes the inequalities (14.5.31) in addition to the independent inequality $\gamma_1 + (c_1^2 - 1)\beta_1 > 0$, thus these are feasible. This idea of independence will also be useful in the general case below.

Let us turn to the general case, i.e., suppose $h, v \geq 2$. First, the $r + h$-th row of $LP(S, (\pi_H \circ \pi_V)^{-1}(c))$ is

$$
(0, \ldots, 0, \overset{h}{\overbrace{1,}} \; \overset{v-1+r+h-1}{\overbrace{0, \ldots, 0}} , 1, 0, \overset{v}{\overbrace{\ldots, 0)}}.
$$

This implies $y_{h+1} = y_{r+2h+v-1} = 0$. If $h = 2$ this implies $y_1 = y_{r+2h+v-2} = 0$; if $h > 3$ this implies $y_h = y_{r+2h+2v-2} = 0$ (the -2 in the $(h + 1)$-th spot in that row is taken care of by the fact $y_{h+1} = 0$ from the previous step), and inductively we obtain $y_{h+1-i} = y_{r+2h+2v-i} = 0$, $i = 1, \ldots, h - 2$, and finally $y_1 = y_{r+h+2v+1} = 0$. Altogether, we have shown $2h$ of the y_i's are zero.

Second, the $r + h + v$-th (last) row is

$$
(0, \ldots, 0, \overset{h+v-1}{\overbrace{1,}} \overset{r+h+v-1}{\overbrace{0, \ldots,}} 0, 1).
$$

This implies $y_{h+v} = y_{r+2h+2v} = 0$. If $v > 2$ this implies $y_{h+v-1} = y_{r+2h+2v-1} = 0$, and inductively we obtain $y_{h+v-i} = y_{r+2h+2v-i} = 0$, $i = 1, \ldots, v - 2$, and finally $y_2 = y_{r+2h+v+1} = 0$. In this step we have shown $2v$ of the y_i's are zero.

So far we have shown $2h + 2v$ of the y_i's are zero using the last $2h + 2v$ rows.

Finally, we consider the first r rows. There are two special rows with possibly positive coefficients $c_r^2 - 1$ and $c_1^2 - 1$, however the corresponding y_1 and y_2 are zero, so as we have the full rank and identity matrix I (with nonnegative coefficients) in $LP(S, (\pi_H \circ \pi_V)^{-1}(c))$ it follows that the remaining r variables y_i are zero, concluding the proof of Theorem 14.5.6. □

References

1. I.A. Cheltsov, Y.A. Rubinstein, Asymptotically log Fano varieties, Adv. Math. **285**, 1241–1300 (2015)
2. G.B. Dantzig, *Linear Programming and Extensions* (Princeton University Press, Princeton, 1963)
3. T. Delzant, Hamiltoniens périodiques et images convexes de l'application moment. Bull. Soc. Math. France **116**, 315–339 (1988)
4. L. Di Cerbo, On Kähler–Einstein surfaces with edge singularities. J. Geom. Phys. **86**, 414–421 (2014)

5. G. Fano, Sulle varietà algebriche a tre dimensioni a curve-sezioni canoniche. Comment. Math. Helv. **14**, 23–64 (1941–1942)
6. W. Fulton, *Introduction to Toric Varieties* (Princeton University Press, Princeton, 1993)
7. P. Griffiths, J. Harris, *Principles of Algebraic Geometry* (Wiley, Hoboken, 1978)
8. M. Gromov, Convex set and Kähler manifolds, in *Advances in Differential Geometry and Topology* (World Scientific Publishing, Singapore, 1990), pp. 1–38
9. H. Guenancia, M. Păun, Conic singularities metrics with prescribed Ricci curvature: the case of general cone angles along normal crossing divisors. J. Diff. Geom. **103**, 15–57 (2016)
10. V.A. Iskovskikh, Yu.G. Prokhorov, Fano varieties, in *Algebraic Geometry, V* (Springer, Berlin, 1999), pp. 1–247
11. T. Jeffres, R. Mazzeo, Y.A. Rubinstein, Kähler–Einstein metrics with edge singularities, (with an appendix by C. Li and Y.A. Rubinstein). Ann. Math. **183**, 95–176 (2016)
12. K. Kaveh, A. Khovanskii, Algebraic equations and convex bodies, in *Perspectives in Analysis, Geometry, and Topology* (Birkhäuser, Basel, 2012), pp. 263–282.
13. R. Lazarsfeld, *Positivity in Algebraic Geometry, I, II* (Springer, Berlin, 2004)
14. R. Lazarsfeld, M. Mustata, Convex bodies associated to linear series. Ann. Sci. Éc. Norm. Supér. **42**(4), 783–835 (2009)
15. H. Maeda, Classification of logarithmic threefolds. Compositio Math. **57**, 81–125 (1986)
16. T. Matsusaka, On canonically polarized varieties, in *Algebraic Geometry* (International Colloquium, Tata Institute of Fundamental Research Studies in Mathematics, Bombay, 1968) (Oxford University Press, 1969), pp. 265–306.
17. R. Mazzeo, Y.A. Rubinstein, The Ricci continuity method for the complex Monge–Ampère equation, with applications to Kähler–Einstein edge metrics. C. R. Math. Acad. Sci. Paris **350**, 693–697 (2012)
18. T. Oda, Convex bodies and algebraic geometry, in *An Introduction to the Theory of Toric Varieties* (Springer, Berlin, 1988)
19. Y.A. Rubinstein, Smooth and singular Kähler–Einstein metrics, in, *Geometric and Spectral Analysis*, ed. by P. Albin et al. Contemporary Mathematics, vol. 630 (American Mathematical Society, Providence and Centre de Recherches Mathématiques, Montreal, 2014), pp. 45–138
20. G. Tian, Kähler–Einstein metrics on algebraic manifolds. Lect. Notes in Math. **1646**, 143–185 (1996)

Chapter 15
Polylog Dimensional Subspaces of ℓ_∞^N

Gideon Schechtman and Nicole Tomczak-Jaegermann

*In memory of Jean Bourgain, the brightest mathematical mind
we have ever encountered*

Abstract We show that a subspace of ℓ_∞^N of dimension $n > (\log N \log \log N)^2$ contains 2-isomorphic copies of ℓ_∞^k where k tends to infinity with $n/(\log N \log \log N)^2$. More precisely, for every $\eta > 0$, we show that any subspace of ℓ_∞^N of dimension n contains a subspace of dimension $m = c(\eta)\sqrt{n}/(\log N \log \log N)$ of distance at most $1 + \eta$ from ℓ_∞^m.

15.1 Introduction

The dichotomy problem of Pisier asks whether a Banach space X either contains, for every n, a subspace K-isomorphic to ℓ_∞^n, for some (equivalently all) $K > 1$, or, for every n, every n-dimensional subspace of X 2-embeds in ℓ_∞^N only if N is exponential in n. This is equivalent to the question of whether for some (equivalently

This research was completed in Fall 2017 while both authors where members of the Geometric Functional Analysis and Application program at MSRI, supported by the National Science Foundation under Grant No. 1440140.

The author "Gideon Schechtman" was supported in part by the Israel Science Foundation.

2010 AMS Subject Classification 46B07, 46B20

G. Schechtman (✉)
Department of Mathematics, Weizmann Institute of Science, Rehovot, Israel
e-mail: gideon@weizmann.ac.il

N. Tomczak-Jaegermann
Department of Mathematics, University of Alberta, Edmonton, AB, Canada
e-mail: nicole.tomczak@ualberta.ca

© Springer Nature Switzerland AG 2020
B. Klartag, E. Milman (eds.), *Geometric Aspects of Functional Analysis*,
Lecture Notes in Mathematics 2266,
https://doi.org/10.1007/978-3-030-46762-3_15

all) absolute $K > 1$ and any sequence $n_N \leq N$ with $n_N / \log N \to \infty$ when $N \to \infty$, every subspace of ℓ_∞^N of dimension n_N contains a subspace of dimension m_N K-isomorphic to $\ell_\infty^{m_N}$ where $m_N \to \infty$ when $N \to \infty$.

We remark in passing that the equivalence between the two versions of the problem ("some $K > 1$" versus "all $K > 1$") is due to the fact proved by R.C. James that, for all $1 < \kappa < K < \infty$, a space which is K isomorphic to ℓ_∞^n contains a subspace κ isomorphic to ℓ_∞^m where $m \to \infty$ as $n \to \infty$. (James proof is essentially included in [5]. A somewhat more precise statement and proof, still due to James, can be read e.g. in [8, p. 283].)

As is exposed in [7], Maurey proved that if X^* has non-trivial type (Equivalently does not contain uniformly isomorphic copies of ℓ_1^n-s. This is a condition stronger than X has non-trivial cotype; equivalently, does not contain uniformly isomorphic copies of ℓ_∞^n-s), then we get the required conclusion: For every n, every n-dimensional subspace of X 2-embeds in ℓ_∞^N only if N is exponential in n.

Another partial result was obtained by Bourgain in [1] where he showed in particular that the conclusion holds if $n_N > (\log N)^4$.

Here we show some improvement over this result of Bourgain: The conclusion holds if $n_N / (\log N \log \log N)^2$ tends to ∞.

Theorem 15.1 *Let n, and N be integers such that $n > (\log N \log \log N)^2$. Then, for some absolute constant $c > 0$ and for every $0 < \eta < 1$, any subspace of ℓ_∞^N of dimension n contains a subspace of dimension $m = c\eta^2 \sqrt{n}/(\log N \log \log N)$ of distance at most $1 + \eta$ from ℓ_∞^m.*

Note that we get some specific estimates for the dimension of the contained subspace $(1 + \eta)$-isomorphic to an ℓ_∞ space of its dimension. Although we are interested in small n-s, the result gives some estimate in the whole range. This is also the case in Bourgain's result: He proved that if $n \geq N^\delta$ than any subspace of ℓ_∞^N of dimension n contains a subspace $(1 + \eta)$-isomorphic to an ℓ_∞ of dimension $m \geq c\eta^5 \delta^2 \sqrt{n}/\log(1/\delta)$. Comparing the two, our result gives better estimates for m when $n \lesssim e^{c(\eta)\sqrt{\log N}}$ and worse when n is larger. Recall also that for n proportional to N, Figiel and Johnson [3] proved earlier that m can be taken of order $\sqrt{N}$ (and no better). This is not recovered by our result.

The general idea of the proof of Theorem 15.1 is the same as in [1] but the technical details are somewhat different. At the end of this note we also speculate that, up to the $(\log \log N)^2$ factor, our result may be best possible.

Our result was essentially achieved a long time ago, circa 1990. Since several people showed interest in it lately we decided to write it up with the hope that more modern methods (and younger minds) may be able to improve it farther.

15.2 Proofs

The main technical tool in the proof of Theorem 15.1 is the following proposition

Proposition 15.1 *Let n, and N be integers such that $n > (\log N)^{3/2} \log\log N$. Let $[a_i(j)]$ be an $n \times N$ matrix with $a_i(j) \geq 0$ for $i = 1, \ldots, n$ and $j = 1, \ldots, N$. Assume that*

$$\sum_{i=1}^{n} a_i(j)^2 \leq 1 \ \text{for} \ \ j = 1, \ldots, N$$

and

$$\sum_{i=1}^{n} a_i(j) \leq 3\sqrt{\log N} \ \text{for} \ \ j = 1, \ldots, N.$$

Moreover, assume that, for some $\gamma > 0$, for every $i = 1, \ldots, n$ there exists $1 \leq j \leq N$ such that $a_i(j) \geq \gamma$. Denote by a_i the i-th row of the matrix. Then, for some positive constants, $c(\gamma)$, $K(\gamma)$ depending only on γ and for every $0 < \eta < 1$, there are disjoint subsets $\sigma_1, \ldots, \sigma_m$ of $\{1, \ldots, n\}$ with $m \geq c(\gamma)\eta^2 n/(\log N)^{3/2} \log\log N$, Such that

$$\|\sum_{r=1}^{m} \sum_{i \in \sigma_r} a_i\|_\infty / \min_{1 \leq r \leq m} \|\sum_{i \in \sigma_r} a_i\|_\infty \leq (1 + K(\gamma)\eta).$$

We first show how to deduce Theorem 15.1 from the proposition above.

Proof of Theorem 15.1 Let X be an n dimensional subspace of ℓ_∞^N. The π_2 norm of the identity on X is equal to $\sqrt{n}$ [4, 9] and by the main theorem of [10] (see [11] for the constant $\sqrt{2}$) this quantity can be computed, up to constant $\sqrt{2}$ on n vectors. This means that there are n vectors $a_i = (a_i(1), \ldots, a_i(N))$, $i = 1, \ldots, n$, in X satisfying

$$\sum_{i=1}^{n} a_i(j)^2 \leq 1, \ \text{for all} \ \ j = 1, \ldots, N$$

and

$$\sum_{i=1}^{n} \|a_i\|_\infty^2 \geq n/2.$$

The first condition implies in particular that $\|a_i\|_\infty^2 \leq 1$ for each i so necessarily for a subset σ' of $\{1, \ldots, n\}$ of cardinality at least $n/4$, $\|a_i\|_\infty \geq 1/2$ for all $i \in \sigma'$. The existence of a subset σ' of $\{1, \ldots, n\}$ of cardinality at least $n/4$ satisfying the two conditions

$$\sum_{i \in \sigma'} a_i(j)^2 \leq 1, \quad \text{for all} \quad j = 1, \ldots, N, \quad \text{and} \quad \|a_i\|_\infty \geq 1/2 \text{ for all } i \in \sigma'$$

(15.1)

is all that we shall use from now on. In Remark 15.1 below we'll show another way to obtain this.

Next we would like to choose a subset σ of σ' of cardinality of order $\sqrt{n \log N}$ such that the matrix $[|a_i(j)|]$, $i \in \sigma$, $j = 1, \ldots, N$, will satisfy the assumptions of Proposition 15.1. So let ξ_i, $i \in \sigma'$, be independent $\{0, 1\}$ valued random variables with $\text{Prob}(\xi_i = 1) = \sqrt{(\log N)/n}$. Since for all j $\sum_{u \in \sigma'} |a_i(j)| \leq \sqrt{n}$, $\mathbb{E} \sum_{u \in \sigma'} |a_i(j)|\xi_i \leq \sqrt{\log N}$. By the most basic concentration inequality, using the fact that $\sum_{i \in \sigma'} a_i(j)^2 \leq 1$, for all j,

$$\text{Prob}(\sum_{i \in \sigma'} |a_i(j)|\xi_i > 3\sqrt{\log N})$$

$$\leq \text{Prob}(\sum_{i \in \sigma'} |a_i(j)|(\xi_i - \mathbb{E}\xi_i) > 2\sqrt{\log N}) \leq e^{-2 \log N} = 1/N^2.$$

It follows that with probability larger than $1 - 1/N$

$$\sum_{i \in \sigma'} |a_i(j)|\xi_i \leq 3\sqrt{\log N}$$

for all j. Since by a similar argument also $\sum_{i \in \sigma'} \xi_i \geq \frac{\sqrt{n \log N}}{16}$ with probability tending to 1 when $N \to \infty$ we get a subset σ of cardinality $n' \geq \frac{\sqrt{n \log N}}{16}$ satisfying

$$\sum_{i \in \sigma} |a_i(j)| \leq 3\sqrt{\log N} \quad \text{for all} \quad j = 1, \ldots, N.$$

Note that the condition $n \geq 256(\log N \log \log N)^2$ implies that $n' \geq (\log N)^{3/2} \log \log N$. It follows that the matrix $[|a_i(j)|]$, $i \in \sigma'$, $j = 1, \ldots, N$ satisfies the conditions of Proposition 15.1 with n' replacing n and $\gamma = 1/2$. We thus get that, for some absolute positive constants c, K, there are disjoint subsets $\sigma_1, \ldots, \sigma_m$ of $\{1, \ldots, n\}$ with

$$m \geq 16cn^2 n'/(\log N)^{3/2} \log \log N \geq cn^2 \sqrt{n}/\log N \log \log N,$$

such that

$$\|\sum_{r=1}^m \sum_{i \in \sigma_r} |a_i|\|_\infty / \min_{1 \leq r \leq m} \|\sum_{i \in \sigma_r} |a_i|\|_\infty \leq (1 + K\eta).$$

Rescaling, we may assume that $\min_{1 \le r \le m} \| \sum_{i \in \sigma_r} |a_i| \|_\infty = 1$. Let j_r denote the label of (one of) the largest coordinates of $\sum_{i \in \sigma_r} |a_i|$. Assume as we may that $\eta < 1/K$. Then no two r's can share the same j_r. Changing the labelling we can also assume $j_r = r$.

Put $x_r = \sum_{i \in \sigma_r} sign(a_i(r)) a_i$. Then for all r, $\|x_r\|_\infty \ge 1$ and for all $j = 1, \ldots, N$,

$$\sum_{r=1}^{m} |x_r(j)| \le 1 + K\eta. \tag{15.2}$$

So the sequence $x_r, r = 1, \ldots, m$, is $(1 + K\eta)$-dominated by the ℓ_∞^m basis; i.e.,

$$\| \sum_{r=1}^{m} \alpha_r x_r \|_\infty \le (1 + K\eta) \max_{1 \le r \le m} |\alpha_r| \text{ for all } \{\alpha_r\}_{r=1}^m.$$

The lower estimate is achieved similarly: Assume $\max_{1 \le r \le m} |\alpha_r| = |\alpha_{r_0}|$ and note that

$$\| \sum_{r=1, r \ne r_0}^{m} \sum_{i \in \sigma_r} |a_i(r_0)| \|_\infty \le K\eta.$$

Then,

$$\| \sum_{r=1}^{m} \alpha_r x_r \|_\infty \ge | \sum_{r=1}^{m} \alpha_r x_r(r_0) |$$

$$\ge |\alpha_{r_0}| \sum_{i \in \sigma_{r_0}} |a_i(r_0)| - \sum_{r=1, r \ne r_0}^{m} |\alpha_r| \sum_{i \in \sigma_r} |a_i(r_0)|$$

$$\ge ((1 - K\eta) \max_{1 \le r \le m} |\alpha_r|.$$

We have thus found a subspace of x of dimension $m \ge c\eta \sqrt{n}/(\log N \log\log N)$ whose distance to ℓ_∞^m is at most $(1 + K\eta)/(1 - K\eta)$. Changing the last quantity to $1 + \eta$, paying by changing c to another absolute constant, is standard. ∎

In the proof of Proposition 15.1 we shall use the following Lemma which follows immediately from Lemma 2 in [2] (but, following the proof of that lemma from [2], is a bit easier to conclude).

Lemma 15.1 *Let ξ_i, $i \in \{1, \ldots, n\}$, be independent $\{0, 1\}$ valued random variables with $Prob(\xi_i = 1) = \delta$. Then for all $q \geq 1$,*

$$(\mathbb{E}(\sum_{i=1}^{n} \xi_i)^q)^{1/q} \leq C(\delta n + q).$$

C is a universal constant.

We now pass to the

Proof of Proposition 15.1 We shall assume as we may that $\eta < \gamma$. We first deal with the small $a_i(j)$-s. Fix $\varepsilon > 0$ to be defined later. Let

$$b_i(j) = \begin{cases} a_i(j) & \text{if } a_i(j) \leq \varepsilon \\ 0 & \text{otherwise.} \end{cases}$$

We will show that for any $\delta > 0$, and with high probability for a random subset $\sigma \subset \{1, \ldots, n\}$ of cardinality $|\sigma| \sim \delta n$

$$\sum_{i \in \sigma} b_i(j) \leq C(\delta \sqrt{\log N} + \varepsilon \log N) \text{ for } j = 1, \ldots, N, \tag{15.3}$$

where C is an absolute constant.

Indeed, set $p = \log N$. Fix $\delta > 0$ and let ξ_i denote selectors with mean δ as in Lemma 15.1. By Chebyshev inequality, (15.3) follows from the estimate

$$\sup_j \left(\mathbb{E}(\sum_{i=1}^{n} \xi_i(\omega) b_i(j))^p \right)^{1/p} \leq C(\delta \sqrt{\log N} + \varepsilon \log N). \tag{15.4}$$

Indeed,

$$\left(\mathbb{E} \sum_{j=1}^{N} (\sum_{i=1}^{n} \xi_i(\omega) b_i(j))^p \right)^{1/p}$$

$$\leq N^{1/\log N} \sup_j \left(\mathbb{E}(\sum_{i=1}^{n} \xi_i(\omega) b_i(j))^p \right)^{1/p} \leq eC(\delta \sqrt{\log N} + \varepsilon \log N).$$

Now apply Chebyshev's inequality.

Fix $1 \leq j \leq N$ and denote $(b_i(j))_i \in \mathbb{R}^n$ by b. Considering the level sets of b we may assume without loss of generality that b is of the form

$$b = \sum_{k=2\log(1/\varepsilon)}^{\infty} 2^{-k/2} \chi_{D_k},$$

(log is $\log_2$) where the sets $D_k \subset \{1, \ldots, n\}$ are mutually disjoint and χ_{D_k} denotes the characteristic function of the set D_k, for $k = 2\log(1/\varepsilon), \ldots$. Thus,

$$\left(\mathbb{E}(\sum_{j=1}^{n} \xi_j(\omega)b_i(j))^p \right)^{1/p}$$

$$\leq \sum_{k=2\log(1/\varepsilon)}^{\infty} 2^{-k/2} \left(\mathbb{E}(\sum_{j\in D_k} \xi_j(\omega))^p \right)^{1/p}$$

$$\leq C \sum_{k=2\log(1/\varepsilon)}^{\infty} 2^{-k/2} \left(\delta|D_k| + p \right) \quad \text{by Lemma 15.1}$$

$$\leq C\delta \sum_{k=2\log(1/\varepsilon)}^{\infty} 2^{-k/2}|D_k| + Cp \sum_{k=2\log(1/\varepsilon)}^{\infty} 2^{-k/2}. \tag{15.5}$$

To estimate the first term in (15.5) note that

$$\sum_{k=2\log(1/\varepsilon)}^{\infty} 2^{-k/2}|D_k| = \|b\|_1 \leq 3\sqrt{\log N}.$$

The second term is clearly smaller than an absolute constant times εp.

Combining the latter two estimates with (15.5) we get (15.4) and hence also (15.3).

To deal with the large coordinates, set, for $j = 1, \ldots, N$,

$$A_j = \{1 \leq i \leq n;\ a_i(j) \geq \varepsilon\}.$$

Since $\sum_{i=1}^{n} a_i(j) \leq 3\sqrt{\log N}$,

$$|A_j| \leq 3\sqrt{\log N}/\varepsilon \quad \text{for } j = 1, \ldots, N. \tag{15.6}$$

An argument similar to the one that proved (15.4) also shows that a random set $\sigma \subset \{1, \ldots, n\}$ of cardinality $|\sigma| \sim \delta n$ satisfies

$$|\sigma \cap A_j| \leq C(\delta\sqrt{\log N}/\varepsilon + \log N). \tag{15.7}$$

Indeed, this follows easily by applying the following inequality with $p = \log N$,

$$\left(\mathbb{E} \left(\sum_{i=1}^{\sqrt{\log N}/\varepsilon} \xi_i \right)^p \right)^{1/p} \leq C(\delta\sqrt{\log N}/\varepsilon + \log N).$$

Moreover, Chebyshev's inequality implies that we can find a set $\sigma \subset \{1, \ldots, n\}$ of cardinality at least $\frac{1}{2}\delta n$ which satisfies (15.3) and (15.7) simultaneously (say, with the same absolute constant C).

Choose now $\delta = 2\eta/\sqrt{\log N}$ and $\varepsilon = \eta/\log N$. Then we get a set $\sigma \subset \{1, \ldots, n\}$ of cardinality at least $\eta n/\sqrt{\log N}$. such that

$$\sum_{i\in\sigma} b_i(j) \leq 3C\eta \quad \text{for} \quad j = 1, \ldots, N, \tag{15.8}$$

and

$$|\sigma \cap A_j| \leq 3C\log N \quad \text{for} \quad j = 1, \ldots, N. \tag{15.9}$$

Define $j_1 \in \{1, \ldots, N\}$ and s_1 by

$$s_1 = \sum_{i\in\sigma\cap A_{j_1}} a_i(j_1) = \max_j \sum_{i\in\sigma\cap A_j} a_i(j).$$

For $r > 1$ define $S_{r-1} = \sigma \setminus (A_{j_1} \cup \cdots \cup A_{j_{r-1}})$ and j_r and s_r by

$$s_r = \sum_{i\in S_{r-1}\cap A_{j_r}} a_i(j_r) = \max_j \sum_{i\in S_{r-1}\cap A_j} a_i(j).$$

By rearranging the columns we may assume $j_r = r$ for all r. Now, (15.9) implies that $|S_r| \geq |\sigma| - 3Cr\log N$ so S_r is not empty for $1 \leq r \leq \frac{\eta n}{3C(\log N)^{3/2}}$. Also,

$$\gamma \leq s_r \leq 3\sqrt{\log N} \leq 3\log N \quad \text{for} \quad 1 \leq r \leq \frac{\eta n}{3C(\log N)^{3/2}}.$$

The sequence s_r is non-increasing, divide it into $(\log((3\log N)/\gamma))/\log(1+\eta)$ intervals such that in each interval $\max s_r/\min s_r$ is at most $1 + \eta$. There is an interval R with $|R| \geq \frac{(\log(1+\eta))\eta n}{3C(\log N)^{3/2}\log((3\log N)/\gamma)} \geq \frac{\eta^2 n}{6C(\log N)^{3/2}\log((3\log N)/\gamma)}$ such that

$$\max_{r\in R} s_r/\min_{r\in R} s_r \leq 1 + \eta.$$

Put $\sigma_r = S_{r-1} \cap A_r$. Since $\min_{r \in R} s_r \geq \gamma > \eta$ we are done in view of (15.8) and the fact that $s_r \geq \sum_{S_{r-1} \cap A_s} a_i(s)$ for $r < s$. ∎

15.3 Remarks

Remark 15.1 Here is an alternative way to get (15.1):

Let X be an n-dimensional normed space which, without loss of generality we assume is in John's position, i.e., the maximal volume ellipsoid inscribed in the unit ball of X is the canonical sphere S^{n-1}. A weak form of the Dvoretzky–Rogers lemma asserts that there are orthonormal vectors $x_1, \ldots, x_n$ such that $\|x_i\|_X \geq c$ for some universal positive constant c. This is proved by a simple volume argument, see for example Theorem 3.4 in [6]. (There it is shown that there are $[n/2]$ such vectors. This is enough for us but it's also easy and well known how to use these $n/2$ orthonormal vectors to get n orthonormal vectors with a somewhat worse lower bound on their norms.)

The map $T : \ell_2^n \to X$ defined by $Te_i = x_i$ is norm one. Note that

$$1 = \|T\| = \sup_{\|x^*\|_{X^*} \leq 1} \left(\sum_{i=1}^n (x^*(x_i))^2 \right)^{1/2}.$$

When X is isometric to a subspace of ℓ_∞^N there are N elements $x_j^* \in B_{X^*}$ such that, for all $x \in X$, $\|x\| = \max_{1 \leq j \leq N} x_j^*(x)$. From this it is easy to deduce that

$$\sup_{\|x^*\|_{X^*} \leq 1} \left(\sum (x^*(x_i))^2 \right)^{1/2} = \max_{1 \leq j \leq N} \left(\sum_{i=1}^n (x_j^*(x_i))^2 \right)^{1/2}.$$

Denoting $a_i(j) = x_j^*(x_i)$ we get (15.1).

Remark 15.2 Here we would like to suggest an approach toward showing that the dichotomy conjecture fails and maybe even that one can't get below the estimate $n > (\log N)^2$ in Theorem 15.1.

Let X and Y be two l dimensional normed spaces. Put $n = l^2$ and $N = 36^l$. Let $\{x_i\}_{i=1}^{6^l}$ be a $1/2$ net in the sphere of X and $\{y_i^*\}_{i=1}^{6^l}$ be a $1/2$ net in the sphere of Y^*. Note that for every $T : X \to Y$,

$$\max_{1 \leq i, j \leq 6^l} y_i^*(Tx_j) \leq \|T\| \leq 4 \max_{1 \leq i, j \leq 6^l} y_i^*(Tx_j).$$

Consequently, $B(X, Y)$, the space of operators from X to Y with the operator norm, 4-embeds into ℓ_∞^N. Note that $\dim(B(X, Y)) = n \sim (\log N)^2$.

(Un)fortunately, $B(X, Y)$ cannot serve as a negative example since it always contains ℓ_∞-s with dimension going to infinity with N. This was pointed out to us by Bill Johnson. Indeed, by Dvoretzky's theorem, ℓ_2^k 2-embeds into Y and into X^*, for some k tending to infinity with n. Let I denote the first embedding and Q be the adjoint of the second embedding. It is then easy to see that $T \to ITQ$ is a 4-embedding of $B(\ell_2^k, \ell_2^k)$ into $B(X, Y)$. Finally, $B(\ell_2^k, \ell_2^k)$ contains isometrically ℓ_∞^k.

However, to get a negative answer to the dichotomy problem, it is enough to find n dimensional X and Y and a subspace Z of $B(X, Y)$ of dimension m with m/n tending to infinity with n which has good cotype, i.e., if Z contains a 2-isomorph of ℓ_∞^k then k is bounded by a universal constant. If one can find such an example with $m \geq cn^2$ for some universal positive constant c then it will even show that one can't get below the estimate $n > (\log N)^2$ in Theorem 15.1.

References

1. J. Bourgain, Subspaces of L_N^∞, arithmetical diameter and Sidon sets, in *Probability in Banach Spaces, V* (Medford, 1984). Lecture Notes in Mathematics, vol. 1153 (Springer, Berlin, 1985), pp. 96–127
2. J. Bourgain, Bounded orthogonal systems and the $\Lambda(p)$-set problem. Acta Math. **162**(3–4), 227–245 (1989)
3. T. Figiel, W.B. Johnson, Large subspaces of l_∞^n and estimates of Gordon-Lewis constant. Israel J. Math. **37**(1–2), 92–112 (1980)
4. D.J.H. Garling, Y. Gordon, Relations between some constants associated with finite dimensional Banach spaces. Israel J. Math. **9**, 346–361 (1971)
5. R.C. James, Uniformly non-square Banach spaces. Ann. Math. **80**(2), 542–550 (1964)
6. V.D. Milman, G. Schechtman, *Asymptotic theory of finite-dimensional normed spaces*. With an appendix by M. Gromov. Lecture Notes in Mathematics, vol. 1200 (Springer, Berlin, 1986), pp. viii+156
7. G. Pisier, Remarques sur un résultat non publié de B. Maurey, (French) [[Remarks on an unpublished result of B. Maurey]] in *Seminar on Functional Analysis*, 1980–1981. (École Polytech., Palaiseau, 1981), Exp. No. V, p. 13
8. G. Schechtman, Euclidean sections of convex bodies, in *Asymptotic Geometric Analysis*. Fields Institute Communications, vol. 68 (Springer, New York, 2013), pp. 271–288
9. M.G. Snobar, p-absolutely summing constants. (Russian) Teor. Funkciĭ Funkcional. Anal. i Priložen. **216**(16), 38–41, (1972)
10. N. Tomczak-Jaegermann, Computing 2-summing norm with few vectors. Ark. Mat. **17**(2), 273–277 (1979)
11. N. Tomczak-Jaegermann, *Banach-Mazur Distances and Finite-Dimensional Operator Ideals*. Pitman Monographs and Surveys in Pure and Applied Mathematics, vol. 38. (Longman Scientific & Technical, Harlow; Wiley, New York, 1989)

Chapter 16
On a Formula for the Volume
of Polytopes

Rolf Schneider

Abstract We carry out an elementary proof of a formula for the volume of polytopes, due to A. Esterov, from which it follows that the mixed volume of polytopes depends only on the product of their support functions.

16.1   Introduction

Esterov [1] has proved the surprising fact that the mixed volume of n convex polytopes in $\mathbb{R}^n$ depends only on the product of their support functions. That an extension of this result to general convex bodies is not true, was pointed out by Kazarnovskiĭ [3, Remark 2]. Esterov deduced his result from a new formula for the volume of a polytope in terms of the nth power of its support function. It is the purpose of this note to carry out an elementary proof of this formula. A motivation will be given after we have stated this formula, in the next section.

16.2   Formulation of the Result

First we fix some terminology. We work in n-dimensional Euclidean space $\mathbb{R}^n$ with scalar product $\langle \cdot, \cdot \rangle$. Its unit sphere is denoted by $\mathbb{S}^{n-1}$. Polytopes are always nonempty, compact, and convex. The volume of a polytope P is denoted by $V_n(P)$. A polyhedral cone is the intersection of a finite family of closed halfspaces with the origin o in their boundaries; equivalently, it is the positive hull of a finite set of vectors. The positive hull of a set $\{v_1, \ldots, v_k\}$ of linearly independent vectors of $\mathbb{R}^n$ is a *simplicial cone* and is denoted by $< v_1, \ldots, v_k >$; this cone is said to be *generated* by $v_1, \ldots, v_k$. A *fan* in $\mathbb{R}^n$ is a finite family $\mathcal{F}$ of polyhedral cones

R. Schneider (✉)
Mathematisches Institut, Albert-Ludwigs-Universität, Freiburg im Breisgau, Germany
e-mail: rolf.schneider@math.uni-freiburg.de

© Springer Nature Switzerland AG 2020
B. Klartag, E. Milman (eds.), *Geometric Aspects of Functional Analysis*,
Lecture Notes in Mathematics 2266,
https://doi.org/10.1007/978-3-030-46762-3_16

with the following properties: every face of a cone in $\mathcal{F}$ is a cone in $\mathcal{F}$, and the intersection of two cones in $\mathcal{F}$ is a face of both. A fan is called *simplicial* if all its cones are simplicial. A fan $\mathcal{F}'$ is a *refinement* of the fan $\mathcal{F}$ if every cone of $\mathcal{F}'$ is contained in a cone of $\mathcal{F}$. Every fan has a simplicial refinement. This is easily seen by induction with respect to the dimension, taking positive hulls with suitable additional rays (or see [2, Thm. 2.6]). For a polytope P and a (nonempty) face F of P, we denote by $N(P, F)$ the normal cone of P at F. The family of all normal cones of P at its faces is a fan, called the *normal fan* of P (see, e.g., Ziegler [5, p. 193], or Ewald [2, p. 17], where it is simply called the 'fan' of P).

If $(v_1, \ldots, v_n)$ is an ordered basis of $\mathbb{R}^n$, we denote by $(v_1^\perp, \ldots, v_n^\perp)$ its Gram–Schmidt orthonormalization. This means that $(v_1^\perp, \ldots, v_n^\perp)$ is orthonormal, the set $\{v_1^\perp, \ldots, v_k^\perp\}$ spans the same subspace as $\{v_1, \ldots, v_k\}$, and $\langle v_k, v_k^\perp \rangle > 0$, for $k = 1, \ldots, n$.

The following is a special case of a more general result of Esterov [1] (with a corrected factor).

Theorem 16.1 (Esterov) *Let $P \subset \mathbb{R}^n$ be a polytope, and let the fan Γ be a simplicial refinement of the normal fan of P. Let $\mathcal{B}(\Gamma)$ be the set of all ordered n-tuples of unit vectors generating cones of Γ. For each n-dimensional cone $C \in \Gamma$, let f_C be the restriction of the nth power of the support function of P to the interior of C. Then*

$$\frac{1}{(n!)^2} \sum_{(v_1,\ldots,v_n)\in\mathcal{B}(\Gamma)} \frac{\partial^n f_{<v_1,\ldots,v_n>}}{\partial v_1^\perp \cdots \partial v_n^\perp} = V_n(P). \tag{16.1}$$

Esterov writes about his result: "We …represent it as a specialization of the isomorphism between two well known combinatorial models of the cohomology of toric varieties." He also gives a brief sketch of an elementary proof, however, for the polytope A under consideration, "assuming for simplicity that the orthogonal complement to the affine span of every (relatively open) face $B \subset A$ intersects B". This is too much of a simplification, since the construction becomes non-trivial if this assumption is not satisfied. Moreover, the statement about the subdivision into "simplices that are in one to one correspondence with the terms of the sum" (the sum in (16.1) is meant) is not correct, since the simplices depend only on the polytope P, whereas the sum gets more terms if the fan Γ is refined. That these extra terms (which are in general not zero) add up to zero, requires an additional argument. The author's statement, "Independence of subdivisions of Γ and linearity follow by definition", seems unjustified.

Since Esterov's surprising result has never been observed in the development of the classical theory of mixed volumes, it might be desirable to have a complete proof along classical lines. Therefore, in the following we carry out Esterov's brief sketch with the necessary details.

16.3   Proof of Theorem 16.1

In the following, we use the common notations lin, pos, aff, conv for, respectively, the linear, positive, affine, or convex hull of a set of vectors or points in $\mathbb{R}^n$. Further, vert P denotes the set of vertices of a polytope P.

First we recall the notion of an *orthoscheme* Δ in $\mathbb{R}^n$. Given a base point z (which will later be the origin, but is needed in greater generality for an induction argument), an ordered orthonormal basis $(u_1, \ldots, u_n)$ of $\mathbb{R}^n$, and a sequence $(a_1, \ldots, a_n)$ of real numbers, it is defined by

$$\Delta := \operatorname{conv}\{z, z_1, \ldots, z_n\}, \quad z_k := z + \sum_{i=1}^{k} a_i u_i \text{ for } k = 1, \ldots, n.$$

This is a simplex, but it may be degenerate, since $a_i = 0$ is allowed.

Let $P \subset \mathbb{R}^n$ be a polytope with interior points. Under the special assumption in Esterov's proof sketch quoted above, P can be decomposed into orthoschemes. Without this assumption, one has to decompose the indicator function of P into the indicator functions of signed orthoschemes.

To prepare this, we denote by $H(Q, u)$ the supporting hyperplane of a polytope $Q \subset \mathbb{R}^n$ with outer normal vector $u \in \mathbb{R}^n \setminus \{o\}$ (for notions from convex geometry that are not explained here, we refer to [4]). By $H^-(Q, u), H^+(Q, u)$ we denote the two closed halfspaces bounded by $H(Q, u)$, where $H^-(Q, u)$ contains Q. In the following, for a point $p \in \mathbb{R}^n$ and a hyperplane $H \subset \mathbb{R}^n$, we denote by $p|H$ the image of p under orthogonal projection to H.

Let $(v_1, \ldots, v_n)$ be a basis of $\mathbb{R}^n$ such that $v_1, \ldots, v_n \in N(P, \{y\})$ for some vertex y of P. Let $(v_1^\perp, \ldots, v_n^\perp)$ be the Gram–Schmidt orthonormalization of $(v_1, \ldots, v_n)$. We define

$$S_1 := P \cap H(P, v_1^\perp), \quad S_2 := S_1 \cap H(S_1, v_2^\perp), \quad \ldots, \quad S_n := S_{n-1} \cap H(S_{n-1}, v_n^\perp).$$

Then $\dim S_k \leq n - k$, so that $S_n = \{y\}$ for the vertex y of P, and $S_k \supseteq S_{k+1}$ for $k = 1, \ldots, n-1$ (equality may hold). We say that $(S_1, \ldots, S_n)$ is *generated by* $(v_1, \ldots, v_n)$. A sequence $(S_1, \ldots, S_n)$ of faces of P is called a *complete tower* of P if $S_1 \supset S_2 \supset \cdots \supset S_n$ and $\dim S_k = n - k$ for $k = 1, \ldots, n$.

Now we define the required orthoschemes. Let $z \in \mathbb{R}^n$. Given $(v_1, \ldots, v_n)$ as above and its generated sequence $(S_1, \ldots, S_n)$ (so that $S_n = \{y\}$), we define a sequence of points by

$$z_1 := z|H(P, v_1^\perp), \quad z_2 := z_1|H(S_1, v_2^\perp), \quad \ldots, \quad z_n := z_{n-1}|H(S_{n-1}, v_n^\perp).$$

We also define a sequence $(a_1, \ldots, a_n)$ of numbers by

$$z_1 = z + a_1 v_1^\perp, \quad z_2 := z_1 + a_2 v_2^\perp, \quad \ldots, \quad z_n = z_{n-1} + a_n v_n^\perp.$$

Here $z_n = y$ and hence $y = z + a_1 v_1^\perp + \cdots + a_n v_n^\perp$. Therefore,

$$a_i = \langle y - z, v_i^\perp \rangle. \tag{16.2}$$

We define the orthoscheme

$$\Delta := \operatorname{conv}\{z, z_1, \ldots, z_n\}.$$

Its volume is given by

$$V_n(\Delta) = \frac{1}{n!}|a_1 \cdots a_n|. \tag{16.3}$$

We denote by $\delta \in \{-1, 0, 1\}$ the sign of $a_1 \cdots a_n$ and call the pair (Δ, δ) a *signed orthoscheme*. It is said to be *induced by* $(v_1, \ldots, v_n)$ (if P and z are given).

We apply this construction from two different starting points. First, we start from the n-dimensional polytope P and associate a signed orthoscheme with each of its complete towers. Let $(S_1, \ldots, S_n)$ be a complete tower of P. Then there is a unique ordered orthonormal basis $(u_1, \ldots, u_n)$ of $\mathbb{R}^n$ such that

$$S_1 := P \cap H(P, u_1), \quad S_2 := S_1 \cap H(S_1, u_2), \quad \ldots \quad S_n := S_{n-1} \cap H(S_{n-1}, u_n). \tag{16.4}$$

We call $(u_1, \ldots, u_n)$ the orthonormal basis *associated with* the complete tower $(S_1, \ldots, S_n)$ of P. Let (Δ, δ) be the signed orthoscheme induced by $(u_1, \ldots, u_n)$. It is also said to be the signed orthoscheme *induced by the complete tower* $(S_1, \ldots, S_n)$.

Definition For given P and z, we denote by $\mathcal{O}(P, z)$ the set of all signed orthoschemes induced by complete towers of P.

Let U be the union of the affine hulls of the facets of all orthoschemes Δ, for $(\Delta, \delta) \in \mathcal{O}(P, z)$. Denoting the indicator function of a set $A \subset \mathbb{R}^n$ by $\mathbb{1}_A$, we state the following

Proposition 16.1

$$\sum_{(\Delta, \delta) \in \mathcal{O}(P, z)} \delta \mathbb{1}_\Delta(x) = \mathbb{1}_P(x) \quad \text{for all } x \in \mathbb{R}^n \setminus U. \tag{16.5}$$

Proof We set

$$g_n(P, z, x) := \sum_{(\Delta, \delta) \in \mathcal{O}(P, z)} \delta \mathbb{1}_\Delta(x)$$

and prove (16.5) by induction with respect to the dimension. The case $n = 1$ is clear. We assume that $n \geq 2$ and that the assertion has been proved in smaller dimensions, for all polytopes and base points. Let P be an n-polytope and z a point in $\mathbb{R}^n$. Let $(\Delta, \delta) \in \mathcal{O}(P, z)$. Then (Δ, δ) is induced by some complete tower $(S_1, \ldots, S_n)$ of P. If $F_1, \ldots, F_m$ are the facets of P, then $S_1 = F_i$ for some $i \in \{1, \ldots, m\}$, and $(S_2, \ldots, S_n)$ is a complete tower of F_i. We have $\Delta = \mathrm{conv}(\Delta' \cup \{z\})$ with some $(\Delta', \delta') \in \mathcal{O}(F_i, z | \mathrm{aff}\, F_i)$ and $\delta = \delta' \sigma(F_i, z)$, where we define, for a facet F of P with outer unit normal vector u,

$$\sigma(F, z) := \left\{ \begin{array}{r} 1 \text{ if } z \in \mathrm{int}\, H^-(P, u), \\ 0 \text{ if } z \in H(P, u), \\ -1 \text{ if } z \in \mathrm{int}\, H^+(P, u). \end{array} \right.$$

For $x \in \mathbb{R}^n \setminus \{z\}$ we define the ray

$$R(z, x) := \{x + \lambda(x - z) : \lambda \geq 0\}.$$

Let $x \in \mathbb{R}^n \setminus U$. Then $x \neq z$. We define $q(x, F_i)$ as the intersection point of $R(z, x)$ and $\mathrm{aff}\, F_i$ if $R(z, x)$ meets $\mathrm{aff}\, F_i$, and as the point x otherwise. Clearly,

$$x \in \Delta \Leftrightarrow R(z, x) \text{ meets } \mathrm{aff}\, F_i \text{ and } q(x, F_i) \in \Delta'$$

and thus $\mathbb{1}_\Delta(x) = \mathbb{1}_{\Delta'}(q(x, F_i))$. This gives

$$g_n(P, z, x) = \sum_{(\Delta, \delta) \in \mathcal{O}(P, z)} \delta \mathbb{1}_\Delta(x)$$

$$= \sum_{i=1}^{m} \sigma(F_i, z) \sum_{(\Delta', \delta') \in \mathcal{O}(F_i, z | \mathrm{aff}\, F_i)} \delta' \mathbb{1}_{\Delta'}(q(x, F_i))$$

$$= \sum_{i=1}^{m} \sigma(F_i, z) g_{n-1}(F_i, z | \mathrm{aff}\, F_i, q(x, F_i))$$

$$= \sum_{i=1}^{m} \sigma(F_i, z) \mathbb{1}\{q(x, F_i) \in F_i\},$$

where the induction hypothesis was applied to $g_{n-1}(F_i, \cdot, \cdot)$. This is possible, since the point $q(x, F_i)$ is not contained in the union of the affine hulls of the $(n-2)$-faces of the orthoschemes Δ', $(\Delta', \delta') \in \mathcal{O}(F_i, z | \mathrm{aff}\, F_i)$, $i = 1, \ldots, m$.

If $x \in P$ (recall that $x \notin U$), there is exactly one index $i \in \{1, \ldots, m\}$ with $q(x, F_i) \in F_i$, and we have $\sigma(F_i, z) = 1$. Therefore, $g_n(P, z, x) = 1$. If $x \notin P$ and some point $q(F_i, x) \in F_i$ exists, then precisely one other index j exists with $q(F_j, x) \in F_j$, and we have $\sigma(F_i, z) = -\sigma(F_j, z)$. This gives $g_n(P, z, x) = 0$. We have proved Eq. (16.5). $\qquad \square$

Integrating (16.5) with respect to Lebesgue measure, we obtain

$$V_n(P) = \sum_{(\Delta,\delta)\in\mathcal{O}(P,z)} \delta V_n(\Delta). \tag{16.6}$$

Our second starting point for constructing signed orthoschemes is the fan Γ. Let (Δ, δ) be the signed orthoscheme induced by $(v_1, \dots, v_n) \in \mathcal{B}(\Gamma)$, where $< v_1, \dots, v_n > \subseteq N(P, \{y\})$ for a vertex y of P. To express the volume of Δ in a suitable way, we note that the restriction of the support function of P to $<v_1, \dots, v_n>$ is given by $u \mapsto \langle y, u \rangle$, hence

$$f_{<v_1,\dots,v_n>}(u) = \langle y, u \rangle^n \quad \text{for } u \in \text{int} <v_1, \dots, v_n>.$$

Writing

$$u = \alpha_1 v_1^\perp + \cdots + \alpha_n v_n^\perp,$$

we have

$$f_{<v_1,\dots,v_n>}(u) = \left(\alpha_1 \langle v_1^\perp, y \rangle + \cdots + \alpha_n \langle v_n^\perp, y \rangle \right)^n$$

and therefore

$$\frac{\partial f_{<v_1,\dots,v_n>}}{\partial v_1^\perp \cdots \partial v_n^\perp} = n! \langle v_1^\perp, y \rangle \cdots \langle v_n^\perp, y \rangle = n! a_1 \cdots a_n = (n!)^2 \delta V_n(\Delta) \tag{16.7}$$

by (16.3), where the numbers $a_1, \dots, a_n$ are those defined by (16.2) with $z = o$.

To utilize this, we need to know which $(v_1, \dots, v_n) \in \mathcal{B}(\Gamma)$ induce signed orthoschemes from $\mathcal{O}(P, o)$. We introduce the following definition.

Definition Let $(v_1, \dots, v_n)$ be an ordered basis of $\mathbb{R}^n$ such that $v_1, \dots, v_n$ are contained in a polyhedral cone C. Then $(v_1, \dots, v_n)$ is called *adapted to* C if there is a sequence $T_1 \subset T_2 \subset \cdots \subset T_n$ where T_k is a k-face of C and $v_1 \in T_1$, $v_k \in T_k \setminus T_{k-1}$ for $k = 2, \dots, n$.

An ordered basis $(v_1, \dots, v_n) \in \mathcal{B}(\Gamma)$ is called *tidy* if it is adapted to the normal cone $N(P, \{y\})$ containing $v_1, \dots, v_n$. The set of all tidy ordered bases $(v_1, \dots, v_n) \in \mathcal{B}(\Gamma)$ is denoted by $\mathcal{T}$, and we set $\mathcal{B}(\Gamma) \setminus \mathcal{T} =: \mathcal{U}$.

From now on, the point $z \in \mathbb{R}^n$ chosen earlier is the origin o.

Proposition 16.2 *The signed orthoscheme induced by $(v_1, \dots, v_n) \in \mathcal{B}(\Gamma)$ belongs to $\mathcal{O}(P, o)$ if and only if $(v_1, \dots, v_n)$ is tidy. Moreover, every signed orthoscheme from $\mathcal{O}(P, o)$ is induced by a unique element of $\mathcal{B}(\Gamma)$.*

Proof We assume first that $(v_1, \dots, v_n) \in \mathcal{B}(\Gamma)$ induces the signed orthoscheme that is induced by the complete tower $(S_1, \dots, S_n)$ of P, with $S_n = \{y\}$. Then the Gram–Schmidt orthonormalization of $(v_1, \dots, v_n)$ is equal to the ordered basis

$(u_1, \ldots, u_n)$ associated with the complete tower $(S_1, \ldots, S_n)$ (that is, defined by (16.4)). By the definition of the Gram–Schmidt orthonormalization, this implies that $v_1 \in N(P, S_1)$ and $v_k \in N(P, S_k) \setminus N(P, S_{k-1})$ for $k = 2, \ldots, n$. The normal cone $N(P, S_k)$ is a k-face of the normal cone $N(P, \{y\})$ (see, e.g., [4, Sect. 2.4], also for the facts on normal cones used below). Thus, $(v_1, \ldots, v_n)$ is adapted to $N(P, \{y\})$, and hence $(v_1, \ldots, v_n)$ is tidy. Conversely, suppose that $(v_1, \ldots, v_n) \in \mathcal{B}(\Gamma)$ is tidy, say that $v_1 \in T_1$ and $v_k \in T_k \setminus T_{k-1}$ for $k = 2, \ldots, n$, where $T_1 \subset T_2 \subset \cdots \subset T_n$ are faces of $N(P, \{y\})$ (for some vertex y of P) with $\dim T_k = k$. Then there is a complete tower $(S_1, \ldots, S_n)$ of P with $N(P, S_k) = T_k$ for $k = 1, \ldots, n$. The signed orthoscheme induced by this tower is induced by $(v_1, \ldots, v_n)$.

A given signed orthoscheme $(\Delta, \delta) \in \mathcal{O}(P, o)$ is induced by a unique complete tower $(S_1, \ldots, S_n)$ of P, say with $S_n = \{y\}$. Then $T_k := N(P, S_k)$ is a k-face of $N(P, \{y\})$ and $T_1 \subset T_2 \subset \cdots \subset T_n$. Since $N(P, \{y\})$ is the union of simplicial cones from Γ, there is a unique cone $C = <v_1, \ldots, v_n> \in \Gamma$ which has k-faces F_k, $k = 1, \ldots, n$, satisfying $F_k \subseteq T_k$. With a suitable (unique) ordering, we then have $v_1 \in F_1$ and $v_k \in F_k \setminus F_{k-1}$ for $k = 2, \ldots, n$. This element $(v_1, \ldots, v_n) \in \mathcal{B}(\Gamma)$ induces (Δ, δ), and it is the only one with this property. This completes the proof of Proposition 16.2. $\qquad\square$

We now see from (16.6) and (16.7) that

$$\frac{1}{(n!)^2} \sum_{(v_1, \ldots, v_n) \in \mathcal{T}} \frac{\partial^n f_{<v_1, \ldots, v_n>}}{\partial v_1^\perp \cdots \partial v_n^\perp} = V_n(P).$$

To complete the proof of (16.1), it remains to show that

$$\sum_{(v_1, \ldots, v_n) \in \mathcal{U}} \frac{\partial^n f_{<v_1, \ldots, v_n>}}{\partial v_1^\perp \cdots \partial v_n^\perp} = 0. \tag{16.8}$$

To prove (16.8), we state a more general version, which can be proved by induction.

Proposition 16.3 *Let $C \subset \mathbb{R}^n$ be an n-dimensional polyhedral cone, and let Γ_C be a simplicial fan such that C is the union of its cones. Let $\mathcal{B}(\Gamma_C)$ be the set of all ordered n-tuples of unit vectors generating cones of Γ_C, and let $\mathcal{U}_C$ be the subset of ordered n-tuples that are not adapted to C. Let $y \in \mathbb{R}^n$ and $f(u) := \langle y, u \rangle^n$ for $u \in C$. Then*

$$\sum_{(v_1, \ldots, v_n) \in \mathcal{U}_C} \frac{\partial^n f_{<v_1, \ldots, v_n>}}{\partial v_1^\perp \cdots \partial v_n^\perp} = 0. \tag{16.9}$$

Proof We abbreviate

$$Df(v_1, \ldots, v_n) := \frac{1}{n!} \frac{\partial^n f_{<v_1, \ldots, v_n>}}{\partial v_1^\perp \cdots \partial v_n^\perp} \quad \text{for } (v_1, \ldots, v_n) \in \mathcal{B}(\Gamma_C),$$

then

$$Df(v_1, \ldots, v_n) = \langle v_1^\perp, y \rangle \cdots \langle v_n^\perp, y \rangle, \qquad (16.10)$$

by (16.7).

We proceed by induction with respect to the dimension, starting with $n = 2$. Let the data be as in the proposition, with $n = 2$. Let $(v_1, v_2) \in \mathcal{U}_C$. Then $v_1 \in \text{int } C$, since otherwise (v_1, v_2) would be adapted to C. Therefore, there exists precisely one cone $< v_1, w_2 > \in \Gamma_C$ with w_2 independent from v_2. The Gram–Schmidt orthonormalizations of (v_1, v_2), (v_1, w_2) are, respectively, $(v_1^\perp, v_2^\perp)$ and $(v_1^\perp, -v_2^\perp)$. It follows from (16.10) that

$$Df(v_1, v_2) + Df(v_1, w_2) = 0.$$

Thus, the elements (v_1, v_2) of $\mathcal{U}_C$ can be grouped into pairs for which the expressions $Df(v_1, v_2)$ sum to 0. Therefore, (16.9) holds for $n = 2$.

Now we assume that $n \geq 3$ and that Proposition 16.3 has been proved in smaller dimensions. Let the data be as in the proposition. We consider two classes of elements $(v_1, \ldots, v_n) \in \mathcal{U}_C$.

Class 1 contains the tuples $(v_1, \ldots, v_n) \in \mathcal{U}_C$ with $\text{pos}\{v_1, \ldots, v_{n-1}\} \cap \text{int } C \neq \emptyset$. For $(v_1, \ldots, v_n)$ in this class, $\text{pos}\{v_1, \ldots, v_{n-1}\}$ is an $(n - 1)$-dimensional face of the cone $< v_1, \ldots, v_n > \in \Gamma_C$ that meets $\text{int } C$ and hence is a face of precisely one other cone $< v_1, \ldots, v_{n-1}, w_n > \in \Gamma_C$. Let $(v_1^\perp, \ldots, v_n^\perp)$ be the Gram–Schmidt orthonormalization of $(v_1, \ldots, v_n)$, and let $(v_1^\perp, \ldots, v_{n-1}^\perp, w_n^\perp)$ be the Gram–Schmidt orthonormalization of $(v_1, \ldots, v_{n-1}, w_n)$. Since v_n and w_n lie in different halfspaces bounded by $\text{lin}\{v_1, \ldots, v_{n-1}\} = \text{lin}\{v_1^\perp, \ldots, v_{n-1}^\perp\}$, we have $v_n^\perp = -w_n^\perp$. It follows from (16.10) that

$$Df(v_1, \ldots, v_n) + Df(v_1, \ldots, v_{n-1}, w_n) = 0.$$

Thus, the elements $(v_1, \ldots, v_n)$ of $\mathcal{U}_C$ in class 1 can be grouped into pairs for which the expressions $Df(v_1, \ldots, v_n)$ sum to 0.

Class 2 contains the tuples $(v_1, \ldots, v_n) \in \mathcal{U}_C$ with $\text{pos}\{v_1, \ldots, v_{n-1}\} \cap \text{int } C = \emptyset$. Let $(v_1, \ldots, v_n)$ be in this class. Then the cone $\text{pos}\{v_1, \ldots, v_{n-1}\}$ is contained in an $(n - 1)$-dimensional face F_i of the cone C. We have $v_n \in C \setminus \text{lin } F_i$, since $v_1, \ldots, v_n$ are linearly independent. If the $(n - 1)$-tuple $(v_1, \ldots, v_{n-1})$ were adapted to the cone F_i, then the n-tuple $(v_1, \ldots, v_n)$ were adapted to the cone C, a contradiction. Thus, $(v_1, \ldots, v_{n-1})$ is not adapted to F_i. We can now apply the inductional hypothesis to the cone F_i and the simplicial fan induced in F_i by Γ_C. This yields

$$\sum_{(v_1, \ldots, v_n) \in \text{ class } 2_i} \langle v_1^\perp, y \rangle \cdots \langle v_{n-1}^\perp, y \rangle = 0, \qquad (16.11)$$

where class 2_i contains the tuples $(v_1, \ldots, v_n) \in \mathcal{U}_C$ with $\mathrm{pos}\{v_1, \ldots, v_{n-1}\} \subseteq F_i$. For each $(v_1, \ldots, v_n) \in$ class 2_i, we have

$$\mathrm{lin}\{v_1^\perp, \ldots, v_{n-1}^\perp\} = \mathrm{lin}\{v_1, \ldots, v_{n-1}\} = \mathrm{lin}\, F_i,$$

and v_n is contained in the open halfspace bounded by $\mathrm{lin}\, F_i$ whose closure contains C. Therefore, $v_n^\perp$ is the same vector for all $(v_1, \ldots, v_n) \in$ class 2_i. Multiplying (16.11) by $\langle v_n^\perp, y \rangle$, we obtain

$$\sum_{(v_1, \ldots, v_n) \in \text{class } 2_i} Df(v_1, \ldots, v_n) = 0.$$

Summing this over $i = 1, \ldots, m$, where $F_1, \ldots, F_m$ are the facets of C, we complete the induction and thus the proof of Proposition 16.3.   □

To prove (16.8), we now apply Proposition 16.3 to the normal cone of each vertex of P and sum over the vertices. This completes the proof of (16.1).

Formula (16.1) is useful if one has to consider a common simplicial refinement of several normal fans, as in the next section. In a volume formula for a single polytope, the superfluous terms may well be omitted. We state an appropriate reformulation of the above result. Let $(S_1, \ldots, S_n)$ be a complete tower of the n-polytope P. We say that *it ends at* the vertex y if $S_n = \{y\}$. The orthonormal basis $(u_1, \ldots, u_n)$ associated with the complete tower $(S_1, \ldots, S_n)$ (defined by (16.4)) is also the unique orthonormal basis defined by

$$u_1 \in N(P, S_1), \quad u_2 \in \mathrm{lin}\{u_1\} + N(P, S_2), \quad \ldots, \quad u_n \in \mathrm{lin}\{u_1, \ldots, u_{n-1}\} + N(P, S_n).$$

For each vertex y of P, we denote by $\mathcal{B}(y)$ the set of all orthonormal bases associated with all complete towers of P ending at y. With these notations, the volume formula obtained from (16.6) and (16.7) can also be written in the form

$$V_n(P) = \frac{1}{n!} \sum_{y \in \text{vert } P} \sum_{(u_1, \ldots, u_n) \in \mathcal{B}(y)} \langle u_1, y \rangle \cdots \langle u_n, y \rangle. \tag{16.12}$$

Note that $\langle u_1, y \rangle \cdots \langle u_n, y \rangle$ is the product of the coordinates of the vertex y with respect to the orthonormal basis $(u_1, \ldots, u_n)$.

16.4   Mixed Volumes

For the extension of Theorem 16.1 to mixed volumes $V(\cdot, \ldots, \cdot)$, it is crucial that the normal fans of finitely many polytopes have a common simplicial refinement.

Theorem 16.2 *Let $P_1, \ldots, P_n \subset \mathbb{R}^n$ be polytopes, and let the fan Γ be a simplicial refinement of the normal fans of $P_1, \ldots, P_n$. Let $\mathcal{B}(\Gamma)$ be the set of all ordered n-tuples of unit vectors generating cones of Γ. For each n-dimensional cone $C \in \Gamma$, let g_C be the restriction of the product of the support functions of $P_1, \ldots, P_n$ to the interior of C. Then*

$$\frac{1}{(n!)^2} \sum_{(v_1,\ldots,v_n) \in \mathcal{B}(\Gamma)} \frac{\partial^n g_{<v_1,\ldots,v_n>}}{\partial v_1^{\perp} \cdots \partial v_n^{\perp}} = V(P_1, \ldots, P_n). \tag{16.13}$$

Proof Let $C \in \Gamma$. To each $i \in \{1, \ldots, n\}$, there is a vertex y_i of P_i such that the support function of P_i on C is given by $\langle \cdot, y_i \rangle$. Let $\lambda_1, \ldots, \lambda_n \geq 0$ and $P = \lambda_1 P_1 + \cdots + \lambda_n P_n$ with polytopes $P_1, \ldots, P_n \subset \mathbb{R}^n$. The support function of P on C is given by $\langle \cdot, \lambda_1 y_1 + \cdots + \lambda_n y_n \rangle$. With f_C defined as in Theorem 16.1 for the polytope P, we have

$$\sum_{i_1,\ldots,i_n=1}^{n} \lambda_{i_1} \cdots \lambda_{i_n} V(P_{i_1}, \ldots, P_{i_n}) = V_n(P)$$

$$= \frac{1}{(n!)^2} \sum_{(v_1,\ldots,v_n) \in \mathcal{B}(\Gamma)} \frac{\partial^n f_{<v_1,\ldots,v_n>}}{\partial v_1^{\perp} \cdots \partial v_n^{\perp}}.$$

Here

$$f_{<v_1,\ldots,v_n>}(u) = \langle u, \lambda_1 y_1 + \cdots + \lambda_n y_n \rangle^n \quad \text{for } u \in \mathrm{pos}\{v_1, \ldots, v_n\},$$

hence

$$\frac{\partial^n f_{<v_1,\ldots,v_n>}}{\partial v_1^{\perp} \cdots \partial v_n^{\perp}} = n! \langle v_1^{\perp}, \lambda_1 y_1 + \cdots + \lambda_n y_n \rangle \cdots \langle v_n^{\perp}, \lambda_1 y_1 + \cdots + \lambda_n y_n \rangle$$

$$= n! \sum_{i_1,\ldots,i_n=1}^{n} \lambda_{i_1} \cdots \lambda_{i_n} \langle v_1^{\perp}, y_{i_1} \rangle \cdots \langle v_n^{\perp}, y_{i_n} \rangle.$$

By comparison we get, in particular,

$$V(P_1, \ldots, P_n) = \frac{1}{n!} \sum_{(v_1,\ldots,v_n) \in \mathcal{B}(\Gamma)} \langle v_1^{\perp}, y_1 \rangle \cdots \langle v_n^{\perp}, y_n \rangle$$

$$= \frac{1}{(n!)^2} \sum_{(v_1,\ldots,v_n) \in \mathcal{B}(\Gamma)} \frac{\partial^n g_{<v_1,\ldots,v_n>}}{\partial v_1^{\perp} \cdots \partial v_n^{\perp}},$$

which completes the proof. $\square$

I thank the referee for correcting an inaccuracy and for several suggestions that improved the presentation.

References

1. A. Esterov, Tropical varieties with polynomial weights and corner loci of piecewise polynomials. Mosc. Math. J. **12**, 55–76 (2012)
2. G. Ewald, *Combinatorial Convexity and Algebraic Geometry*. Graduate Texts in Mathematics, vol. 168 (Springer, New York, 1996)
3. B.Ya. Kazarnovskiĭ, On the action of the complex Monge–Ampère operator on piecewise linear functions. (Russian) Funktsional. Anal. i Prilozhen. **48**, 19–29 (2014). English translation in Funct. Anal. Appl. **48**, 15–23 (2014)
4. R. Schneider, *Convex Bodies: The Brunn–Minkowski Theory*. 2nd edn. Encyclopedia of Mathematics and Its Applications, vol. 151 (Cambridge University Press, Cambridge, 2014)
5. G.M. Ziegler, *Lectures on Polytopes*. Graduate Texts in Mathematics, vol. 152 (Springer, New York, 1995)

LECTURE NOTES IN MATHEMATICS   🐎 Springer

Editors in Chief: J.-M. Morel, B. Teissier;

Editorial Policy

1. Lecture Notes aim to report new developments in all areas of mathematics and their applications – quickly, informally and at a high level. Mathematical texts analysing new developments in modelling and numerical simulation are welcome.

   Manuscripts should be reasonably self-contained and rounded off. Thus they may, and often will, present not only results of the author but also related work by other people. They may be based on specialised lecture courses. Furthermore, the manuscripts should provide sufficient motivation, examples and applications. This clearly distinguishes Lecture Notes from journal articles or technical reports which normally are very concise. Articles intended for a journal but too long to be accepted by most journals, usually do not have this "lecture notes" character. For similar reasons it is unusual for doctoral theses to be accepted for the Lecture Notes series, though habilitation theses may be appropriate.

2. Besides monographs, multi-author manuscripts resulting from SUMMER SCHOOLS or similar INTENSIVE COURSES are welcome, provided their objective was held to present an active mathematical topic to an audience at the beginning or intermediate graduate level (a list of participants should be provided).

   The resulting manuscript should not be just a collection of course notes, but should require advance planning and coordination among the main lecturers. The subject matter should dictate the structure of the book. This structure should be motivated and explained in a scientific introduction, and the notation, references, index and formulation of results should be, if possible, unified by the editors. Each contribution should have an abstract and an introduction referring to the other contributions. In other words, more preparatory work must go into a multi-authored volume than simply assembling a disparate collection of papers, communicated at the event.

3. Manuscripts should be submitted either online at www.editorialmanager.com/lnm to Springer's mathematics editorial in Heidelberg, or electronically to one of the series editors. Authors should be aware that incomplete or insufficiently close-to-final manuscripts almost always result in longer refereeing times and nevertheless unclear referees' recommendations, making further refereeing of a final draft necessary. The strict minimum amount of material that will be considered should include a detailed outline describing the planned contents of each chapter, a bibliography and several sample chapters. Parallel submission of a manuscript to another publisher while under consideration for LNM is not acceptable and can lead to rejection.

4. In general, **monographs** will be sent out to at least 2 external referees for evaluation.

   A final decision to publish can be made only on the basis of the complete manuscript, however a refereeing process leading to a preliminary decision can be based on a pre-final or incomplete manuscript.

   Volume Editors of **multi-author works** are expected to arrange for the refereeing, to the usual scientific standards, of the individual contributions. If the resulting reports can be forwarded to the LNM Editorial Board, this is very helpful. If no reports are forwarded or if other questions remain unclear in respect of homogeneity etc, the series editors may wish to consult external referees for an overall evaluation of the volume.

5. Manuscripts should in general be submitted in English. Final manuscripts should contain at least 100 pages of mathematical text and should always include

   – a table of contents;
   – an informative introduction, with adequate motivation and perhaps some historical remarks: it should be accessible to a reader not intimately familiar with the topic treated;
   – a subject index: as a rule this is genuinely helpful for the reader.
   – For evaluation purposes, manuscripts should be submitted as pdf files.

6. Careful preparation of the manuscripts will help keep production time short besides ensuring satisfactory appearance of the finished book in print and online. After acceptance of the manuscript authors will be asked to prepare the final LaTeX source files (see LaTeX templates online: https://www.springer.com/gb/authors-editors/book-authors-editors/manuscriptpreparation/5636) plus the corresponding pdf- or zipped ps-file. The LaTeX source files are essential for producing the full-text online version of the book, see http://link.springer.com/bookseries/304 for the existing online volumes of LNM). The technical production of a Lecture Notes volume takes approximately 12 weeks. Additional instructions, if necessary, are available on request from lnm@springer.com.

7. Authors receive a total of 30 free copies of their volume and free access to their book on SpringerLink, but no royalties. They are entitled to a discount of 33.3 % on the price of Springer books purchased for their personal use, if ordering directly from Springer.

8. Commitment to publish is made by a *Publishing Agreement*; contributing authors of multiauthor books are requested to sign a *Consent to Publish form*. Springer-Verlag registers the copyright for each volume. Authors are free to reuse material contained in their LNM volumes in later publications: a brief written (or e-mail) request for formal permission is sufficient.

Addresses:

Professor Jean-Michel Morel, CMLA, École Normale Supérieure de Cachan, France
E-mail: moreljeanmichel@gmail.com

Professor Bernard Teissier, Equipe Géométrie et Dynamique,
Institut de Mathématiques de Jussieu – Paris Rive Gauche, Paris, France
E-mail: bernard.teissier@imj-prg.fr

Springer: Ute McCrory, Mathematics, Heidelberg, Germany,
E-mail: lnm@springer.com

Printed in the United States
By Bookmasters